T0291567

Elementary Differential Geometry

The link between the physical world and its visualisation is geometry. This easy-to-read, generously illustrated textbook presents an elementary introduction to differential geometry with emphasis on geometric results. Avoiding formalism as much as possible, the author harnesses basic mathematical skills in analysis and linear algebra to solve interesting geometric problems, which prepare students for more advanced study in mathematics and other scientific fields such as physics and computer science.

The wide range of topics includes curve theory, a detailed study of surfaces, curvature, variation of area and minimal surfaces, geodesics, spherical and hyperbolic geometry, the divergence theorem, triangulations, and the Gauss–Bonnet theorem. The section on cartography demonstrates the concrete importance of elementary differential geometry in applications. Clearly developed arguments and proofs, colour illustrations, and over 100 exercises and solutions make this book ideal for courses and self-study. The only prerequisites are one year of undergraduate calculus and linear algebra.

Christian Bär is Professor of Geometry in the Institute for Mathematics at the University of Potsdam, Germany.

Elementary Differential Geometry

Christian Bär
Universität Potsdam, Germany

CAMBRIDGE
UNIVERSITY PRESS

University Printing House, Cambridge CB2 8BS, United Kingdom

One Liberty Plaza, 20th Floor, New York, NY 10006, USA

477 Williamstown Road, Port Melbourne, VIC 3207, Australia

314-321, 3rd Floor, Plot 3, Splendor Forum, Jasola District Centre, New Delhi - 110025, India

79 Anson Road, #06-04/06, Singapore 079906

Cambridge University Press is part of the University of Cambridge.

It furthers the University's mission by disseminating knowledge in the pursuit of
education, learning and research at the highest international levels of excellence.

www.cambridge.org
Information on this title: www.cambridge.org/9780521721493

Originally published in German as Elementare Differentialgeometrie by Walter de Gruyter 2001

First published in English by Cambridge University Press 2010

Reprinted with corrections 2011

A catalogue record for this publication is available from the British Library

Library of Congress Cataloging in Publication data
Bär, Christian.
Elementary differential geometry / Christian Bär.
p. cm.
ISBN 978-0-521-89671-9 (Hardback) – ISBN 978-0-521-72149-3 (Pbk.)
1. Geometry, Differential–Textbooks. I. Title.
QA641.B325 2010
516.3′6–dc22

2010001343

ISBN 978-0-521-89671-9 Hardback
ISBN 978-0-521-72149-3 Paperback

Contents

Preface

This book evolved from courses about elementary differential geometry which I have taught in Freiburg, Hamburg and Potsdam. The word "elementary" should not be understood as "particularly easy", but indicates that the development of formalism, which would be necessary for a deeper study of differential geometry, is avoided as much as possible. We will instead approach geometrically interesting problems using tools from the standard fundamental courses in analysis and linear algebra. It is possible to raise interesting questions even about objects as "simple" as plane curves. The proof of the four-vertex theorem, for example, is anything but trivial.

The book is suitable for students from the second year of study onwards and can be used in lectures, seminars, or for private study.

The first chapter is interesting mostly for historical reasons. The reader can here find out how geometric results have been obtained from axioms for thousands of years, since Euclid. In particular, the controversy about the parallel axiom will be explained. In this chapter we will mostly follow Hilbert's presentation of plane geometry, since it is rather close to Euclid's formulation of the axioms and yet meets today's requirements for mathematical rigour. In the mean time the axiomatic system has been simplified significantly [2]. A presentation with only seven axioms can be found in [30].

Anyone who is only interested in differential geometry can begin with the second chapter. The theory of curves is developed here, with particular focus on curves in the plane and in three-dimensional space. Curvature, a central notion in differential geometry, appears for the first time. Of particular interest are global results, i.e. statements about the overall shape of closed curves. The above-mentioned four-vertex theorem and the theorems of Fenchel and Fáry–Milnor fall into this category. They tell us how much a space curve needs to curve so that it can close up (Fenchel) and how much a curve needs to curve to become knotted (Fáry–Milnor).

We begin to study surfaces in three-dimensional space in the third chapter. The necessary concepts are introduced, e.g. different notions of curvature, and some important classes of surfaces are studied in more detail. One of them is the class of minimal surfaces, which appear in nature as soap films. Some examples are illustrated in colour as well.

In the fourth chapter we change our point of view and concentrate on geometric quantities that can be obtained using measurements taken on the surface itself only. We study the shortest connecting curves between two points on a given surface, for example. This stance suggests the introduction of general Riemannian metrics, which allows us to construct new important geometries. The most prominent example is the hyperbolic plane, which, as Hilbert showed, cannot be realised as a "classical" surface. One reason why the hyperbolic plane is so important is that it ended the controversy about the parallel axiom. It is therefore often referred to as a non-Euclidean geometry. We devote ourselves to hyperbolic and spherical geometry and derive the most important trigonometric laws. Spherical geometry is used to discuss applications in cartography. We conclude the chapter with a comparison of different models of hyperbolic geometry illustrated by a woodcut of Dutch artist M. C. Escher.

In the fifth chapter we derive Gauss's divergence theorem and deduce that the total Gauss curvature of a closed surface does not depend on the Riemannian metric. The total curvature is thus a "topological invariant" of the surface.

The last chapter is dedicated to the topological interpretation of this quantity. We show that every compact surface can be triangulated, i.e. that it can be cut into triangles in a suitable way. The Gauss–Bonnet theorem then tells us that the total curvature can be found by counting vertices, edges and triangles. We conclude with the outlook and recommendations for further study.

Three appendices follow: first hints for solutions to the exercises, then a collection of useful formulae concerned with the inner geometry of surfaces and the most important trigonometric laws, and finally, the mathematical symbols used in this book are listed, to make it easier to look them up. As is the custom, the book ends with the references and the index.

The enumeration of theorems, lemmas, examples and so forth is done using three numbers, where the first one denotes the chapter and the second one the section. The numerous exercises are enumerated by two numbers, the first one being the chapter. They are interspersed in the text and mainly discuss examples, which can be used to practice the material treated so far. The further logical arguments do not build on most exercises, but reading and doing the exercises is recommended to establish the necessary familiarity with the introduced concepts.

At this point it is my pleasure to thank all those that made this book a success, e.g. by pointing out mistakes and making suggestions, including B. Ammann, M. Aubel, F. Auer, L. Außenhofer, P. Ghanaat, H. Karcher, A. Kreuzer, D. Lengeler, F. Pfäffle, C. Pries, W. Reichel, E. Schröder,

C. Schulz, T. Seidel, U. Semmelmann, H. Wendland, U. Witting, and U. Wöske. I am of course solely responsible for any mistakes in this book, which it will inevitably contain. I would be very grateful for a note telling me about them which can be sent to baer@math.uni-potsdam.de. Sincere thanks also go to Cambridge University Press, in particular C. Dennison, for her always pleasant and trusting collaboration.

This book has been typeset using LaTeX with the PSTricks package to produce the drawings. The coloured illustrations have been created with povray, the maps in the section on cartography with the Generic Mapping Tools for Unix. The transformations of Escher's woodcut to the Klein model of hyperbolic geometry on page 220 and to the half-plane mode on page 222 have been carried out with gimp using the mathmap plugin. I am very grateful to the developers of all this great open source software. The illustrations on pages 142, 206, and 256 have been created with Maple.

At last, a special thank you to A. Hornecker who contributed the coloured illustrations and many of the drawings and to P. Meerkamp who wrote a first translation of the German edition into English.

Notation

We here introduce some terms and conventions that will appear in this book again and again.

The cardinality of a set A is denoted using absolute value bars:

$$|A| = \text{number of elements of } A.$$

For the difference of two sets we write

$$A - B = \{x \in A \mid x \notin B\}.$$

$\mathbb{R}^n$ denotes the vector space of all column vectors with n real entries:

$$\begin{pmatrix} x^1 \\ \vdots \\ x^n \end{pmatrix} = (x^1, \dots, x^n)^\top \in \mathbb{R}^n.$$

The individual entries usually have their indices at the top. For a subset $A \subset \mathbb{R}^n$ the expression $\bar{A}$ denotes the closure, ∂A the boundary and $\mathring{A}$ the interior.

The Euclidean standard scalar product on $\mathbb{R}^n$ is written using angle brackets:

$$\left\langle (x^1, \dots, x^n)^\top, (y^1, \dots, y^n)^\top \right\rangle = \sum_{j=1}^{n} x^j y^j.$$

For a subspace $V \subset \mathbb{R}^n$

$$V^\perp = \{x \in \mathbb{R}^n \mid \langle x, y \rangle = 0 \text{ for all } y \in V\}$$

is the orthogonal complement.

The vector product on $\mathbb{R}^3$ is given by

$$\begin{pmatrix} x^1 \\ x^2 \\ x^3 \end{pmatrix} \times \begin{pmatrix} y^1 \\ y^2 \\ y^3 \end{pmatrix} = \begin{pmatrix} x^2y^3 - x^3y^2 \\ -x^1y^3 + x^3y^1 \\ x^1y^2 - x^2y^1 \end{pmatrix}.$$

For the real part of a complex number z we write $\Re(z)$ and for the natural logarithm of a positive real number x we write $\ln(x)$.

A *smooth* map denotes one that is infinitely often differentiable. For the differential or its Jacobian matrix at a point p we write

$$D_pF = \begin{pmatrix} \frac{\partial F^1}{\partial x^1}(p) & \cdots & \frac{\partial F^1}{\partial x^n}(p) \\ \vdots & & \vdots \\ \frac{\partial F^m}{\partial x^1}(p) & \cdots & \frac{\partial F^m}{\partial x^n}(p) \end{pmatrix},$$

where $F = (F^1, \ldots, F^m)^\top : \mathbb{R}^n \to \mathbb{R}^m$. For functions $f : \mathbb{R}^n \to \mathbb{R}$ specifically

$$\operatorname{grad} f = Df^\top = \left(\frac{\partial f}{\partial x^1}, \ldots, \frac{\partial f}{\partial x^n}\right)^\top$$

denotes the gradient.

The group of invertible real $n \times n$ matrices is denoted by $\mathrm{GL}(n)$, the subgroup of orthogonal matrices by $\mathrm{O}(n)$:

$$\mathrm{O}(n) = \{A \in \mathrm{GL}(n) \mid A^\top A = \mathrm{Id}\},$$

and the subgroup of special orthogonal matrices by $\mathrm{SO}(n)$:

$$\mathrm{SO}(n) = \{A \in \mathrm{O}(n) \mid \det(A) = 1\}.$$

Here $A^\top$ denotes the transpose of A.

1 Euclidean geometry

We familiarise ourselves with the axioms of Euclidean geometry in the plane and derive some geometric implications, among them the existence of the parallel line. We briefly discuss the historical importance of the parallel axiom. The existence of the Cartesian model shows the consistency of the Euclidean axioms. The Cartesian model is used to investigate Euclidean trigonometry.

1.1 The axiomatic approach

Geometry is one of the oldest of all sciences. Remarkable geometric knowledge was already present in the advanced oriental cultures of the fifth–third centuries BCE. Practical problems from metrology, architecture, astronomy and navigation were considered on an abstract level and led to geometric laws. For instance, the Egyptians used the formula for the area of a triangle

$$\text{area} = \frac{\text{length of base line} \times \text{height}}{2}$$

and the approximate formula for the area of a circle

$$\text{area} = \left(\text{diameter} - \frac{\text{diameter}}{9}\right)^2.$$

The latter corresponds to an approximation of π by $\frac{256}{81} \approx 3.1605$. No difference was made between exact and approximate formulae in principle. Mathematical knowledge was there in the form of laws, justifications or proofs were not given.

This changed in Greece between 350 and 200 BCE. Aspects of usefulness were then superseded by the desire for understanding. Mathematicians not only wanted to know certain laws, but also why they hold. This was the starting point of the axiomatisation of geometry. At first only a few intuitively evident axioms were laid down, from which it was thought that everything else could be derived logically in a rigorous manner. In what follows we will become acquainted with the axioms of Euclidean plane geometry and go through some rather simple implications as illustrations of the axiomatic proof. We will mainly follow the formulation of the axioms presented by Hilbert in [13].

The axioms can be classified into five groups. We begin with the incidence axioms. To formulate these, we need two sets $\mathscr{P}$ and $\mathscr{G}$, whose elements we call points and straight lines respectively. Further assume that for every point $p \in \mathscr{P}$ and every straight line $L \in \mathscr{G}$ the statement "p is contained in L", in symbols "$p \in L$", is either true or false. Note that the symbol "$\in$" does not denote a set-theoretic inclusion in this case, since the straight lines L are for now not sets, but abstract elements of $\mathscr{G}$. We nevertheless want to use this suggestive notation. Let us now move on to the first axioms.

Incidence axioms These axioms make some statements about the containedness of points in straight lines.

Axiom I$_1$ *For any two points there exists a straight line that goes through both of them,*

$$\forall p, q \in \mathscr{P} \quad \exists L \in \mathscr{G}: \quad p \in L \text{ and } q \in L.$$

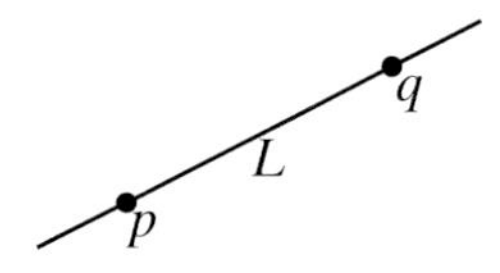

Axiom I$_2$ *There is at most one straight line through any two distinct points,*

$$\forall p, q \in \mathscr{P},\ p \neq q, \quad \forall L, M \in \mathscr{G},\ p \in L,\ q \in L,\ p \in M,\ q \in M: \quad L = M.$$

For any two distinct points p and q, there is by those first two axioms exactly one straight line that goes through both of them, we will from now on denote it by $L(p,q)$.

Axiom I$_3$ *Every straight line contains at least two distinct points,*

$$\forall L \in \mathscr{G} \quad \exists p, q \in \mathscr{P},\ p \neq q: \quad p \in L \text{ and } q \in L.$$

Axiom I$_4$ *There exist three points that do not lie on a straight line,*

$$\exists p, q, r \in \mathscr{P}: \quad \nexists L \in \mathscr{G} \text{ with } p \in L,\ q \in L,\ r \in L.$$

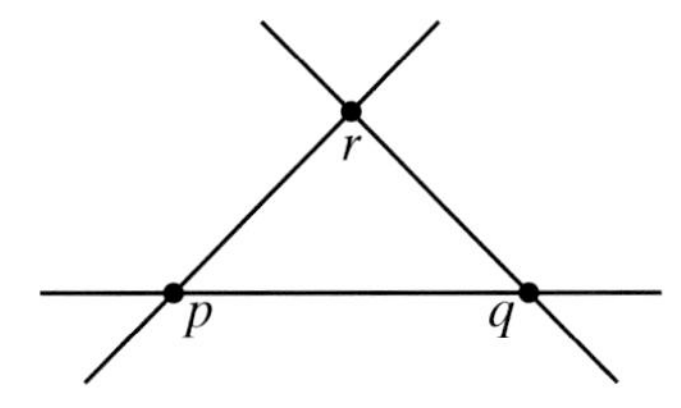

Axiom I$_4$ expresses that our geometry has at least two dimensions.

Axioms of order For the formulation of these axioms we need, in addition to the notions of $\mathscr{P}$, $\mathscr{G}$ and $\in$, that for every triple (p, q, r) of points the statement "q lies between p and r" must be either true or false. The following axioms must be satisfied.

AXIOM A$_1$ *If q lies between p and r, then p, q and r are three pairwise distinct points on a straight line.*

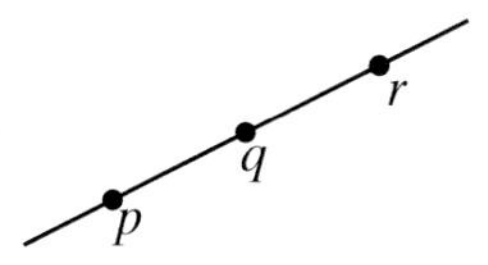

AXIOM A$_2$ *If q lies between p and r, then q lies between r and p.*

For two points p and q we call the set of all points that lie between p and q the ***line segment*** from p to q and write $\overline{pq}$. Axiom A$_2$ therefore implies $\overline{pr} = \overline{rp}$.

AXIOM A$_3$ *For any two distinct points p and q there exists a point r, such that q lies between p and r.*

Attention This axiom does not say that for any two given points, there exists another point between them. We will first have to *prove* this, see theorem 1.1.1.

AXIOM A$_4$ *Given three points, at most one of them lies between the two others.*

If ***two straight lines*** L and M have a point p in common, $p \in L$ and $p \in M$, then we sometimes say that L and M ***intersect***, in symbols $L \cap M \neq \emptyset$. We say that a ***line segment*** $\overline{pr}$ and a ***straight line*** L ***intersect*** if there exists a point q with $q \in L$ between p and r.

AXIOM A$_5$ *Let p, q and r be three points that do not lie on a straight line, let L be a straight line that does not contain any of these three points. If L intersects the line segment $\overline{pq}$, then L intersects precisely one of the other two line segments $\overline{pr}$ or $\overline{qr}$.*

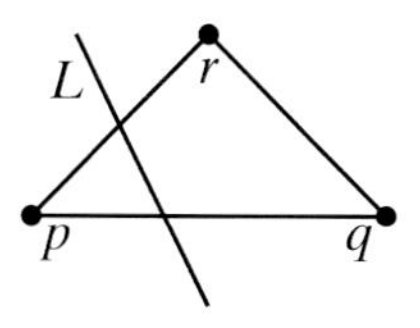

This means that a straight line which enters a triangle must leave it through one of the other two sides. It also illustrates that our geometry does not have more than two dimensions. In three dimensions axiom A$_5$ would not be valid:

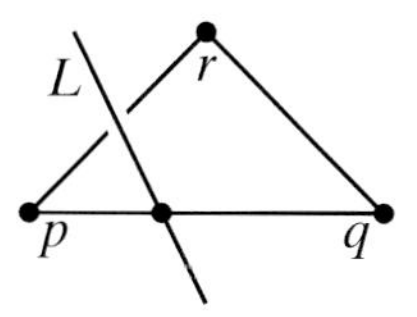

Let us now prove a first theorem, using the axioms stated so far.

Theorem 1.1.1 *For any two distinct points p and q there exists a point r which lies between p and q, i.e. the line segment $\overline{pq}$ is not empty.*

Proof Let p and q be two points. By axiom I_4 there exists a point s that does not lie on the straight line $L(p,q)$. By axiom A_3 there is a point t such that s lies between p and t. Another application of axiom A_3 gives a point u such that q lies between t and u. The straight line $L := L(s,u)$ intersects the line segment $\overline{pt}$ at s.

The point t does not lie on the straight line $L(p,q)$, since otherwise s would by axiom A_1 also lie on that straight line, contradicting the choice of s. We can therefore apply axiom A_5 to the straight line L and the three points p, q and t. As L intersects the line segment $\overline{pt}$, L must by axiom A_5 also intersect one of the two line segments $\overline{pq}$ or $\overline{tq}$, unless it contains one of the three points p, q or t.

First case *L contains p or t.*

Then L agrees with the straight line $L(p,t)$ by axiom I_2. Hence u lies on $L(p,t)$ and axiom A_1 implies that q lies on $L(p,t)$ as well. Hence p, q and t do lie on a straight line, a contradiction.

Second case *L contains q.*

Several applications of axiom I_2 show that the points s, u, q, t and p lie on a straight line, a contradiction.

Third case *L intersects the line segment $\overline{tq}$ at a point v.*

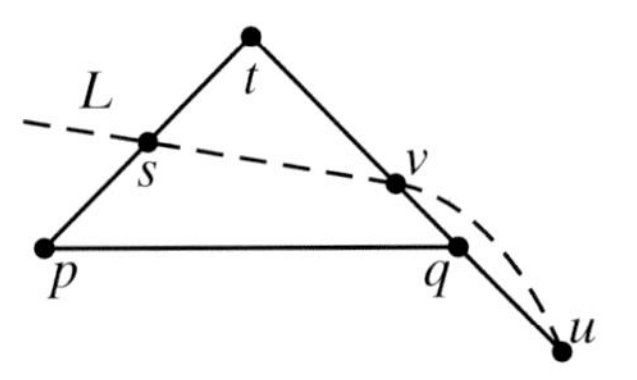

Then L and $L(t,q)$ have the two points u and v in common. If $u = v$, then u would lie between t and q while q lies between t and u, a contradiction to axiom A_4. Thus L and $L(t,q)$ have two distinct points in common, and axiom I_2 implies that $L = L(t,q)$. But then both s and p lie on L, i.e. p, q and t lie on a straight line, a contradiction.

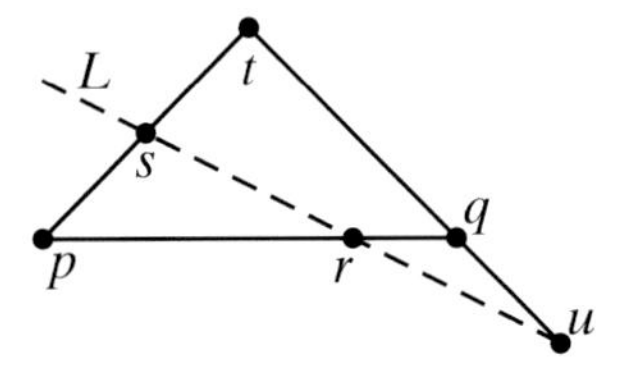

Thus L and $\overline{pq}$ must intersect at a point r. In particular, $\overline{pq}$ cannot be empty. □

Exercise 1.1 Let p, q, r and s be points on a straight line. Show that if q lies between p and s and if additionally r lies between q and s, then r lies between p and s as well.

Using this exercise and theorem 1.1.1, it is now easy to solve the following exercise.

Exercise 1.2 Show the following: there is an infinite number of distinct points between two points.

Definition 1.1.2 Let L be a straight line, $p \in L$. Let q and r be two points on L, neither of which is p. We say that q and r ***lie on the same side of the point*** p if p does not lie between q and r.

Exercise 1.3 Let L be a straight line and let $p \in L$ be a point on L. Show that the relation "q_1 lies on the same side of p as q_2" defines an equivalence relation on the set $\{q \in L \mid q \neq p\}$.

An equivalence class of points on L that do not equal p can then be referred to as a ***side of*** p ***on*** L.

Exercise 1.4 Show that there are exactly two sides of p on L.

Definition 1.1.3 Let L be a straight line and let p and q be two points that do not lie on L. We say that p and q ***lie on the same side of a straight line*** L if the line segment $\overline{pq}$ does not intersect the straight line L.

Exercise 1.5 Let L be a straight line. Show that the relation "q_1 lies on the same side of L as q_2" defines an equivalence relation on the set $\{q \mid q \notin L\}$.

Again we can call an equivalence class of points not on L a ***side of*** L.

Exercise 1.6 Show that there are exactly two sides of L.

Congruence axioms To formulate the third group of axioms, the congruence axioms, we need in addition to the previous notions that for every pair $(\overline{pq}, \overline{p_1q_1})$ of line segments the formal statement "$\overline{pq}$ is congruent to $\overline{p_1q_1}$" is either true or false.

Axiom K_1 (reproduction of lengths) *Let $\overline{pq}$ be a line segment, let L_1 be a straight line, let $p_1, r_1 \in L_1$, $r_1 \neq p_1$. Then there is a point $q_1 \in L_1$ on the same side of p_1 as r_1 such that $\overline{pq}$ is congruent to $\overline{p_1q_1}$.*

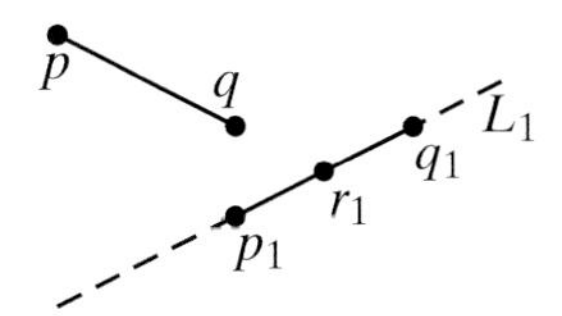

In this axiom only the existence of a congruent line segment is required. Its uniqueness needs to be proved later with the aid of other axioms.

AXIOM K_2 *If the line segments* $\overline{p_1q_1}$ *and* $\overline{p_2q_2}$ *are both congruent to the line segment* $\overline{pq}$, *then* $\overline{p_1q_1}$ *is congruent to* $\overline{p_2q_2}$ *as well.*

Four more congruence axioms will follow. We can nevertheless already prove a first implication.

Lemma 1.1.4 *The congruence of line segments defines an equivalence relation on the set of line segments.*

Proof (a) Let $\overline{pq}$ be a line segment. We show that $\overline{pq}$ is congruent to itself. Let L be a straight line that contains p, $p \in L$. By axiom K_1 there exists a point r on L such that $\overline{pq}$ and $\overline{pr}$ are congruent. Then $\overline{p_1q_1} := \overline{p_2q_2} := \overline{pq}$ is congruent to $\overline{pr}$ and hence $\overline{pq}$ is by axiom K_2 congruent to itself.

(b) (symmetry) Let $\overline{pq}$ be congruent to $\overline{p_1q_1}$. We show that then $\overline{p_1q_1}$ is congruent to $\overline{pq}$. It follows from (a) that $\overline{p_1q_1}$ is congruent to $\overline{p_1q_1}$. Axiom K_2 now gives that $\overline{p_1q_1}$ is congruent to $\overline{pq}$.

(c) (transitivity) If $\overline{p_1q_1}$ is congruent to $\overline{p_2q_2}$ and $\overline{p_2q_2}$ is congruent to $\overline{p_3q_3}$, then we need to show that $\overline{p_1q_1}$ is congruent to $\overline{p_3q_3}$. This follows directly from axiom K_2 together with (b). □

We will from now on sometimes denote "$\overline{p_1q_1}$ is congruent to $\overline{p_2q_2}$" by "$\overline{p_1q_1} \equiv \overline{p_2q_2}$".

AXIOM K_3 (additivity of line segments) *Let* L *and* L_1 *be straight lines, let* $p,q,r \in L$ *be three pairwise distinct points on* L *and* $p_1,q_1,r_1 \in L_1$ *likewise on* L_1. *Assume that the line segments* $\overline{pq}$ *and* $\overline{qr}$ *do not have any common points,* $\overline{pq} \cap \overline{qr} = \emptyset$. *Analogously let* $\overline{p_1q_1} \cap \overline{q_1r_1} = \emptyset$.

If now $\overline{pq} \equiv \overline{p_1q_1}$ *and* $\overline{qr} \equiv \overline{q_1r_1}$, *then* $\overline{pr} \equiv \overline{p_1r_1}$.

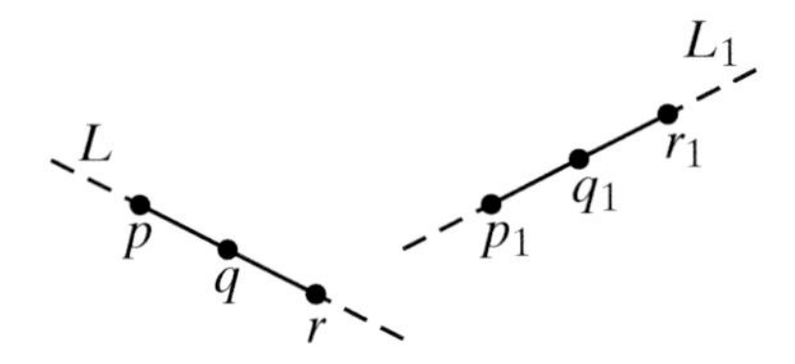

We need the concept of the *angle* for the formulation of the three other congruence axioms.

Definition 1.1.5 An ***angle*** is an equivalence class of triples of points p, q and r that do not lie on a straight line, where two triples (p,q,r) and (p_1,q_1,r_1) are equivalent if

(i) $q = q_1$,
(ii) $L(p,q) = L(p_1,q)$ and p and p_1 lie on the same side of q,
(iii) $L(r,q) = L(r_1,q)$ and r and r_1 lie on the same side of q,

or if

(i) $q = q_1$,
(ii) $L(p,q) = L(r_1,q)$ and p and r_1 lie on the same side of q,
(iii) $L(r,q) = L(p_1,q)$ and r and p_1 lie on the same side of q.

For the equivalence class of (p,q,r) we write $\angle(p,q,r)$. The point q is then called the ***vertex*** of the angle $\angle(p,q,r)$.

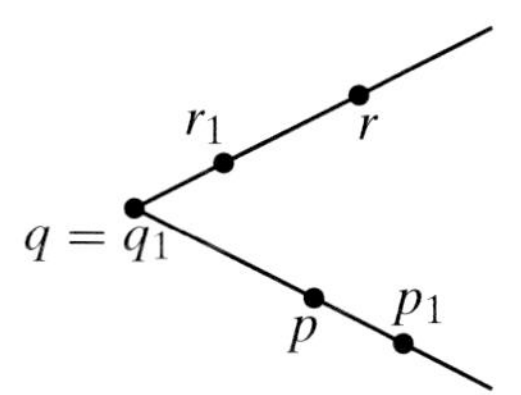

We now additionally require that for any two angles $\angle(p,q,r)$ and $\angle(p_1,q_1,r_1)$ the formal statement "$\angle(p,q,r)$ is congruent to $\angle(p_1,q_1,r_1)$" is either true or false. Again, we write "$\angle(p,q,r) \equiv \angle(p_1,q_1,r_1)$" if $\angle(p,q,r)$ is congruent to $\angle(p_1,q_1,r_1)$.

AXIOM K_4 *The congruence of angles induces an equivalence relation on the set of angles.*

AXIOM K_5 (reproduction of angles) *Let p, q, r be points that do not lie on a straight line, and let p_1, q_1, s_1 be another set of points that do not lie on a straight line. Then there exists a point r_1 on the same side of $L(p_1,q_1)$ as s_1 such that the angle $\angle(p_1,q_1,r_1)$ is congruent to the angle $\angle(p,q,r)$.*

If r_2 is another point with the same properties as r_1, i.e. r_2 also lies on the same side of $L(p_1,q_1)$ as s_1 and if $\angle(p_1,q_1,r_2) \equiv \angle(p,q,r)$, then $\angle(p_1,q_1,r_1) = \angle(p_1,q_1,r_2)$.

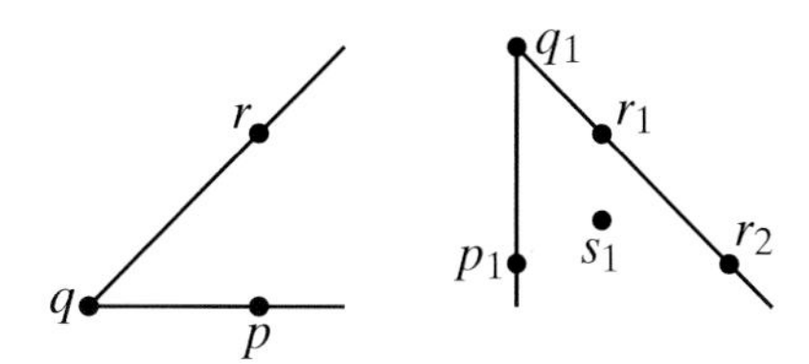

Axiom K_5 says that we can reproduce a given angle in a unique way if we are given the vertex, one line adjacent to the angle and the side of the other line next to the angle.

The last congruence axiom relates the congruence of line segments to that of angles. Until now the two notions of congruence existed entirely separately.

AXIOM K_6 *Let (p,q,r) be a triple of points that do not lie on a straight line and (p_1,q_1,r_1) likewise. If $\overline{pq} \equiv \overline{p_1q_1}$, $\overline{pr} \equiv \overline{p_1r_1}$ and $\angle(q,p,r) \equiv \angle(q_1,p_1,r_1)$, then*

$$\angle(p,q,r) \equiv \angle(p_1,q_1,r_1).$$

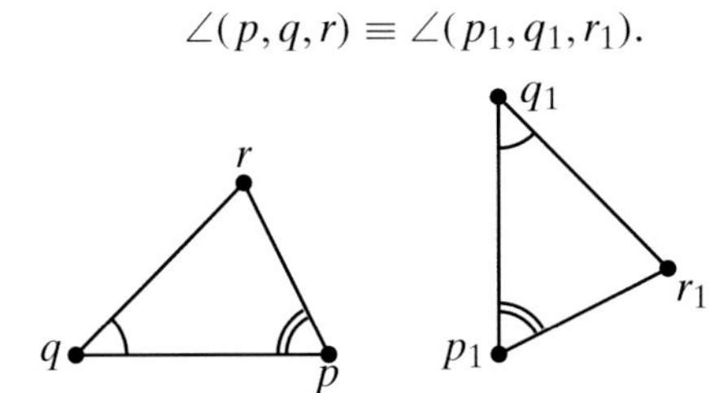

Let us make some inferences using the axioms introduced so far. We will first extend the statement of the last axiom.

Theorem 1.1.6 *Let (p,q,r) be a triple of points that do not lie on a straight line, (p_1,q_1,r_1) likewise. If $\overline{pq} \equiv \overline{p_1q_1}$, $\overline{pr} \equiv \overline{p_1r_1}$ and $\angle(q,p,r) \equiv \angle(q_1,p_1,r_1)$, then*

$$\angle(p,q,r) \equiv \angle(p_1,q_1,r_1), \qquad \angle(p,r,q) \equiv \angle(p_1,r_1,q_1), \qquad \overline{qr} \equiv \overline{q_1r_1}.$$

Proof The angle congruences follow directly from axiom K_6, in the second case after renaming the variables. It remains to show that $\overline{qr} \equiv \overline{q_1r_1}$. By axiom K_1 we can find a point s_1 on the straight line $L(q_1,r_1)$ which is on the same side as r_1 and satisfies $\overline{qr} \equiv \overline{q_1s_1}$.

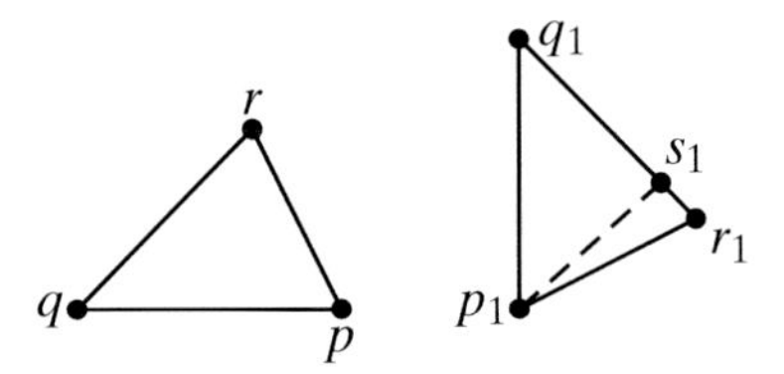

We apply axiom K_6 to (p,q,r) and (p_1,q_1,s_1) and conclude that

$$\angle(q,p,r) \equiv \angle(q_1,p_1,s_1).$$

On the other hand we have $\angle(q,p,r) \equiv \angle(q_1,p_1,r_1)$, and by the uniqueness from axiom[1] K_5

[1] For an application of the uniqueness statement from axiom K_5 it must be ensured that r_1 and s_1 lie on the same side of $L(p_1,q_1)$. This is left as an exercise for the reader.

$$\angle(q_1, p_1, s_1) = \angle(q_1, p_1, r_1).$$

If we now had $r_1 \neq s_1$, then we would conclude that p_1 and q_1 both lie on the straight line $L(r_1, s_1)$, i.e. p_1, q_1 and r_1 would lie on a straight line, which contradicts the assumption.

Thus $r_1 = s_1$ and hence $\overline{qr} \equiv \overline{q_1 r_1}$. □

Theorem 1.1.7 (Congruence of adjacent angles) *Suppose that the pairwise distinct points p, q and s lie on a straight line L, while $r \notin L$. Analogously, let $p_1, q_1, s_1 \in L_1$ be pairwise distinct, $r_1 \notin L_1$. If $\angle(p, q, r)$ and $\angle(p_1, q_1, r_1)$ are congruent, then the same is true for $\angle(s, q, r)$ and $\angle(s_1, q_1, r_1)$.*

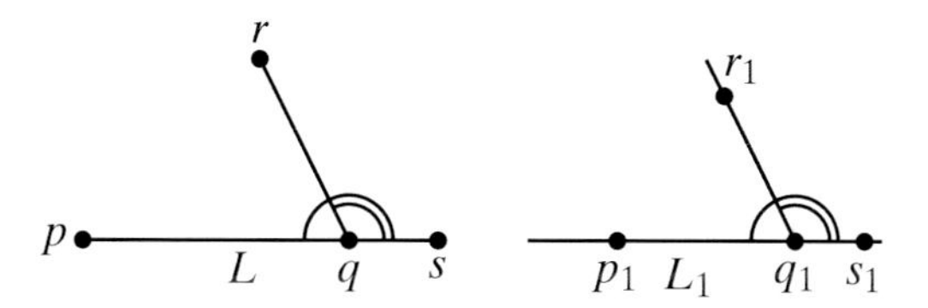

The angle $\angle(s, q, r)$ is sometimes called the ***adjacent angle*** of $\angle(p, q, r)$. The theorem thus says that adjacent angles of congruent angles are congruent as well.

Proof The points p_1, r_1 and s_1 can by axiom K_1 be assumed to have been chosen in such a way that $\overline{pq} \equiv \overline{p_1 q_1}$, $\overline{rq} \equiv \overline{r_1 q_1}$ and $\overline{sq} \equiv \overline{s_1 q_1}$.

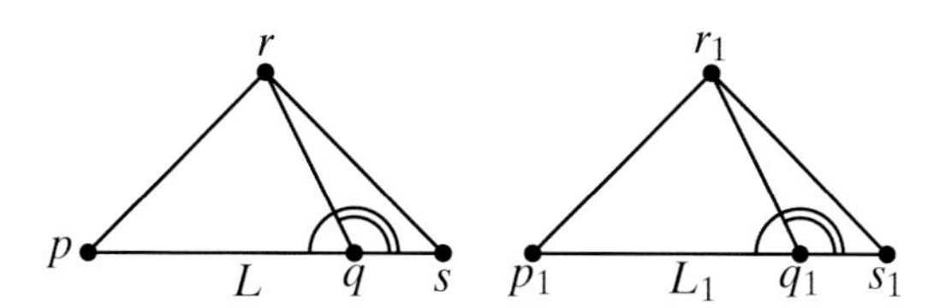

From theorem 1.1.6, applied to (p, q, r) and (p_1, q_1, r_1), it follows that $\overline{pr} \equiv \overline{p_1 r_1}$. By axiom K_3 we then have $\overline{ps} \equiv \overline{p_1 s_1}$. Applying theorem 1.1.6 once again, this time to (r, p, s) and (r_1, p_1, s_1), we obtain $\overline{rs} \equiv \overline{r_1 s_1}$ and $\angle(q, s, r) \equiv \angle(q_1, s_1, r_1)$. Axiom K_6 then says for (q, s, r) and (q_1, s_1, r_1) that $\angle(s, q, r) \equiv \angle(s_1, q_1, r_1)$. □

Theorem 1.1.8 (Congruence of vertical angles) *Let L and M be two distinct straight lines that intersect at p. Let $r, q \in L$ lie on two distinct sides of p and let $s, t \in M$ lie on two distinct sides of p as well. Then*

$$\angle(q, p, s) \equiv \angle(r, p, t).$$

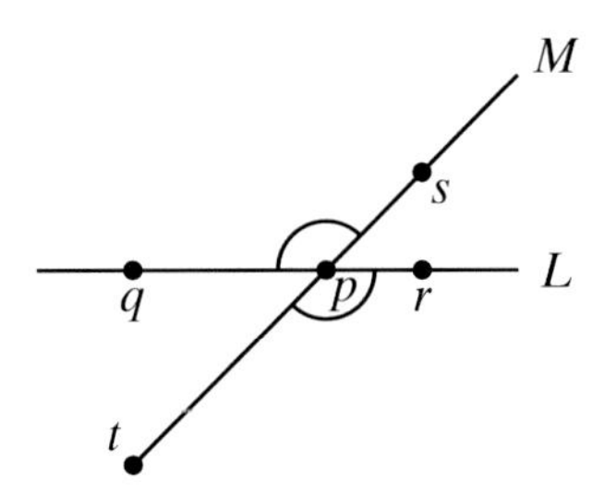

Proof Both $\angle(q,p,s)$ and $\angle(r,p,t)$, are adjacent angles of $\angle(q,p,t)$. The angle $\angle(q,p,t)$ is congruent to itself by axiom K4. The claim therefore follows by theorem 1.1.7. □

After these preparations we now come to the first truly interesting geometric theorem.

Theorem 1.1.9 (Existence of a parallel) *Let L be a straight line, p a point, $p \notin L$. There then exists a straight line M that contains p and that does not intersect L.*

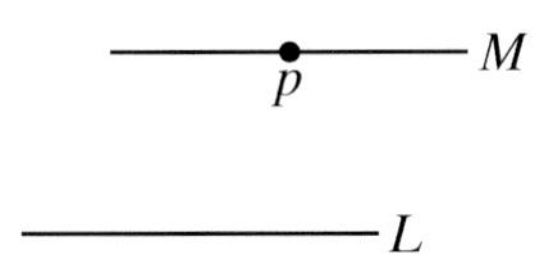

We then say that M is a ***parallel*** to L through p. The theorem says that such parallels always exist.

Proof Let L be a straight line and p a point that does not lie on L. We will first construct the line M and then show that it has the desired properties.

For the construction we choose a point $q \in L$ and add the straight line $N := L(p,q)$. We choose another point $r \in L$, $r \neq q$. Then $r \notin N$, since otherwise $p \in L(p,q) = L(q,r) = L$. We reproduce angle $\angle(r,q,p)$ as in axiom K5 on the straight line N at the point p, i.e. we find points $s \in N$ and $t \notin N$ on the same side of N as r, such that the angle $\angle(t,p,s)$ is congruent to the angle $\angle(r,q,p)$. We now set $M := L(p,t)$.

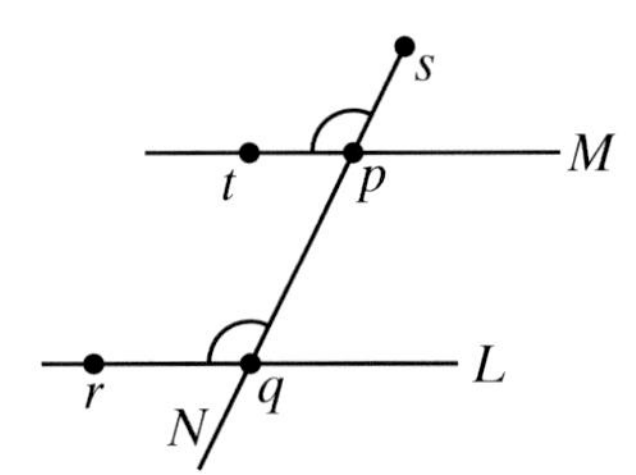

It remains to show that L and M do not intersect. Suppose that L and M did intersect at a point u. We restrict ourselves to the case that u lies on the same side of N as r and t. The other case is dealt with in a similar way.

We reproduce the line segment $\overline{uq}$ on the straight line M as in axiom K_1, beginning at the point p, and we do *not* do this on the same side as u. This means that $\overline{uq} \equiv \overline{pv}$, with a point $v \in M$ such that p lies between u and v.

The angles $\angle(u,p,s)$ and $\angle(q,p,v)$ are congruent by theorem 1.1.8. Hence the angles $\angle(u,q,p)$ and $\angle(q,p,v)$ are congruent as well. We apply axiom K_6 to (u,q,p) and (v,p,q), and see that $\angle(q,p,u) \equiv \angle(p,q,v)$.

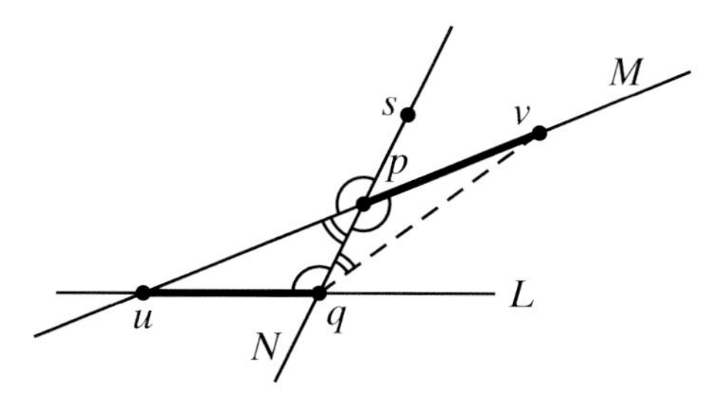

But $\angle(q,p,u)$ is an adjacent angle of $\angle(u,p,s)$, so by theorem 1.1.7 and by the uniqueness of the reproduction of angles $\angle(p,q,v)$ must be an adjacent angle of $\angle(u,q,p)$. Hence v lies on the straight line L, and thus the same is true for p, which contradicts the assumption. □

Parallel axiom We have seen that the existence of parallels can be proved using the axioms introduced so far. What about their uniqueness? This is the subject of the parallel axiom.

AXIOM P (parallel axiom) *Let L be a straight line, p a point, $p \notin L$. Then there is at most one straight line that contains p and does not intersect L.*

There has been controversy about the necessity of this axiom for thousands of years. Many mathematicians believed that uniqueness could, as existence, be deduced from the other axioms. There were many proof attempts. Carl Friedrich Gauss (1777–1855) was probably the first person to truly believe in the independence of the parallel axiom. However, he never published his views about it. The debate was ended by the Russian Nikolai Iwanowitsch Lobatschewski (1792–1856) and by the Hungarian Janos Bolyai (1802–1860), who, independently of each other, found a geometry that satisfies all axioms except for the parallel axiom. Hence the parallel axiom cannot be derived from the other axioms. Bolyai's father, a teacher of mathematics himself, was so worried about his son's result that he wrote a letter to Gauss, asking him for his opinion. In his answer Gauss was enthusiastic about the work of the younger Bolyai, but added the remark that he could not praise him, since this would be self-praise as he, Gauss, had known this for many years. Janos Bolyai never again published work on mathematics.

The afore-mentioned geometry is hyperbolic geometry, also known as non-Euclidean geometry. We will cover it in more detail later.

Completeness axioms When real numbers are introduced in analysis, one becomes acquainted with the completeness axioms, which distinguish the real numbers from the rational numbers and other ordered number fields. To determine the Euclidean geometry completely, we need the appropriate axioms.

AXIOM $\mathbf{V_1}$ (Archimedes's axiom) *Let $\overline{pq}$ and $\overline{rs}$ be two line segments. Then there exists a natural number n such that the line segment $\overline{r_1 s_n}$, which results from the n-fold reproduction of the line segment $\overline{rs}$ on the straight line $L(p,q)$, starting at p in direction q, contains the line segment $\overline{pq}$.*

$p = r_1$ $s_2 = r_3$ q

$s_1 = r_2$ r_n s_n

Before formulating the last axiom, we summarise those notions that we need for the axiomatic formulation of Euclidean geometry. We have

- a set $\mathscr{P}$, whose elements are called points,
- a set $\mathscr{G}$, whose elements we call straight lines,
- a relation $\in$ between $\mathscr{P}$ and $\mathscr{G}$,
- a three-figure relation "between" on $\mathscr{P}$,
- a relation $\equiv_1$ on the set of straight lines,
- a relation $\equiv_2$ on the set of angles.

One could now formally define a Euclidean geometry as a 6-tuple $(\mathscr{P}, \mathscr{G}, \in,$ between$, \equiv_1, \equiv_2)$ that consists of such notions, and that satisfies axioms I_1–I_4, A_1–A_5, K_1–K_6, P, V_1 as well as V_2, which still is to be formulated.

An ***extension*** of our geometry is a second 6-tuple $(\mathscr{P}', \mathscr{G}', \in',$ between$', \equiv_1', \equiv_2')$ such that $\mathscr{P} \subset \mathscr{P}'$, $\mathscr{G} \subset \mathscr{G}'$ and the relations $\in'$, between$'$, $\equiv_1'$ as well as $\equiv_2'$ agree with the corresponding relations $\in, \ldots$ after a restriction to $\mathscr{P}$ and $\mathscr{G}$.

AXIOM $\mathbf{V_2}$ (maximality) *Let $(\mathscr{P}', \mathscr{G}', \in',$ between$', \equiv_1', \equiv_2')$ be an extension of our geometry. Then $\mathscr{P}' = \mathscr{P}$ and $\mathscr{G}' = \mathscr{G}$.*

Our geometry is therefore assumed to be non-extensible, i.e. maximal.

For a further discussion of axiomatic geometry the reader is referred to Hilbert's book [13]. Proofs of numerous geometric theorems, a discussion of the uniqueness of Euclidean geometry and of the necessity of Archimedes's axiom, straightedge-and-compass constructions and spatial Euclidean geometry can be found there. On a final note, some axioms can be weakened slightly, e.g. axioms K_4 and V_2.

1.2 The Cartesian model

While the axiomatic method ensures logical clarity in the structural layout of geometry, it is also annoyingly clumsy. The proofs of even relatively simple geometric facts can easily become quite laborious. In addition, dealing with geometric objects that are not composed of line segments, circular lines, and so forth is relatively involved.

We therefore now follow the ideas of the Frenchman René Descartes (1596–1650) and characterise points by coordinates, which give the positions of the points in the plane. This allows us to use methods from algebra and infinitesimal calculus in geometry, extending our mathematical toolkit considerably.

We hence make the definition

$$\mathscr{P} := \mathbb{R}^2.$$

Straight lines are defined as sets of points of the form

$$L = L_{p,v} := \{x \in \mathbb{R}^2 \mid x = p + t \cdot v,\ t \in \mathbb{R}\},$$

where $p, v \in \mathbb{R}^2$, $v \neq 0$, are fixed. The set of straight lines is then

$$\mathscr{G} := \{L_{p,v} \mid p, v \in \mathbb{R}^2, v \neq 0\}.$$

We say that a point p is ***contained*** in a straight line L if $p \in L$ in the set-theoretic sense.

Exercise 1.7 Verify the validity of axioms I_1–I_4.

A point $q \in \mathbb{R}^2$ lies ***between*** p and $r \in \mathbb{R}^2$, $p \neq r$, if there exists a $t \in (0, 1)$ such that $q = t \cdot p + (1 - t) \cdot r$.

Exercise 1.8 Show that axioms A_1–A_4 are valid.

It remains to define the congruence relations. This is done using a suitable group of maps, which acts on the set of points.

Definition 1.2.1 Let $A \in \mathrm{O}(n)$ be an orthogonal matrix, i.e. it satisfies $A^\top \cdot A = \mathrm{Id}$, where $A^\top$ is the transpose of A. Let $b \in \mathbb{R}^n$.

Then the map

$$F_{A,b} : \mathbb{R}^n \to \mathbb{R}^n, \qquad F_{A,b}(x) = Ax + b,$$

is called a ***Euclidean motion***. The vector b is sometimes called the ***translational part***. For a fixed dimension n, we call the set of all Euclidean motions,

$$\mathrm{E}(n) := \{F_{A,b} \mid A \in \mathrm{O}(n), b \in \mathbb{R}^n\},$$

the ***Euclidean motion group***.

Exercise 1.9 Show the following: composition of maps turns $\mathrm{E}(n)$ into a group with neutral element Id. Products and inverses are given by $F_{A,b} \circ F_{B,c} = F_{AB,Ac+b}$ and $F_{A,b}^{-1} = F_{A^{-1},-A^{-1}b}$ respectively.

Exercise 1.10 Show that axiom A_5 is valid.

Hint An application of a Euclidean motion maps the straight line L to the x-axis $L' = \{(x,0)^\top \mid x \in \mathbb{R}\}$. Then argue that two points $(x,y)^\top$ and $(x',y')^\top$ from $\mathbb{R}^2 - L'$ lie on the same side of L' if and only if y and y' are both positive or both negative.

Definition 1.2.2 Two line segments $\overline{pq}$ and $\overline{rs}$ are said to be ***congruent*** if there exists a Euclidean motion $F \in \mathrm{E}(2)$ such that

$$\overline{F(p)F(q)} = \overline{rs}.$$

Analogously, the angle $\angle(p,q,r)$ is ***congruent*** to $\angle(p_1,q_1,r_1)$ if there is an $F \in \mathrm{E}(2)$ such that

$$\angle(F(p),F(q),F(r)) = \angle(p_1,q_1,r_1).$$

Exercise 1.11 Show the validity of axioms K_1–K_6.

Exercise 1.12 Show the validity of the parallel axiom P.

Exercise 1.13 Show the validity of Archimedes's axiom V_1.

The proof of the maximality axiom V_2 requires more thought than those of the other axioms. We therefore want to show it here.

Theorem 1.2.3 *With the definitions made above the maximality axiom V_2 is valid.*

Proof Let $\mathcal{P}' \supset \mathcal{P}$ and $\mathcal{G}' \supset \mathcal{G}$ be respectively the set of points and the set of straight lines of an extension of our Cartesian model of Euclidean geometry. We will call the points from $\mathcal{P}$ *old points*, and those from $\mathcal{P}' - \mathcal{P}$ *new points*. We use analogous notation when dealing with straight lines. It needs to be shown that there cannot be new points and straight lines at all, $\mathscr{P} = \mathscr{P}'$ and $\mathscr{G} = \mathscr{G}'$.

The proof is accomplished in three steps. We first show that the old straight lines do not contain new points, then that there are no new points, and finally that there are no new straight lines.

(a) *Old straight lines do not contain new points, i.e. for $L \in \mathscr{G}$ and $p \in \mathscr{P}'$ with $p \in' L$ we have $p \in \mathscr{P}$.*

One must not become confused at this point. According to our definition, old straight lines are sets of old points. In the *set-theoretic sense* old straight lines only contain old points anyway. Nevertheless, it is a priori possible that new points are contained in old straight lines for the extension $\in'$ of the set-theoretic inclusion $\in$. It now needs to be shown that this is not the case.

For this purpose let $L \in \mathscr{G}$ be an old straight line. Suppose that L contains a new point n. We choose an old point $p_1 \in L$. As L is of the form $L = L_{p,v}$, we can write $p_1 = p + t_1 v =: c(t_1)$, $t_1 \in \mathbb{R}$. If we reproduce the line segment $\overline{(0,0)^\top (0,1)^\top}$ sufficiently often on L, starting at p_1 in direction n, then we obtain by Archimedes's axiom a second point $q_1 \in L$ such that n lies between p_1 and q_1. Reproducing the line segment $\overline{(0,0)^\top (0,1)^\top}$ on an old straight line starting at an old point always gives another old point. Thus q_1 is an old point. We write $q_1 = c(s_1)$, $s_1 \in \mathbb{R}$. Without loss of generality let $t_1 < s_1$.

We now decompose the real numbers into two disjoint subsets, $\mathbb{R} = T \sqcup S$, where $T = \{t \in \mathbb{R} \mid c(t) \text{ lies on the same side of } n \text{ as } c(t_1)\}$ and $S = \{t \in \mathbb{R} \mid c(t) \text{ lies on the same side of } n \text{ as } c(s_1)\}$. This decomposition of the real numbers constitutes a Dedekind cut. By the completeness of $\mathbb{R}$ either the subset T must have a maximum or S a minimum. We deal with the case that T has a maximum t_2. The other case can be treated analogously.

We set $p_2 := c(t_2)$. As $t_2 \in T$, the points p_1 and p_2 must lie on the same side of n. By Archimedes's axiom, there is a natural number k such that k-fold reproduction of the line segment $\overline{p_2 n}$ on L from p_2 in direction q_1 contains the line segment $\overline{p_2 q_1}$. We set $p_3 := c(t_3)$ with $t_3 = t_2 + (s_1 - t_2)/k$. Then the line segment $\overline{p_2 p_3}$ has the property that its k-fold reproduction on L from p_2 in direction q_1 gives the line segment $\overline{p_2 q_1}$.

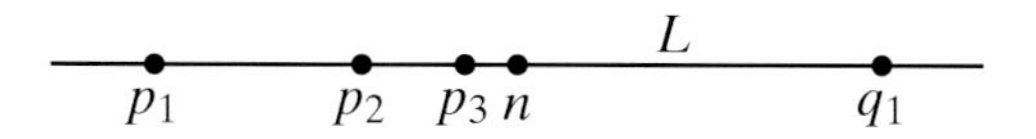

As the line segment $\overline{p_2 q_1}$ is contained in the k-fold reproduced line segment $\overline{p_2 n}$, the line segment $\overline{p_2 p_3}$ must be contained in $\overline{p_2 n}$. Thus p_3 lies between p_2 and n. It follows that $t_3 \in T$ and $t_3 > t_2$. This contradicts the maximality of t_2.

(b) *There are no new points, i.e. $\mathscr{P} = \mathscr{P}'$.*

Let $p \in \mathscr{P}'$ be a point. We choose an old point $q \in \mathscr{P}$ and consider the straight line L through p and q. We now choose three old points $r, s, t \in \mathscr{P}$ that do not lie on a straight line and none of which lies on L, such that L and the line segment $\overline{rs}$ intersect at q.

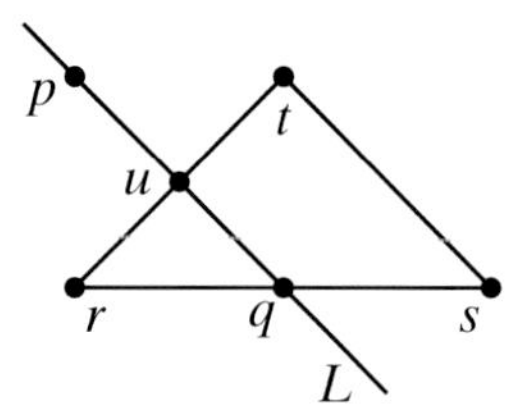

It follows from axiom A_5 that L intersects the line segment $\overline{rt}$ or $\overline{ts}$ at a point u. As the two old straight lines $L(r,t)$ and $L(s,t)$ do not contain any new points by (a), u must be a new point. Thus $L = L(u,q)$ is an old straight line and contains, again by (a), no new points. Hence p is an old point, $p \in \mathscr{P}$.

(c) *There are no new straight lines, i.e.* $\mathscr{G} = \mathscr{G}'$.

Let $L \in \mathscr{G}'$ be a straight line. It follows from axiom I_3 that L contains two distinct points p and q. They must be old points by (b). Hence $L = L(p,q)$ is an old straight line, $L \in \mathscr{G}$. $\square$

Exercise 1.14 Give another proof for part (a) of theorem 1.2.3 by considering Cauchy sequences instead of Dedekind cuts.

The axioms of Euclidean plane geometry are thus valid for the Cartesian model. In particular, we see that the axioms are *consistent*.

Doing geometry in the Cartesian model has the advantage that we now have the whole mathematical machinery of differential and integral calculus to hand. This makes the treatment of Euclidean trigonometry relatively easy.

We will as of now use angle brackets to denote the standard scalar product on $\mathbb{R}^n$:

$$\langle x, y\rangle = \sum_{i=1}^{n} x^i y^i$$

for $x = (x^1, \ldots, x^n)^\top, y = (y^1, \ldots, y^n)^\top \in \mathbb{R}^n$.

If x, y and z are three points in $\mathbb{R}^2$ which do not lie on a straight line, then $|\langle y-x, z-x\rangle| \le \|y-x\| \cdot \|z-x\|$ by the Cauchy–Schwarz inequality. Hence $|\langle y-x, z-x\rangle/(\|y-x\| \cdot \|z-x\|)| \le 1$. Since $\cos : [0,\pi] \to [-1,1]$ is bijective we can define their ***interior angle*** as the unique number $\gamma \in [0,\pi]$ such that $\cos(\gamma) = \langle y-x, z-x\rangle/(\|y-x\| \cdot \|z-x\|)$.

Exercise 1.15 Show that if x, y, z, and x', y', z' represent the same angle in the sense of definition 1.1.5, $\angle(y,x,z) = \angle(y',x',z')$, then they have the same interior angle, $\gamma = \gamma'$.

Exercise 1.16 Show that two angles are congruent if and only if they have the same interior angle.

This is analogous to the fact that two line segments are congruent if and only if they have equal length.

Theorem 1.2.4 (cosine rule for Euclidean geometry) *Let $p, q, r \in \mathbb{R}^2$. Let $a = \|p - q\|$, $b = \|p - r\|$ and $c = \|q - r\|$ be the sides of a triangle with vertices p, q and r. Let γ be the interior angle at the vertex p. Then*

$$c^2 = a^2 + b^2 - 2ab\cos(\gamma).$$

Proof Euclidean motions do not change the side and angle ratios. We can therefore after the application of a suitable Euclidean motion assume that $p = (0,0)^\top$, $q = (a,0)^\top$ and $r = (x,y)^\top$.

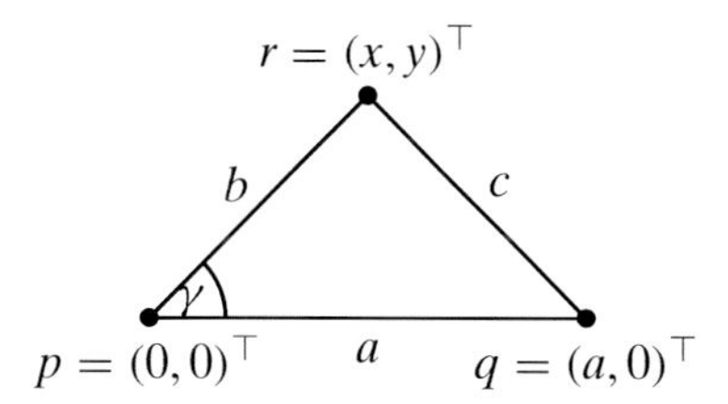

Hence

$$c^2 = (x-a)^2 + y^2 = a^2 - 2ax + x^2 + y^2 = a^2 + b^2 - 2ax.$$

Further

$$\cos(\gamma) = \frac{\langle q, r\rangle}{ab} = \frac{xa + y \cdot 0}{ab} = \frac{ax}{ab}$$

and thus $ax = ab\cos(\gamma)$. The claim follows. □

Corollary 1.2.5 (Pythagoras's theorem) *Using the notation from theorem* 1.2.4 *and letting the angle be a right angle, $\gamma = \pi/2$, the following equality holds:*

$$a^2 + b^2 = c^2.$$ □

Theorem 1.2.6 (sine rule for Euclidean geometry) *If β denotes the interior angle at vertex q and α the one at vertex r, then the following equality holds:*

$$\frac{a}{b} = \frac{\sin(\alpha)}{\sin(\beta)}.$$

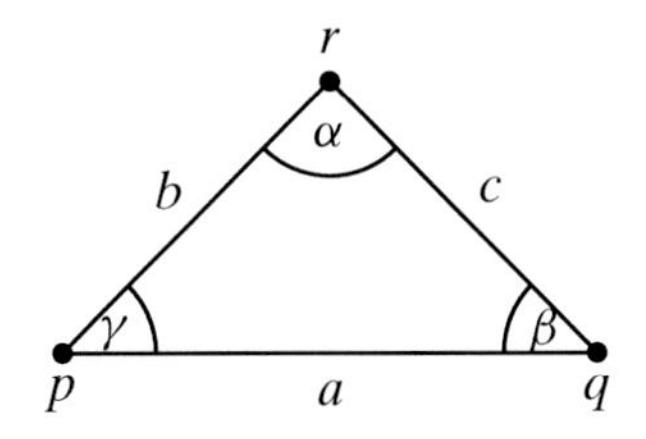

Proof By the cosine rule the angle at vertex r satisfies

$$-2bc\cos(\alpha) = a^2 - (b^2 + c^2)$$

and hence

$$4b^2c^2\cos^2(\alpha) = \left(-a^2 + b^2 + c^2\right)^2.$$

Analogously,

$$4a^2c^2\cos^2(\beta) = \left(a^2 - b^2 + c^2\right)^2.$$

We therefore obtain

$$\begin{aligned}
\frac{\sin^2(\alpha)}{\sin^2(\beta)} &= \frac{4a^2b^2c^2\left(1 - \cos^2(\alpha)\right)}{4a^2b^2c^2\left(1 - \cos^2(\beta)\right)} \\
&= \frac{4a^2b^2c^2 - a^2\left(-a^2 + b^2 + c^2\right)^2}{4a^2b^2c^2 - b^2\left(a^2 - b^2 + c^2\right)^2} \\
&= \frac{a^2}{b^2} \cdot \frac{4b^2c^2 - \left(a^4 + b^4 + c^4 - 2a^2b^2 - 2a^2c^2 + 2b^2c^2\right)}{4a^2c^2 - \left(a^4 + b^4 + c^4 - 2a^2b^2 + 2a^2c^2 - 2b^2c^2\right)} \\
&= \frac{a^2}{b^2}.
\end{aligned}$$

□

Theorem 1.2.7 (angle sum in the Euclidean triangle) *The sum of the interior angles in the Euclidean triangle satisfies*

$$\alpha + \beta + \gamma = \pi.$$

Proof (a) We first prove the statement for right-angle triangles. Let $\gamma = \pi/2$. Then

$$\sin(\alpha + \beta + \pi/2) = \cos(\alpha + \beta) = \cos(\alpha)\cos(\beta) - \sin(\alpha)\sin(\beta). \tag{1.1}$$

Using the cosine rule 1.2.4 and Pythagoras's theorem we obtain

$$\begin{aligned}\cos(\alpha)\cos(\beta) &= \frac{-a^2+b^2+c^2}{2bc}\cdot\frac{a^2-b^2+c^2}{2ac}\\ &= \frac{2b^2\cdot 2a^2}{4abc^2}\\ &= \frac{ab}{c^2}.\end{aligned} \tag{1.2}$$

The sine rule 1.2.6 together with the cosine rule and Pythagoras's theorem gives

$$\begin{aligned}\sin(\alpha)\sin(\beta) &= \frac{b}{a}\sin^2(\alpha)\\ &= \frac{b}{a}\cdot\left(1-\cos^2(\alpha)\right)\\ &= \frac{b}{a}\cdot\left(1-\frac{\left(-a^2+b^2+c^2\right)^2}{4b^2c^2}\right)\\ &= \frac{b}{a}\cdot\left(1-\frac{4b^4}{4b^2c^2}\right)\\ &= \frac{b}{a}\cdot\frac{c^2-b^2}{c^2}\\ &= \frac{ab}{c^2}.\end{aligned} \tag{1.3}$$

Substituting (1.2) and (1.3) into (1.1), we obtain $\sin(\alpha+\beta+\pi/2)=0$, i.e. $\alpha+\beta+\pi/2$ is an integral multiple of π,

$$\alpha+\beta+\pi/2 = k\cdot\pi,$$

$k\in\mathbb{Z}$. As all angles in a right-angle triangle are >0 and $\leq\pi/2$, we have $\pi/2<\alpha+\beta+\pi/2<3\pi/2$ and thus $k=1$. It follows that $\alpha+\beta+\pi/2=\pi$, which proves the claim for right-angle triangles.

(b) In a general triangle let a be the longest side and r the vertex opposite to it. We drop a perpendicular from r to a and thus divide the triangle into two right-angle triangles.

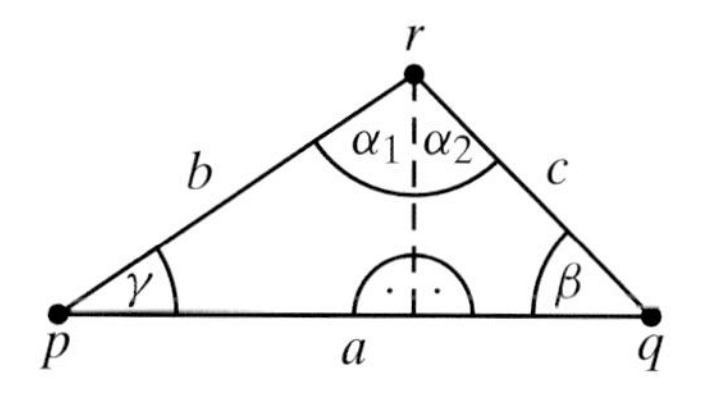

We know from part (a) of this proof that in right-angle triangles $\alpha_1 + \gamma + \pi/2 = \pi$ and $\alpha_2 + \beta + \pi/2 = \pi$. Addition of those two equations gives $\alpha + \beta + \gamma = \pi$. □

Exercise 1.17 Prove the *half-angle theorem of Euclidean geometry*:

$$\tan^2\left(\frac{\alpha}{2}\right) = \frac{(a-b+c)(a+b-c)}{(a+b+c)(-a+b+c)}.$$

Hint Check and use the formula $\tan^2(t) = (1-\cos(2t))/(1+\cos(2t))$.

For comparison, we conclude this section by looking at another important classical geometry, i.e. ***spherical geometry***. The points of this geometry are not in the plane, but on the surface of a sphere. We set

$$\mathscr{P} := S^2 := \{(x,y,z)^\top \in \mathbb{R}^3 \mid x^2+y^2+z^2 = 1\}.$$

We call S^2 the ***two-dimensional sphere***.

This raises the question of what the "straight lines" are in this geometry. The straight lines in the plane can be characterised by the fact that each of them is the shortest curve connecting any two of its points. On the sphere the great circles have this property. A ***great circle*** is the subset of S^2 in which a two-dimensional vector subspace of $\mathbb{R}^3$, i.e. a plane that passes through 0, intersects the surface of the sphere.

We therefore set

$$\mathscr{G} := \{S^2 \cap E \mid E \text{ is a two-dimensional vector subspace of } \mathbb{R}^3\}.$$

The incidence relation $\in$ must here be interpreted in the set-theoretic sense again.

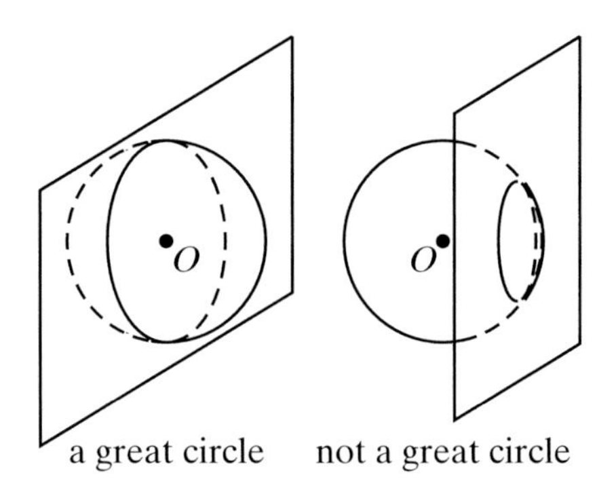

a great circle   not a great circle

Exercise 1.18 Which of the axioms I_1–I_4 are valid?

Given three points on a great circle, it is not possible to say in a meaningful way which of them lies between the two others. Let us try with the following definition: if $p, q, r \in S^2$, then q lies ***between*** p and r if and only if p, q and r are three pairwise distinct points on a great circle.

Exercise 1.19 What is a line segment? Which of the axioms A_1–A_5 are valid?

The congruence relations on the other hand can easily be defined in a geometrically meaningful way. We simply exchange the Euclidean motion group E(2) for the orthogonal group O(3). One needs to pay attention to the fact that for an orthogonal matrix $A \in O(3)$ and $p \in S^2$ the point Ap should lie on S^2 again, i.e. A indeed maps S^2 to itself only.

Exercise 1.20 Show that axioms K_1–K_6 are valid.

There are no parallels since any two great circles intersect. Theorem 1.1.9 therefore does not hold. This is one of the ways to see that not all incidence and ordering axioms can hold.

Because of its importance in navigation, amony other things, spherical geometry has been studied for a very long time. We will investigate it in more detail in sections 4.9 and 4.10.

2 Curve theory

We analyse curves in n*-dimensional space with a special focus on plane curves and space curves. Length, curvature and torsion are introduced. We prove Hopf's Umlaufsatz for simple closed curves, characterise convex curves and derive the four-vertex theorem. The isoperimetric inequality, which compares the length of a simple closed plane curve with the enclosed area, is proved using the Fourier series. We show that for a given curvature and torsion the resulting space curve is unique up to a Euclidean motion. We investigate how much a space curve needs to curve if it is closed and make the result even stronger in the case that the space curve is knotted.*

2.1 Curves in $\mathbb{R}^n$

We now want to use the tools of differentiation and integration to describe curves in n-dimensional space. We usually graphically imagine a curve as a bent line in space. Mathematically we express this as follows:

Definition 2.1.1 Let $I \subset \mathbb{R}$ be an interval. A ***parametrised curve*** is a map $c : I \to \mathbb{R}^n$ that can be differentiated infinitely often. A parametrised curve is said to be ***regular*** if its velocity vector does not vanish anywhere; $\dot{c}(t) \neq 0$ for all $t \in I$.

The interval I from the definition may be open, closed or half-open; furthermore, I can be bounded or unbounded. The condition $\dot{c}(t) \neq 0$ ensures that the point $c(t)$ on the curve moves at $t \in I$. In particular, this excludes the constant map $c(t) = c_0$. This certainly makes sense, as the image of this map only consists of the point c_0; not exactly what we have in mind when thinking of a curve. Let us look at some examples.

Example 2.1.2 A ***straight line*** can be described as a regular parametrised curve:

$$c : \mathbb{R} \to \mathbb{R}^n,$$
$$c(t) = c_0 + t \cdot v,$$

where $c_0 \in \mathbb{R}^n$ and $v \in \mathbb{R}^n - \{0\}$. This obviously satisfies the condition $\dot{c}(t) = v \neq 0$.

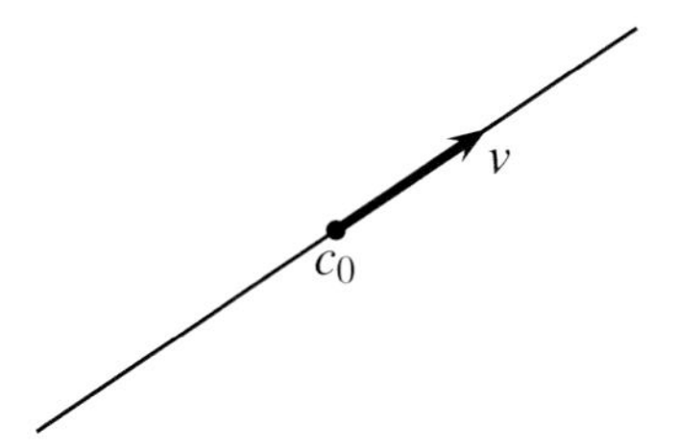

Example 2.1.3 A ***circular curve*** in the plane around the origin $(0, 0)$ with radius $r > 0$ looks as follows:

$$c : \mathbb{R} \to \mathbb{R}^2,$$
$$c(t) = \begin{pmatrix} r \cdot \cos(t) \\ r \cdot \sin(t) \end{pmatrix}.$$

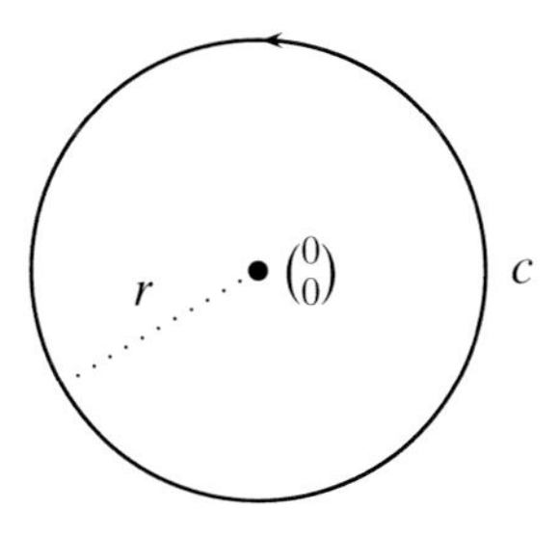

The arrow in the sketch shows the direction in which the curve traverses the image. This example shows that a regular parametrised curve is not necessarily injective. Because of the periodicity $c(t + 2\pi) = c(t)$ the curve runs infinitely often through every point that is in the image. We could, of course, restrict the domain of c to an interval with a length of exactly one period, but we shall discuss this later.

Example 2.1.4 A ***helix*** in three-dimensional space can be parametrised as follows:

$$c : \mathbb{R} \to \mathbb{R}^3,$$
$$c(t) = \begin{pmatrix} r \cdot \sin(t) \\ r \cdot \cos(t) \\ h \cdot t \end{pmatrix},$$

where $r > 0$ and $h > 0$.

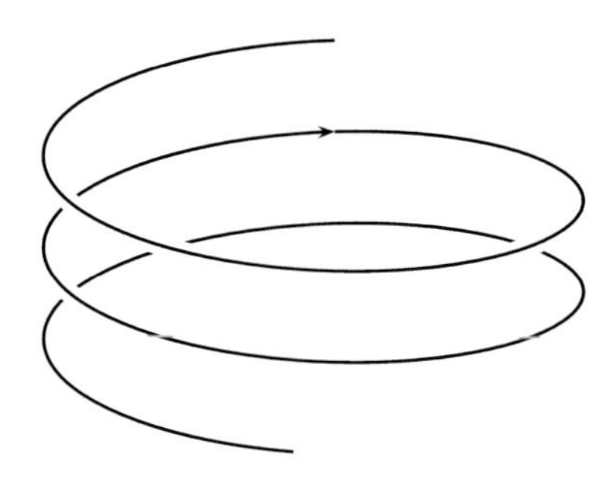

Example 2.1.5 The following regular parametrised curve is called a ***tractrix***:

$$c : (0, \pi/2) \to \mathbb{R}^2,$$
$$c(t) = \begin{pmatrix} \sin(t) \\ \cos(t) + \ln\tan(t/2) \end{pmatrix}.$$

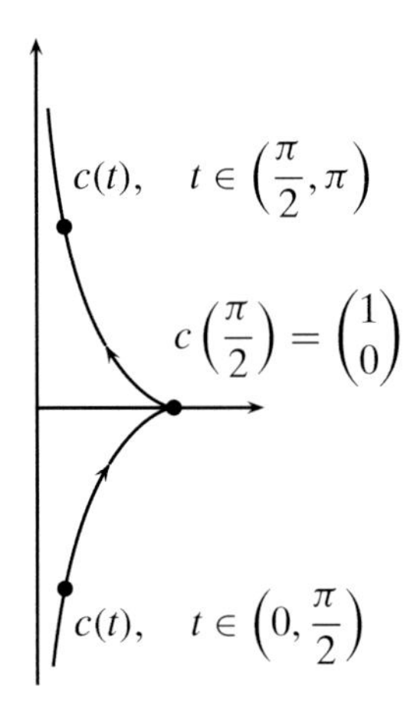

With the given rule for the computation of $c(t)$ from t one could define c on all of $(0, \pi)$. But for $t = \pi/2$ one obtains $\dot{c}\,(\pi/2) = (0, 0)^\top$. Then c would no longer be regular on the whole of $(0, \pi)$. It can be seen in the illustration that c has a cusp for $c\,(\pi/2) = (1, 0)^\top$. This is excluded by our definition of a regular parametrised curve.

Exercise 2.1 Show that the tractrix has the following property: for each point on the curve the line segment of the tangent line from the curve point to the y-axis has length 1.

This allows the following interpretation: if you walk along the y-axis while dragging a stone (or tired dog) on a rope of length 1, then the stone (or unhappy dog) will follow the tractrix.

Example 2.1.6 The ***logarithmic spiral*** is given by

$$c : \mathbb{R} \to \mathbb{R}^2,$$
$$c(t) = \begin{pmatrix} e^{t/10} \cdot \cos(t) \\ e^{t/10} \cdot \sin(t) \end{pmatrix}.$$

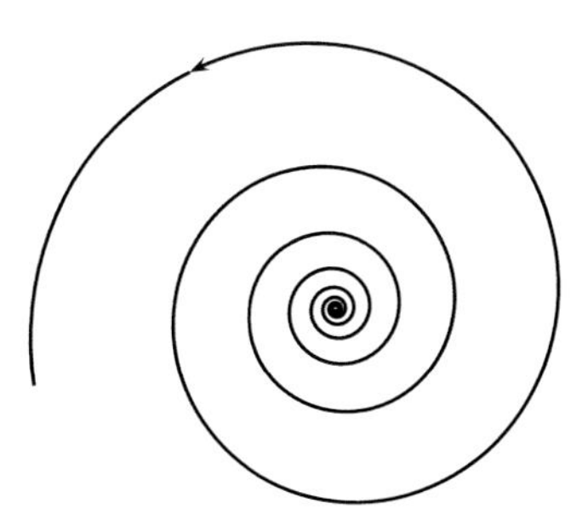

A parametrised curve is more than just the set of points on the curve in $\mathbb{R}^n$, i.e. more than just the image $c(I)$ of c. It is also specified in which direction the curve traverses the image. One often wants to change this *parametrisation* while leaving the image as it is. For this we use the following:

Definition 2.1.7 Let $c : I \to \mathbb{R}^n$ be a parametrised curve. A ***parameter transformation*** of c is a bijective map $\varphi : J \to I$, where $J \subset \mathbb{R}$ is another interval such that both φ and $\varphi^{-1} : I \to J$ can be differentiated infinitely often. The parametrised curve $\tilde{c} = c \circ \varphi : J \to \mathbb{R}^n$ is called a ***reparametrisation*** of c.

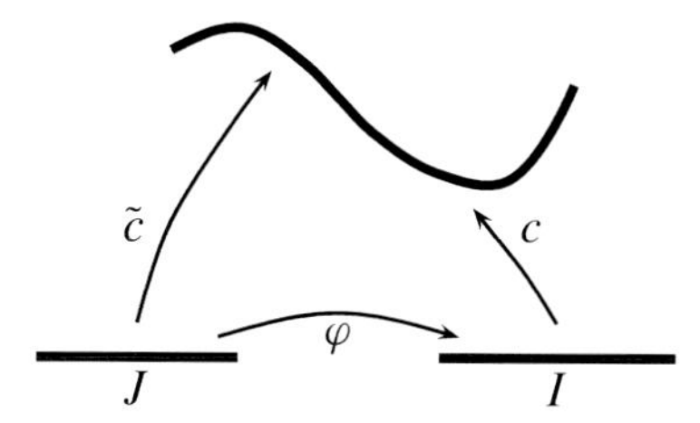

Since $c = \tilde{c} \circ \varphi^{-1}$ one can get c back from $\tilde{c}$. One should note here that the derivative of a parameter transformation φ cannot vanish anywhere, since according to the chain rule

$$\left(\varphi^{-1}\right)^{\cdot}(\varphi(t)) \cdot \dot{\varphi}(t) = \left(\varphi^{-1} \circ \varphi\right)^{\cdot}(t) = 1.$$

This also ensures that a reparametrisation of a regular parametrised curve is again regular:

$$\dot{\tilde{c}}(t) = \dot{c}(\varphi(t)) \cdot \dot{\varphi}(t) \neq 0.$$

A parameter transformation can either reverse or preserve the direction in which the curve traverses the image. The trivial parameter transformation

$\varphi(t) = t$, for example, does not change the parametrised curve, while the parameter transformation $\psi(t) = -t$ reverses the orientation.

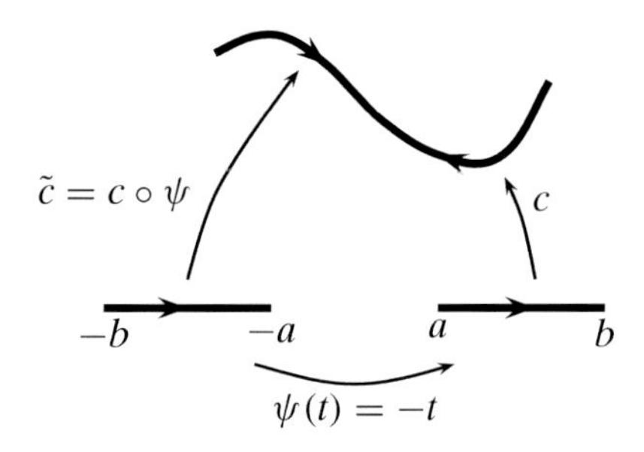

Definition 2.1.8 A parameter transformation φ is called ***orientation-preserving*** if $\dot{\varphi}(t) > 0$ for all t. The parameter transformation φ is said to be ***orientation-reversing*** if $\dot{\varphi}(t) < 0$ for all t.

Every parameter transformation is either orientation-preserving or orientation-reversing. This can easily be seen as follows: suppose that there exists a $t_1 \in I$ with $\dot{\varphi}(t_1) < 0$ and a $t_2 \in I$ with $\dot{\varphi}(t_2) > 0$, then by the intermediate value theorem there would also exist a t_3 between t_1 and t_2 with $\dot{\varphi}(t_3) = 0$. But this is impossible, as we have seen before.

We imagine a curve as a parametrised curve, while the actual choice of parametrisation is regarded as irrelevant. This is made more precise through the following:

Definition 2.1.9 A ***curve*** is an equivalence class of regular parametrised curves, where those curves that are reparametrisations of each other are regarded as equivalent.

The regular parametrised curves from examples 2.1.2–2.1.6 are different curves since they have distinct images in $\mathbb{R}^n$ and hence cannot be obtained from each other by reparametrisations.

The regular parametrised curves

$$c_1 : \mathbb{R} \to \mathbb{R}^2, \quad c_1(t) = (t, t)$$

and

$$c_2 : (0, \infty) \to \mathbb{R}^2, \quad c_2(t) = (\ln t, \ln t),$$

however, are equivalent, since $c_1 = c_2 \circ \varphi$ with $\varphi(t) = e^t$, and hence represent the same curve.

If a curve is represented by a regular parametrised curve $c : I \to \mathbb{R}^n$, then the image $c(I)$ is said to be the ***trace*** of the curve.

Curves do not have an intrinsic orientation, since it can be reversed by a parameter transformation. To be able to define an orientation we make use of the following definition:

Definition 2.1.10 An ***oriented curve*** is an equivalence class of parametrised curves, where those that originate from each other through *orientation-preserving* reparametrisations are regarded as equivalent.

Every oriented curve determines exactly one curve. Every curve has exactly two ***orientations***, i.e. there are exactly two oriented curves that determine a given curve.

Definition 2.1.11 A ***unit speed curve*** or ***curve parametrised by arc-length*** is a regular parametrised curve $c : I \to \mathbb{R}^n$ with $\|\dot{c}(t)\| = 1$ for all $t \in I$.

Curves parametrised by arc-length are exactly those that traverse their image in $\mathbb{R}^n$ with constant velocity 1. Slightly more generally we also say:

Definition 2.1.12 A ***curve parametrised proportional to arc-length*** is a regular parametrised curve $c : I \to \mathbb{R}^n$, for which $\|\dot{c}\|$ is constant (but not necessarily equal to 1).

Curves parametrised by arc-length are for many purposes particularly convenient. But do they exist?

Proposition 2.1.13 *For every regular parametrised curve c there exists an orientation-preserving parameter transformation φ such that the reparametrisation $c \circ \varphi$ is parametrised by arc-length.*

Proof Let $c : I \to \mathbb{R}^n$ be a regular parametrised curve. We choose $t_0 \in I$ and set

$$\psi(s) := \int_{t_0}^{s} \|\dot{c}(t)\| \, dt.$$

Since $\psi'(s) = \|\dot{c}(s)\| > 0$ the map ψ is strictly monotonically increasing and hence injective. Thus

$$\psi : I \to J := \psi(I)$$

is an orientation-preserving parameter transformation. We denote the inverse map by $\varphi := \psi^{-1} : J \to I$. Then φ and ψ can be differentiated infinitely often, and for the first derivative of φ we have the formula

$$\dot{\varphi}(t) = \frac{1}{\psi'(\varphi(t))} = \frac{1}{\|\dot{c}(\varphi(t))\|}.$$

It follows by the chain rule that

$$\left\|(c \circ \varphi)^{\cdot}(t)\right\| = \|\dot{c}(\varphi(t)) \cdot \dot{\varphi}(t)\| = \left\|\frac{\dot{c}(\varphi(t))}{\|\dot{c}(\varphi(t))\|}\right\| = 1.$$

Hence $c \circ \varphi$ is a curve parametrised by arc-length. □

So curves can be reparametrised by arc-length. To what extent are these reparametrisations unique?

Lemma 2.1.14 *Let $c_1 : I_1 \to \mathbb{R}^n$ and $c_2 : I_2 \to \mathbb{R}^n$ be different parametrisations by arc-length of the same curve, then the corresponding parameter transformation $\varphi : I_1 \to I_2$ with $c_1 = c_2 \circ \varphi$ is of the form*

$$\varphi(t) = t + t_0$$

for a $t_0 \in \mathbb{R}$ if c_1 and c_2 have the same orientation. If c_1 and c_2 have opposite orientations, then it is of the form

$$\varphi(t) = -t + t_0.$$

Proof We have

$$1 = \|\dot{c}_1(t)\| = \|\dot{c}_2(\varphi(t)) \cdot \dot{\varphi}(t)\| = \|\dot{c}_2(\varphi(t))\| \cdot |\dot{\varphi}(t)| = |\dot{\varphi}(t)|.$$

Hence $\varphi(t) = \pm t + t_0$. □

Exercise 2.2 Let $c : I \to \mathbb{R}^n$ be a parametrised curve and let $F \in \mathrm{E}(n)$ be a Euclidean motion. Show that if c is a curve parametrised by arc-length, then $F \circ c$ is also a curve parametrised by arc-length. Also show that if c is parametrised proportional to arc-length, then so is $F \circ c$.

To see what parametrisations by arc-length have to do with lengths, we first have to define the length of curves.

Definition 2.1.15 Let $c : [a, b] \to \mathbb{R}^n$ be a parametrised curve. Then

$$L[c] := \int_a^b \|\dot{c}(t)\| \; dt$$

is called the ***length*** of c.

Lemma 2.1.16 *The length of a parametrised curve is not changed by reparametrisation.*

Proof This can be deduced using the substitution rule. If $\tilde{c} = c \circ \varphi$ is a reparametrisation of c, $\varphi : [a', b'] \to [a, b]$, then we have

$$\begin{aligned} L[\tilde{c}] &= \int_{a'}^{b'} \left\| (c \circ \varphi)^{\cdot}(t) \right\| dt \\ &= \int_{a'}^{b'} \| \dot{c}(\varphi(t)) \| \cdot |\dot{\varphi}(t)| \, dt \\ &= \int_{a}^{b} \| \dot{c}(s) \| \, ds \\ &= L[c]. \end{aligned}$$

□

Lemma 2.1.16 says that we can talk about the length of curves, not only of the length of parametrised curves. The length does not depend on the particular parametrisation. Indeed, the length of a road does not depend on the speed at which you have driven along it.

Now we also understand why parametrisations by arc-length are so useful, since if $c : [a, b] \to \mathbb{R}^n$ is a curve parametrised by arc-length, then

$$L[c|_{[a,s]}] = \int_a^s 1 \, dt = s - a$$

for every $s \in [a, b]$. A curve parametrised by arc-length is exactly as long as the parameter interval.

The definition of length in terms of an integral may at first seem a bit arbitrary. A different approach looks as follows:

Definition 2.1.17 A ***polygon*** in $\mathbb{R}^n$ is a tuple $P = (a_0, \ldots, a_k)$ of vectors $a_i \in \mathbb{R}^n$ such that $a_{i+1} \neq a_i$ for all $i = 0, \ldots, k-1$.

We first imagine the vectors a_i as the vertices of the polygon and two successive vertices a_i and a_{i+1} to be connected by the corresponding segment of a straight line.

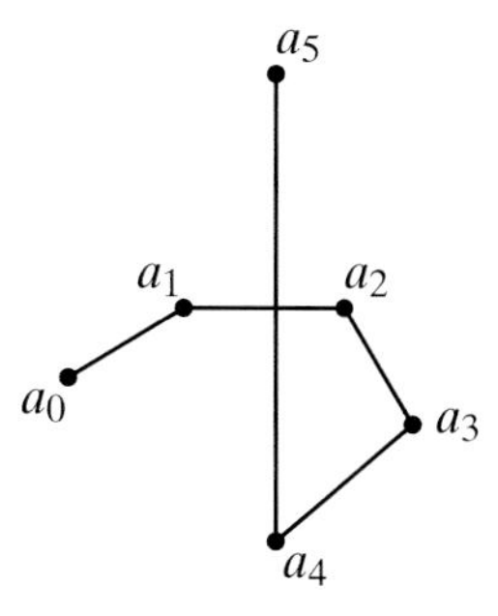

The condition $a_{i+1} \neq a_i$ that two successive vertices must be distinct could have been omitted in this section, in which we only considered the lengths of curves and polygons. We only make it in order not to encounter problems when defining the angle at vertex a_i later, cf. definition 2.3.10.

Since they have vertices, polygons are not (regular parametrised) curves according to our definition. But it is clear what the length of such a polygon is, namely the sum of the lengths of the line segments:

$$L[P] := \sum_{i=0}^{k-1} \|a_{i+1} - a_i\| .$$

To define the length of a parametrised curve one could also approximate the curve by polygons and describe its length as the limit of the lengths of the polygons, if it exists. The following proposition says that this approach leads to the same concept of length.

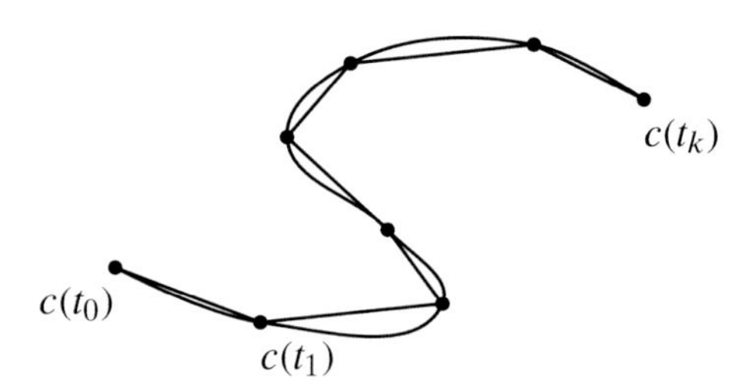

Proposition 2.1.18 (Approximation of length with polygons) *Let $c : [a,b] \to \mathbb{R}^n$ be a parametrised curve. Then for every (however small) $\varepsilon > 0$ there exists a $\delta > 0$ such that for every partition $a = t_0 < t_1 < \cdots < t_k = b$ of the interval making up the domain with mesh smaller than δ (i.e. $t_{i+1} - t_i < \delta$ for all i) the following holds:*

$$|L[c] - L[P]| < \varepsilon,$$

where[1] $P = (c(t_0), c(t_1), \ldots, c(t_k))$.

Proof Let $\varepsilon > 0$ be given. We choose an $\varepsilon' \in \big(0, \varepsilon/(1 + \sqrt{n}(b-a))\big)$. According to the theorem about Riemann sums [18, p. 106, theorem 2.7] applied to the integral

$$L[c] = \int_a^b \|\dot{c}(t)\| \, dt,$$

[1] Should, inconsistent with our definition of polygons, two or more successive vertices of P be the same, $c(t_{i+1}) = c(t_i)$, then we simply omit the repeated vertices in P. Identical successive vertices would not contribute to the length of a polygon anyway.

there exists for ε' a $\delta_0 > 0$ such that for every partition $a = t_0 < t_1 < \cdots < t_k = b$ with mesh smaller than δ_0

$$\left| L[c] - \sum_{i=0}^{k-1} \|\dot{c}\left(t_{i+1}\right)\| \cdot \left(t_{i+1} - t_i\right) \right| < \varepsilon'. \tag{2.1}$$

The components $\dot{c}^j : [a,b] \to \mathbb{R}$ of the derivative $\dot{c}$ of c are continuous functions on the *compact* interval $[a,b]$, hence even uniformly continuous on $[a,b]$. Thus there exists $\delta_j > 0$ such that

$$\left| \dot{c}^j(t) - \dot{c}^j(s) \right| < \varepsilon',$$

whenever $|t - s| < \delta_j$, $t, s \in [a,b]$.

We set $\delta := \min\{\delta_0, \delta_1, \ldots, \delta_n\}$. Now let there be a partition $a = t_0 < t_1 < \cdots < t_k = b$ with mesh smaller than δ. According to the mean value theorem there exist $\tau_{i,j} \in (t_i, t_{i+1})$ such that

$$c^j(t_{i+1}) - c^j(t_i) = \dot{c}^j(\tau_{i,j}) \cdot (t_{i+1} - t_i).$$

Further

$$\begin{aligned}
&\left| \|c(t_{i+1}) - c(t_i)\| - \|\dot{c}(t_{i+1})\|(t_{i+1} - t_i) \right| \\
&= \left| \left\| \begin{pmatrix} \dot{c}^1(\tau_{i,1}) \\ \vdots \\ \dot{c}^n(\tau_{i,n}) \end{pmatrix} \right\| (t_{i+1} - t_i) - \left\| \dot{c}(t_{i+1}) \right\| (t_{i+1} - t_i) \right| \\
&= \left| \left\| \begin{pmatrix} \dot{c}^1(\tau_{i,1}) \\ \vdots \\ \dot{c}^n(\tau_{i,n}) \end{pmatrix} \right\| - \left\| \begin{pmatrix} \dot{c}^1(t_{i+1}) \\ \vdots \\ \dot{c}^n(t_{i+1}) \end{pmatrix} \right\| \right| \cdot (t_{i+1} - t_i) \\
&\le \left\| \begin{pmatrix} \dot{c}^1(\tau_{i,1}) - \dot{c}^1(t_{i+1}) \\ \vdots \\ \dot{c}^n(\tau_{i,n}) - \dot{c}^n(t_{i+1}) \end{pmatrix} \right\| \cdot (t_{i+1} - t_i) \\
&= \sqrt{\sum_{j=1}^{n} (\dot{c}^j(\tau_{i,j}) - \dot{c}^j(t_{i+1}))^2} \cdot (t_{i+1} - t_i) \\
&\le \sqrt{n} \cdot \varepsilon' \cdot (t_{i+1} - t_i).
\end{aligned}$$

The first inequality above holds because of the inverse "triangle inequality" $|\|x\| - \|y\|| \le \|x - y\|$. Summation over i gives

$$\begin{aligned}
&\left| L[P] - \sum_{i=0}^{k-1} \|\dot{c}(t_{i+1})\|(t_{i+1} - t_i) \right| \\
&= \left| \sum_{i=0}^{k-1} \|c(t_{i+1}) - c(t_i)\| - \sum_{i=0}^{k-1} \|\dot{c}(t_{i+1})\|(t_{i+1} - t_i) \right| \\
&\leq \sqrt{n} \cdot \varepsilon' \cdot (b - a).
\end{aligned} \tag{2.2}$$

The claim follows, since

$$\begin{aligned}
|L[P] - L[c]| &\leq \left| L[P] - \sum_{i=0}^{k-1} \|\dot{c}(t_{i+1})\|(t_{i+1} - t_i) \right| \\
&\quad + \left| \sum_{i=0}^{k-1} \|\dot{c}(t_{i+1})\|(t_{i+1} - t_i) - L[c] \right| \\
&\overset{(2.1)(2.2)}{\leq} \sqrt{n} \cdot \varepsilon' \cdot (b - a) + \varepsilon' \\
&< \varepsilon.
\end{aligned}$$

□

Exercise 2.3 Let $c : [a, b] \to \mathbb{R}^n$ be a parametrised curve. Let P be a polygon inscribed in c as in proposition 2.1.18. Show that

$$L[P] \leq L[c].$$

Further show that

$$L[c] = \sup_P L[P],$$

where the supremum is taken over all inscribed polygons P.

Hint When an inscribed polygon P' results from P by refining the partition, then the triangle inequality gives

$$L[P] \leq L[P'].$$

Definition 2.1.19 A parametrised curve $c : \mathbb{R} \to \mathbb{R}^n$ is called ***periodic with period*** L if for all $t \in \mathbb{R}$ we have $c(t + L) = c(t)$, $L > 0$, and there is no $0 < L' < L$ such that $c(t + L') = c(t)$ for all $t \in \mathbb{R}$ as well. A curve is called ***closed*** if it has a periodic regular parametrisation.

The circular line from example 2.1.3 is a periodic parametrised curve with period $L = 2\pi$. Hence the curve represented by this parametrisation is closed.

Not every parametrisation of a closed curve is periodic. One can, for example, reparametrise a periodic parametrised curve in such a way that it becomes slower during each traverse of the curve. Then the time interval needed for the traverse will increase. This reparametrisation is then no longer periodic. However, the statement in the following exercise holds.

Exercise 2.4 Show that if $c : \mathbb{R} \to \mathbb{R}^n$ is a parametrisation by arc-length of a closed curve, then c is periodic.

Definition 2.1.20 A closed curve is a ***simple closed curve*** if it has a periodic regular parametrisation c with period L such that $c|_{[0,L)}$ is injective.

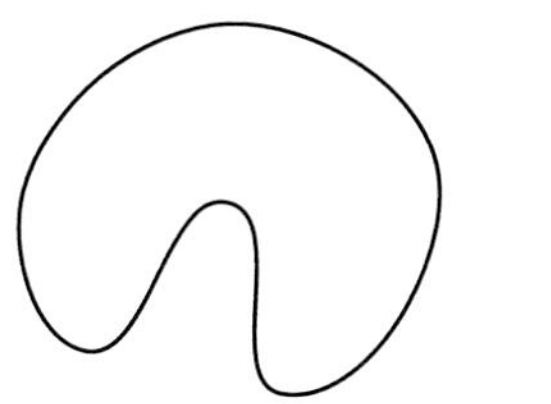

a simple closed curve

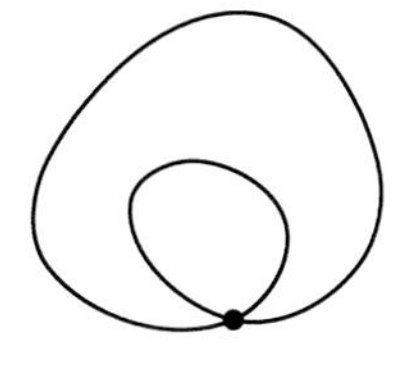

a closed curve, but not simple closed

This condition says that the curve does not have any self-intersections, apart from the point where it "closes itself". As parameter transformations are bijective, the injectivity condition holds not only for *one* periodic parametrisation of the closed curve, but automatically for *all* of its periodic parametrisations. The choice of the periodic parametrisation of the closed curve does not play a role in determining whether it is a simple closed curve or not.

Exercise 2.5 We let a disc of radius 1 roll on the x-axis in the x–y plane. Consider a second disc of radius $r > 0$ with the same centre and rigidly connected with the first.

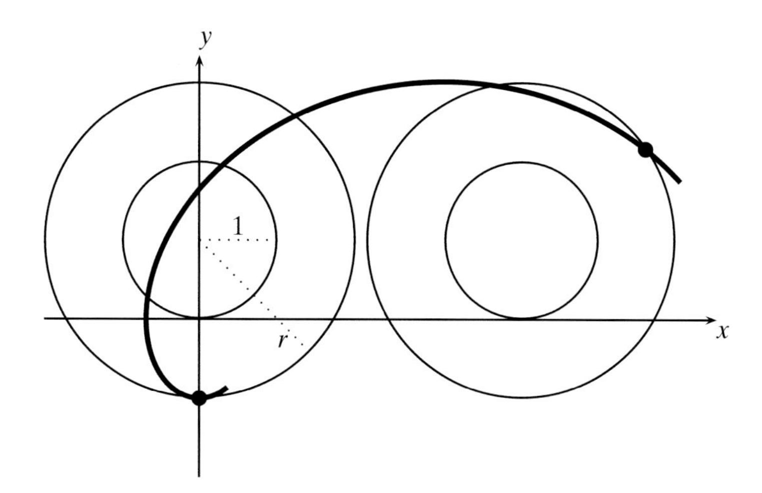

(a) Describe the path on which a point on the rim of the second disc moves by a parametrised curve.
This curve is called ***cycloid***.
(b) Sketch the curve for $r < 1$, $r = 1$ and $r > 1$.
(c) In which cases is the parametrised curve regular?
(d) Calculate the length of the curve for one turn of the disc for the case $r = 1$.

Exercise 2.6 Let $x, y \in \mathbb{R}^n$. Show that the unique shortest curve that connects x with y is the corresponding segment of a straight line:

(a) using proposition 2.1.18;
(b) without using polygons.

Exercise 2.7 Show that the two regular parametrised curves c_1 and c_2 have the same trace, but are nevertheless not equivalent, where

$$c_1 : [0, 2\pi] \to \mathbb{R}^2, \quad c_1(t) = \begin{pmatrix} \cos(t) \\ \sin(t) \end{pmatrix},$$

$$c_2 : [0, 2\pi] \to \mathbb{R}^2, \quad c_2(t) = \begin{pmatrix} \cos(2t) \\ \sin(2t) \end{pmatrix}.$$

Remark This example shows that in general curves are not fully determined by their trace. The following exercise shows that this is different for curves without self-intersections.

Exercise 2.8 Let c_1 and c_2 be regular parametrised curves with the same trace and with compact parameter intervals. Show that c_1 and c_2 are equivalent if c_1 and c_2 are injective.

2.2 Plane curves

In this section we will focus on those curves that lie in the plane, i.e. take values in $\mathbb{R}^2$.

Definition 2.2.1 A parametrised curve $c : I \to \mathbb{R}^2$ is called a ***parametrised plane curve***. Analogously we define ***plane regular parametrised curves***, ***plane curves*** and ***oriented plane curves***.

A specific feature of a plane curve is the possibility of defining its normal field. For this purpose let $c : I \to \mathbb{R}^2$ be a unit speed curve. We define the ***normal field*** by

$$n(t) := \begin{pmatrix} 0 & -1 \\ 1 & 0 \end{pmatrix} \cdot \dot{c}(t).$$

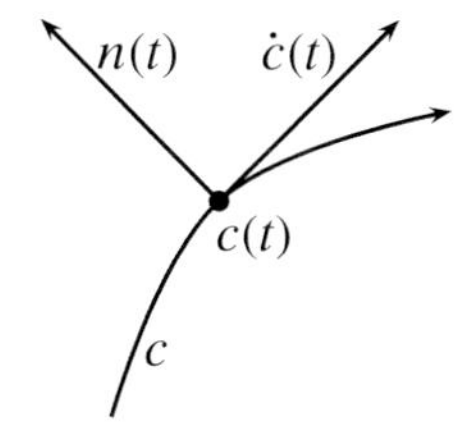

This definition is made in such a way that $(\dot c(t), n(t))$ always form a positively oriented orthonormal basis of $\mathbb{R}^2$. In other words, we rotate the velocity vector by 90 degrees anti-clockwise.

Since c is a unit speed curve we have

$$\langle \dot c, \dot c \rangle \equiv 1.$$

Differentiating this equation gives

$$0 \equiv \langle \ddot c, \dot c \rangle + \langle \dot c, \ddot c \rangle = 2\langle \ddot c, \dot c \rangle .$$

Thus $\dot c(t)$ and $\ddot c(t)$ are perpendicular to each other. Hence $\ddot c(t)$ is a multiple of the normal vector $n(t)$:

$$\ddot c(t) = \kappa(t) \cdot n(t).$$

Definition 2.2.2 The function $\kappa : I \to \mathbb{R}$ is called the ***curvature*** of c.

The curvature is a measure of how much a curve deviates from a straight line. If c is a unit speed curve, then c is exactly a straight line when $\ddot c \equiv 0$, i.e. when $\kappa \equiv 0$. Curvature is positive if the curve bends in the direction of its normal vector, i.e. in the direction in which the image is traversed to the left, and negative when it bends to the right.

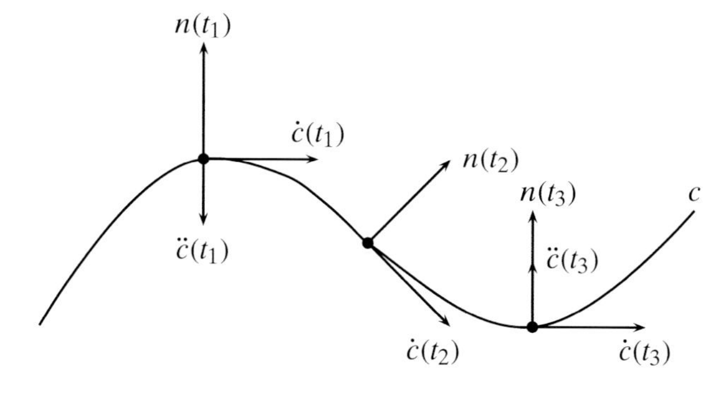

$$\kappa(t_1) < 0 \qquad \kappa(t_2) = 0 \qquad \kappa(t_3) > 0$$

Example 2.2.3 Let us consider the circular curve $c : \mathbb{R} \to \mathbb{R}^2$ of radius $r > 0$, parametrised by arc-length $c(t) = (r \cdot \cos(t/r), r \cdot \sin(t/r))^\top$. Then $\dot c(t) =$

$(-\sin(t/r), \cos(t/r))^\top$ and $\ddot{c}(t) = (1/r)(-\cos(t/r), -\sin(t/r))^\top = (1/r)n(t)$. Hence $\kappa \equiv 1/r$.

Exercise 2.9 Let c be a unit speed curve. Let $F \in \mathrm{E}(2)$ be an orientation-preserving Euclidean motion, $F(x) = Ax + b$ with $A \in \mathrm{SO}(2)$ and $b \in \mathbb{R}^2$. Show that $F \circ c$ has the same curvature as c.

How does a Euclidean motion F that is not orientation-preserving affect the curvature of c?

Exercise 2.10 Let c be a plane regular parametrised curve (not necessarily a unit speed curve). Show that the curvature of the curve is given by the formula

$$\kappa(t) = \frac{\det(\dot{c}(t), \ddot{c}(t))}{\|\dot{c}(t)\|^3}.$$

More precisely this means the following: if $\tilde{c} = c \circ \varphi$ is an orientation-preserving reparametrisation by arc-length with curvature $\tilde{\kappa}$, then $\tilde{\kappa} = \kappa \circ \varphi$.

Exercise 2.11 Let $c : I \to \mathbb{R}^2$ be a unit speed curve. The curve lies on a disc of radius R, i.e. $\|c(t)\| \leq R$ for all $t \in I$. Suppose that the curve touches the boundary of the disc at $t_0 \in \mathring{I}$, i.e. $\|c(t_0)\| = R$. Show that

$$|\kappa(t_0)| \geq \frac{1}{R}.$$

Hint Consider $(d/dt)\big|_{t=t_0} \|c(t)\|^2$ and $(d^2/dt^2)\big|_{t=t_0} \|c(t)\|^2$.

Proposition 2.2.4 (Frenet formulae) *Let $c : I \to \mathbb{R}^2$ be a plane unit speed curve. We set $v := \dot{c}$. Let κ be the curvature of c and let n be the normal vector. Then*

$$(\dot{v}(t), \dot{n}(t)) = (v(t), n(t)) \begin{pmatrix} 0 & -\kappa(t) \\ \kappa(t) & 0 \end{pmatrix}.$$

Proof The equation $\dot{v} = \kappa \cdot n$ is exactly the definition of the curvature. By differentiating the equation $\langle n, n\rangle \equiv 1$ we conclude as above that $\dot{n}(t)$ is perpendicular to $n(t)$ and hence must be a multiple of $v(t)$, $\dot{n}(t) = \alpha(t) \cdot v(t)$. We differentiate $\langle n, v\rangle \equiv 0$ and obtain

$$\begin{aligned} 0 &= \langle \dot{n}, v\rangle + \langle n, \dot{v}\rangle \\ &= \langle \alpha \cdot v, v\rangle + \langle n, \kappa \cdot n\rangle \\ &= \alpha + \kappa. \end{aligned}$$

Thus $\alpha = -\kappa$ and hence $\dot{n} = -\kappa \cdot v$. □

Lemma 2.2.5 *Let $c : [a,b] \to \mathbb{R}^2$ be a unit speed curve. Then there exists a C^∞-function $\vartheta : [a,b] \to \mathbb{R}$ such that*

$$\dot{c}(t) = \begin{pmatrix} \cos(\vartheta(t)) \\ \sin(\vartheta(t)) \end{pmatrix}.$$

If ϑ_1 and ϑ_2 are two such functions, then they differ only by an integer multiple of 2π, $\vartheta_1 = \vartheta_2 + 2k\pi$ with $k \in \mathbb{Z}$ constant. In particular $\vartheta(b) - \vartheta(a)$ is uniquely determined by c.

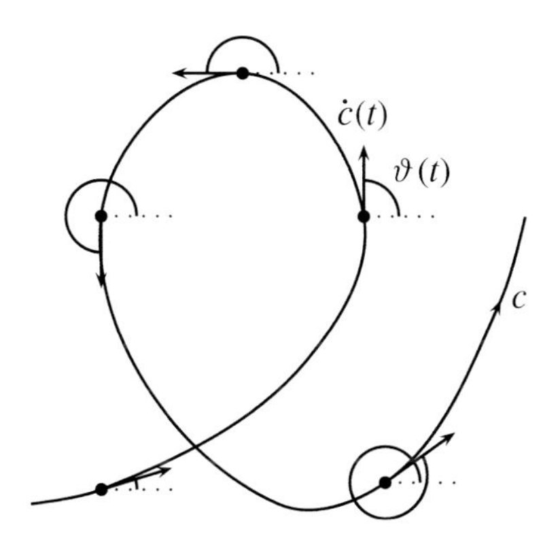

The number $\vartheta(t)$ measures the angle between the velocity vector $\dot{c}(t)$ and the x-axis. However, this angle is unique only up to integer multiples of 2π. Every unit vector can be written in the form $(\cos(\vartheta), \sin(\vartheta))^\top$. The important statement of this lemma is that the angle ϑ can be chosen as a continuous and even smooth function of t. Of course, one could uniquely determine the angle by requiring that it is in the interval $[0, 2\pi)$. But then the function ϑ would have jump discontinuities at those points where the velocity vector has completed one turn.

Proof (a) We first consider the case that the image $\dot{c}([a,b])$ is fully contained in one of the following four semicircles:

$$S_R := \{(x,y)^\top \in S^1 \subset \mathbb{R}^2 | \, x > 0\}, \quad S_L := \{(x,y)^\top \in S^1 \subset \mathbb{R}^2 | \, x < 0\},$$
$$S_O := \{(x,y)^\top \in S^1 \subset \mathbb{R}^2 | \, y > 0\}, \quad S_U := \{(x,y)^\top \in S^1 \subset \mathbb{R}^2 | \, y < 0\}.$$

Here S_R stands for the right semicircle, S_L for the left one, S_O for the top semicircle and S_U for the lower one. For instance, suppose the image is in S_R. This means the condition $\dot{c}^1 > 0$ holds for the first coordinate of $\dot{c}$. Hence our function ϑ satisfies

$$\frac{\dot{c}^2(t)}{\dot{c}^1(t)} = \frac{\sin(\vartheta(t))}{\cos(\vartheta(t))} = \tan(\vartheta(t)).$$

Thus

$$\vartheta(t) = \arctan\left(\frac{\dot{c}^2(t)}{\dot{c}^1(t)}\right) + 2k\pi$$

with $k \in \mathbb{Z}$. Here k is constant, since ϑ would otherwise not be continuous. From the formula it is obvious that ϑ is even smooth. If the initial value $\vartheta(a)$ is given, then k and thus also ϑ are uniquely determined. The other three cases are handled similarly.

(b) We now drop the condition that the image $\dot{c}([a, b])$ needs to be fully contained in a semicircle. We divide the compact interval $[a, b]$, $a = t_0 < t_1 < \cdots < t_m = b$ such that every $\dot{c}([t_i, t_{i+1}])$ is contained in one of the four semicircles. We can choose $\vartheta(a)$ and, according to (a), obtain a unique smooth $\vartheta : [a, t_1] \to \mathbb{R}$ with the desired properties. Hence $\vartheta(t_1)$ is determined and we obtain, again according to (a), a unique smooth continuation $\vartheta : [a, t_2] \to \mathbb{R}$. We continue inductively and finally obtain a smooth $\vartheta : [a, b] \to \mathbb{R}$.

The only choice that we made was the initial value $\vartheta(a)$. It is unique only up to integer multiples of 2π. Hence ϑ is also unique up to integer multiples of 2π. □

Definition 2.2.6 Let $c : \mathbb{R} \to \mathbb{R}^2$ be a plane unit speed curve, periodic with period L. Let $\vartheta : \mathbb{R} \to \mathbb{R}$ be as in Lemma 2.2.5. Then

$$n_c := \frac{1}{2\pi}(\vartheta(L) - \vartheta(0))$$

is called the ***winding number*** of c.

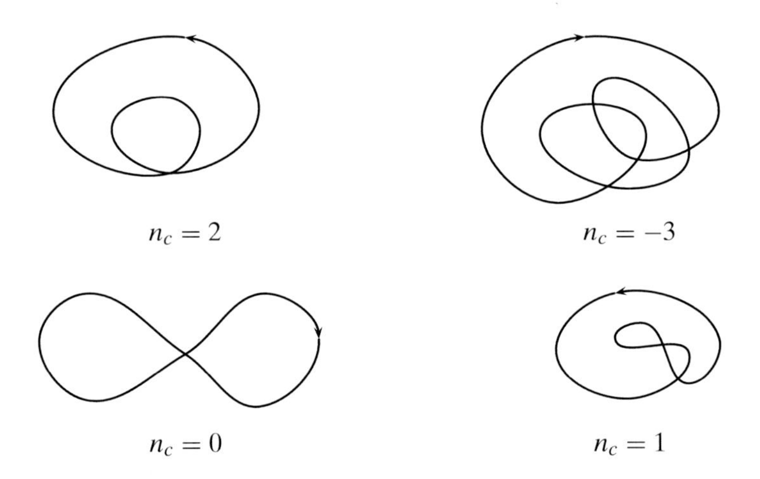

That the trigonometric function from lemma 2.2.5 is unique only up to a constant summand $2k\pi$ does not matter for the definition of the winding number since this summand cancels in the difference $\vartheta(L) - \vartheta(0)$.

Example 2.2.7 The circle of radius $r > 0$ has the parametrisation by arc-length $c(t) = (r \cdot \cos(t/r), r \cdot \sin(t/r))^\top$ with period $L = 2\pi r$. For the velocity vector we obtain

$$\dot{c}(t) = \begin{pmatrix} -\sin(t/r) \\ \cos(t/r) \end{pmatrix}$$
$$= \begin{pmatrix} \cos(t/r + \pi/2) \\ \sin(t/r + \pi/2) \end{pmatrix}.$$

Hence we obtain the trigonometric function $\vartheta(t) = t/r + \pi/2$ and thus the winding number

$$n_c = \frac{1}{2\pi}(\vartheta(2\pi r) - \vartheta(0)) = 1.$$

Lemma 2.2.8 *Let $c_1, c_2 : \mathbb{R} \to \mathbb{R}^2$ be two plane curves parametrised by arc-length, both periodic with period L. If c_2 results from c_1 in an orientation-preserving reparametrisation, then*

$$n_{c_1} = n_{c_2}.$$

If c_2 results from c_1 in a reparametrisation that is not orientation-preserving, then

$$n_{c_1} = -n_{c_2}.$$

Proof According to lemma 2.1.14 the following holds for the parameter transformation φ with $c_1 = c_2 \circ \varphi$:

$$\varphi(t) = \pm t + t_0,$$

where the sign depends on whether the parameter transformation preserves or reverses the orientation. If ϑ_2 is a trigonometric function for c_2 as in lemma 2.2.5, then, in the orientation-preserving case, $\vartheta_1 := \vartheta_2 \circ \varphi$ is a such function for c_1, since

$$\dot{c}_1(t) = \dot{c}_2(t + t_0)$$
$$= (\cos(\vartheta_2(t + t_0)), \sin(\vartheta_2(t + t_0)))^\top.$$

But if ϑ_1 is a trigonometric function for c_1, then the same holds for $\tilde{\vartheta}_1$, where $\tilde{\vartheta}_1(t) := \vartheta_1(t + L)$ and L is the period of c_1. It follows that

$$
\begin{aligned}
2\pi(n_{c_2} - n_{c_1}) &= (\vartheta_2(L) - \vartheta_2(0)) - (\vartheta_1(L) - \vartheta_1(0)) \\
&= \vartheta_1(L - t_0) - \vartheta_1(-t_0) - \vartheta_1(L) + \vartheta_1(0) \\
&= \left(\tilde{\vartheta}_1(-t_0) - \tilde{\vartheta}_1(0)\right) - (\vartheta_1(-t_0) - \vartheta_1(0)) \\
&= 0,
\end{aligned}
$$

thus $n_{c_1} = n_{c_2}$.

In the orientation-reversing case one observes that for a trigonometric function ϑ_2 for c_2 the function $\vartheta_1(t) := \vartheta_2(-t + t_0) + \pi$ is a trigonometric function for c_1, since then $\varphi(t) = -t + t_0$ and hence

$$
\begin{aligned}
\dot{c}_1(t) &= -\dot{c}_2(-t + t_0) \\
&= -(\cos(\vartheta_2(-t + t_0)), \sin(\vartheta_2(-t + t_0)))^\top \\
&= (\cos(\vartheta_2(-t + t_0) + \pi), \sin(\vartheta_2(-t + t_0) + \pi))^\top .
\end{aligned}
$$

We conclude that

$$
\begin{aligned}
2\pi(n_{c_2} + n_{c_1}) &= (\vartheta_2(L) - \vartheta_2(0)) + (\vartheta_1(L) - \vartheta_1(0)) \\
&= \vartheta_1(-L + t_0) - \vartheta_1(t_0) + \vartheta_1(L) - \vartheta_1(0) \\
&= (\vartheta_1(-L + t_0) - \vartheta_1(0)) - \left(\tilde{\vartheta}_1(-L + t_0) - \tilde{\vartheta}_1(0)\right) \\
&= 0,
\end{aligned}
$$

hence $n_{c_1} = -n_{c_2}$. □

The lemma shows that the winding number of an oriented closed plane curve is well defined. When the orientation is reversed the sign of the winding number is changed.

Further we can establish that the winding number is always a *whole* number. From $\cos(\vartheta(L)) = \cos(\vartheta(0))$ and $\sin(\vartheta(L)) = \sin(\vartheta(0))$, it follows that $e^{i\vartheta(L)} = e^{i\vartheta(0)}$ and hence $\vartheta(L) - \vartheta(0) \in 2\pi\mathbb{Z}$.

The reader may already expect there to be a connection between winding number and curvature. If a curve curves long enough to the left ($\kappa > 0$), then it should at some point complete a winding and should make a positive contribution to the winding number. Correspondingly curvature to the right ($\kappa < 0$) should make a negative contribution to the winding number. This is formalised in the following theorem.

Theorem 2.2.9 *Let $c : \mathbb{R} \to \mathbb{R}^2$ be a plane unit speed curve with period L. Let $\kappa : \mathbb{R} \to \mathbb{R}$ be the curvature of c. Then*

$$
n_c = \frac{1}{2\pi} \int_0^L \kappa(t)dt.
$$

Proof We write as in lemma 2.2.5 $\dot{c}(t) = (\cos(\vartheta(t)), \sin(\vartheta(t)))^\top$. Differentiation gives $\ddot{c}(t) = (-\sin(\vartheta(t)) \cdot \dot{\vartheta}(t), \cos(\vartheta(t)) \cdot \dot{\vartheta}(t))^\top$. On the other hand $\ddot{c}(t) = \kappa(t) \cdot n(t) = \kappa(t) \cdot (-\sin(\vartheta(t)), \cos(\vartheta(t)))^\top$. Thus

$$\dot{\vartheta}(t) = \kappa(t). \tag{2.3}$$

So the curvature gives exactly the change in the angle of the velocity vector $\dot{c}$ with a fixed axis. The fundamental theorem of calculus gives as desired

$$n_c = \frac{1}{2\pi}(\vartheta(L) - \vartheta(0)) = \frac{1}{2\pi}\int_0^L \dot{\vartheta}(t)dt = \frac{1}{2\pi}\int_0^L \kappa(t)dt. \qquad \square$$

We expect that a closed plane curve whose winding number has an absolute value of at least 2 must intersect itself. If a curve performs two or more turns before it closes, then it must intersect itself. This impression is correct, as the following theorem says.

Theorem 2.2.10 (Hopf's Umlaufsatz)[2] *A simple closed plane curve has winding number* 1 *or* -1.

To prove Hopf's Umlaufsatz we first need a generalisation of lemma 2.2.5.

Definition 2.2.11 Let $X \subset \mathbb{R}^n$ and $x_0 \in X$. Then X is called ***star-shaped*** with respect to x_0 if for every point $x \in X$ the whole straight line segment between x and x_0 is fully contained in X, i.e. for every $t \in [0, 1]$ we have $tx + (1-t)x_0 \in X$.

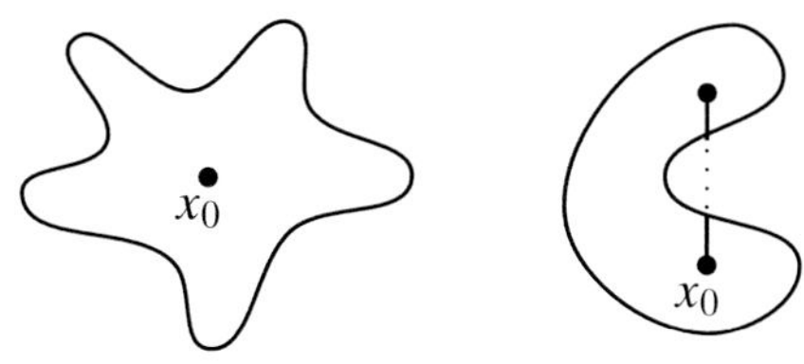

star-shaped w.r.t. x_0 not star-shaped w.r.t. x_0

Lemma 2.2.12 (lifting lemma) *Let $X \subset \mathbb{R}^n$ be star-shaped with respect to x_0. Let $e : X \to S^1 \subset \mathbb{R}^2$ be a continuous map. Then there exists a continuous map $\vartheta : X \to \mathbb{R}$ such that*

$$e(x) = \begin{pmatrix} \cos(\vartheta(x)) \\ \sin(\vartheta(x)) \end{pmatrix}$$

for all $x \in X$. The map ϑ is uniquely determined if $\vartheta(x_0) = \vartheta_0$ is given.

2 "Hopf's winding theorem".

Proof of the lifting lemma (a) Let us first consider the case

$$n = 1, \quad X = [0,1] \quad \text{and} \quad x_0 = 0.$$

This is essentially lemma 2.2.5, except that $\dot{c}$ is replaced by e. Since in this case e is only assumed to be continuous, ϑ is, of course, also only continuous and generally not C^∞. Let us quickly remind ourselves of the proof:

> we divide the interval $[0,1]$ into smaller intervals which are each mapped by e into one of the four semicircles. There the function ϑ can be written down explicitly using the inverse tangent or the inverse cotangent. It is uniquely determined by the initial value $\vartheta(0)$ and the continuity condition.

(b) Now let $X \subset \mathbb{R}^n$ be a general star-shaped set, $x_0 \in X$. Let $x \in X$. Because X is star-shaped, the straight line segment from x_0 to x is fully contained in X and we can define the map

$$e_x : [0,1] \to S^1, \qquad e_x(t) := e(tx + (1-t)x_0).$$

According to (a) there is exactly one continuous $\vartheta_x : [0,1] \to \mathbb{R}$ with $\vartheta_x(0) = \vartheta_0$ and $e_x(t) = (\cos(\vartheta_x(t)), \sin(\vartheta_x(t)))^\top$. If there exists a map ϑ as in the claim of lemma 2.2.12, then the uniqueness of ϑ_x gives

$$\vartheta_x(t) = \vartheta(tx + (1-t)x_0).$$

In particular $\vartheta(x) = \vartheta_x(1)$. This proves the uniqueness of ϑ.

(c) To prove existence we set $\vartheta(x) := \vartheta_x(1)$. Then

$$\begin{pmatrix} \cos(\vartheta(x)) \\ \sin(\vartheta(x)) \end{pmatrix} = \begin{pmatrix} \cos(\vartheta_x(1)) \\ \sin(\vartheta_x(1)) \end{pmatrix} = e_x(1) = e(x)$$

and $\vartheta(x_0) = \vartheta_0$. It remains to check that $\vartheta : X \to \mathbb{R}$ is continuous.

For this purpose let $x \in X$ and $\varepsilon > 0$. Let $0 = t_0 < t_1 < \cdots < t_N = 1$ be a partition of the interval $[0,1]$ such that e_x, restricted to the subinterval $[t_i, t_{i+1}]$, always lies in one of the four open semicircles. Because of the continuity of e we have that for all $y \in X$ close enough to x

$$\|e_x(t) - e_y(t)\| < \varepsilon$$

for all $t \in [0,1]$. If ε is small enough, then e_y, restricted to an subinterval $[t_i, t_{i+1}]$, is always fully contained in the same one of the four open semicircles as e_x. This also means that we can take the same partition of $[0,1]$ for e_y

as for e_x. By induction over the number N of the necessary subintervals it can easily be shown that in the case of the left or the right semicircle

$$\vartheta_x(t) = \arctan\left(\frac{e_x^2(t)}{e_x^1(t)}\right) + 2k\pi$$

and

$$\vartheta_y(t) = \arctan\left(\frac{e_y^2(t)}{e_y^1(t)}\right) + 2k\pi,$$

where $k \in \mathbb{Z}$ is the same for x and y. We write $e = (e^1, e^2)^\top$. For the upper and lower semicircles we have the corresponding formulae:

$$\vartheta_x(t) = \operatorname{arccot}\left(\frac{e_x^1(t)}{e_x^2(t)}\right) + 2k\pi$$

and

$$\vartheta_y(t) = \operatorname{arccot}\left(\frac{e_y^1(t)}{e_y^2(t)}\right) + 2k\pi.$$

In particular we have, depending on the semicircle,

$$\vartheta(x) - \vartheta(y) = \arctan\left(\frac{e^2(x)}{e^1(x)}\right) - \arctan\left(\frac{e^2(y)}{e^1(y)}\right)$$

and

$$\vartheta(x) - \vartheta(y) = \operatorname{arccot}\left(\frac{e^1(x)}{e^2(x)}\right) - \operatorname{arccot}\left(\frac{e^1(y)}{e^2(y)}\right)$$

respectively. The continuity of e, arctan and arccot gives continuity of ϑ. □

Remark If the map $e : X \to S^1$ is not surjective, then the map $\vartheta : X \to \mathbb{R}$ can be obtained much more easily. For example, suppose that $(\cos(\varphi), \sin(\varphi))^\top$ is not in the image of e. For every $k \in \mathbb{Z}$ the map

$$\Psi_k : (\varphi + 2\pi(k-1), \varphi + 2\pi k) \to S^1 - (\cos(\varphi), \sin(\varphi))^\top, \quad \Psi_k(t) = (\cos(t), \sin(t))^\top,$$

is a homeomorphism. Then $\vartheta := \Psi_k^{-1} \circ e : X \to (\varphi + 2\pi(k-1), \varphi + 2\pi k) \subset \mathbb{R}$ is continuous and satisfies

$$e(x) = \begin{pmatrix} \cos(\vartheta(x)) \\ \sin(\vartheta(x)) \end{pmatrix}.$$

The k is then uniquely determined by the condition $\vartheta(x_0) = \vartheta_0$. In particular, all $x_1, x_2 \in X$ satisfy that $|\vartheta(x_1) - \vartheta(x_2)| < 2\pi$.

After these technical preparations we can now prove Hopf's Umlaufsatz.

Proof of Hopf's Umlaufsatz (a) Let c be a periodic parametrisation with period L. Let $x_0 := \max\{c^1(t) \mid t \in \mathbb{R}\}$. Since the trace of c is compact, the maximum will actually be attained. The straight line $\{(x,y)^\top \in \mathbb{R}^2 \mid x = x_0\}$ intersects the trace of c at a point p. With a parameter transformation of the form $t \mapsto t + t_0$ we can achieve that $c(0) = p$.

Let G be the straight line that is parametrised by $s \mapsto p + s\cdot(1,0)^\top$. On the half-line for $s > 0$ there are now no points of c left. If necessary, we reparametrise with the map $t \mapsto -t$ in order to achieve that $\dot{c}(0) = (0,1)^\top$. This transformation reverses the orientation and hence changes the sign of the winding number, but this is not relevant for the claim.

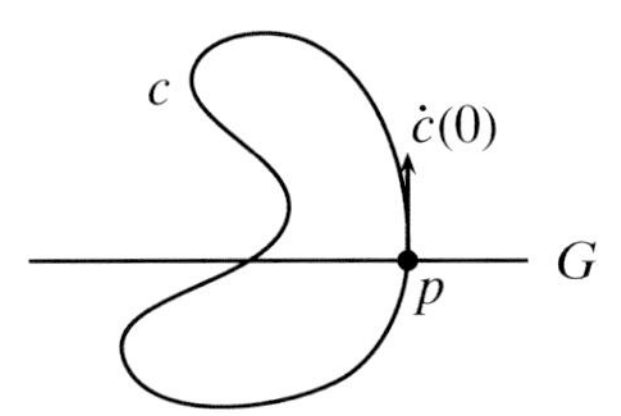

(b) We set $X := \{(t_1,t_2)^\top \in \mathbb{R}^2 \mid 0 \le t_1 \le t_2 \le L\}$. Then X is a star-shaped set w.r.t. $(0,0)^\top$.

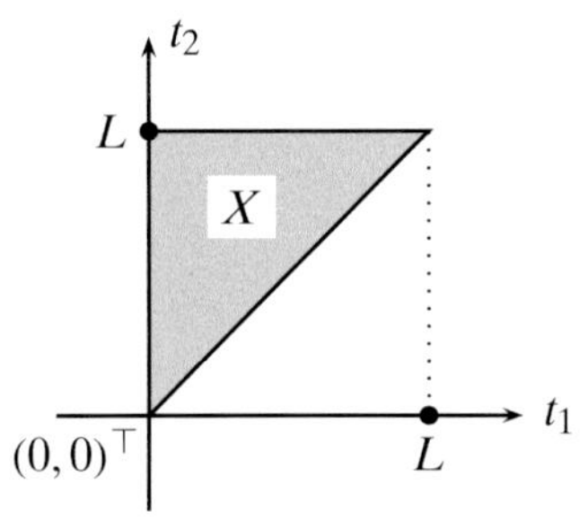

We consider the continuous map

$$e : X \to S^1,$$

$$e(t_1,t_2) = \begin{cases} \dfrac{c(t_2) - c(t_1)}{\|c(t_2) - c(t_1)\|}, & t_2 > t_1 \text{ and } (t_1,t_2) \neq (0,L), \\ \dot{c}(t), & t_2 = t_1 = t, \\ -\dot{c}(0), & (t_1,t_2) = (0,L). \end{cases}$$

Note that e is only well defined because c was assumed to be a *simple* closed curve. Otherwise $c(t_1) = c(t_2)$ could have appeared in the expression for the first case in the above definition.

We now choose a function $\vartheta : X \to \mathbb{R}$ for e as in the lifting lemma 2.2.12. $t \mapsto \vartheta(t,t)$ is a trigonometric function as in lemma 2.2.5 since $e(t,t) = \dot{c}(t)$. For the winding number one obtains

$$2\pi n_c = \vartheta(L,L) - \vartheta(0,0) = \vartheta(L,L) - \vartheta(0,L) + \vartheta(0,L) - \vartheta(0,0). \tag{2.4}$$

(c) If a $t \in (0, L)$ satisfied $e(0, t) = (1, 0)^\top$, then $c(t)$ would be on the right half-line of G, contradicting (a). In addition, $(1, 0)^\top$ is perpendicular to $\dot{c}(0) = e(0, 0) = -e(0, L)$. Hence $(1, 0)^\top$ is not in the image of the map $t \mapsto e(0, t)$. Because of the remark after lemma 2.2.12, the image of $t \mapsto \vartheta(0, t)$ will be in an interval of the form $(2\pi k, 2\pi(k+1))$, $k \in \mathbb{Z}$. From $e(0, L) = -\dot{c}(0) = (0, -1)^\top$ it follows that $\vartheta(0, L) = 3\pi/2 + 2\pi k$ and since $e(0, 0) = \dot{c}(0) = (0, 1)$ we have shown that $\vartheta(0, 0) = \pi/2 + 2\pi k$. Hence

$$\vartheta(0, L) - \vartheta(0, 0) = \pi.$$

Analogously $(-1, 0)^\top$ is not in the image of the map $t \mapsto e(t, L)$ and we obtain

$$\vartheta(L, L) - \vartheta(0, L) = \pi.$$

From (2.4) follows that

$$2\pi n_c = \pi + \pi = 2\pi.$$

□

Definition 2.2.13 A plane curve is called ***convex*** if each of its points is such that the curve lies entirely on one side of the tangent through this point.

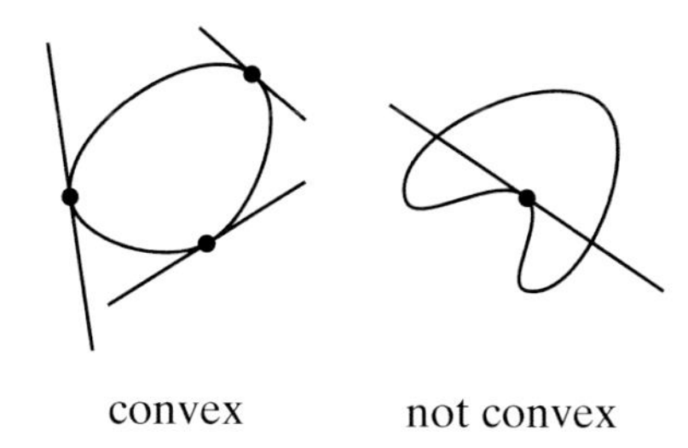

convex not convex

If c is a plane curve parametrised by arc-length and n the normal field along c, then the convexity condition states for the point $c(t_0)$ that

$$\langle c(t) - c(t_0), n(t_0) \rangle \geq 0 \tag{2.5}$$

for all t or

$$\langle c(t) - c(t_0), n(t_0) \rangle \leq 0 \tag{2.6}$$

for all t.

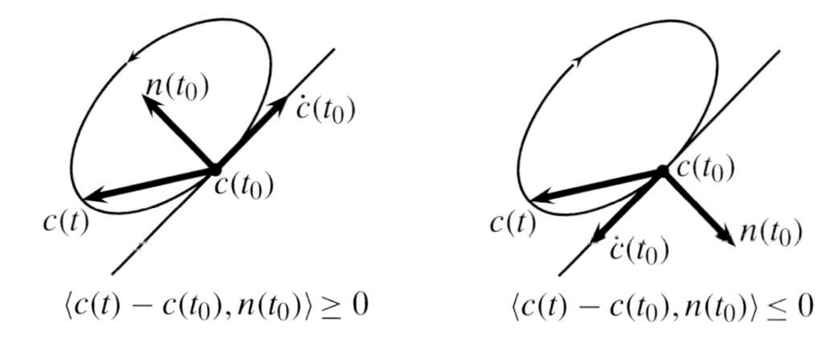

$\langle c(t) - c(t_0), n(t_0) \rangle \geq 0$ $\langle c(t) - c(t_0), n(t_0) \rangle \leq 0$

A priori one could for a given curve need condition (2.5) for some t_0 and (2.6) for other t_0. In reality this does not occur. This is because if a convex curve satisfies $\langle c(t) - c(t_1), n(t_1)\rangle \geq 0$ for all t and $\langle c(t) - c(t_2), n(t_2)\rangle \leq 0$ for all t, then continuity implies that there exists a t_3 between t_1 and t_2 such that $\langle c(t) - c(t_3), n(t_3)\rangle = 0$ for all t. This implies that c is a straight line and thus both (2.5) and (2.6) are satisfied for all t_0. We have proved:

Lemma 2.2.14 *Let $c : I \to \mathbb{R}^2$ be a plane curve parametrised by arc-length with normal field n. Then c is convex if and only if for all $t, t_0 \in I$*

$$\langle c(t) - c(t_0), n(t_0)\rangle \geq 0$$

or for all $t, t_0 \in I$

$$\langle c(t) - c(t_0), n(t_0)\rangle \leq 0.$$

Intuitively it seems clear that a convex curve always curves in the same sense, i.e. always to the left or always to the right. This is formalised in the following theorem.

Theorem 2.2.15 *Let $c : \mathbb{R} \to \mathbb{R}^2$ be a simple closed plane curve parametrised by arc-length. Let $\kappa : \mathbb{R} \to \mathbb{R}$ be its curvature. The curve is convex if and only if $\kappa(t) \geq 0$ for all $t \in \mathbb{R}$ or $\kappa(t) \leq 0$ for all $t \in \mathbb{R}$.*

Proof Let $\vartheta : \mathbb{R} \to \mathbb{R}$ be a trigonometric function as in lemma 2.2.5. We already know by (2.3) that $\dot{\vartheta} = \kappa$.

(a) Now let c be convex. We have to show that κ does not change sign. According to lemma 2.2.14 we can assume that

$$\langle c(t) - c(t_0), n(t_0)\rangle \geq 0$$

for all $t, t_0 \in \mathbb{R}$. The case $\langle c(t) - c(t_0), n(t_0)\rangle \leq 0$ is handled analogously and leads to the opposite sign of κ. In our case we show that $\kappa(t_0) \geq 0$ for all t_0. The Taylor series of c is

$$c(t) = c(t_0) + \dot{c}(t_0)(t - t_0) + \tfrac{1}{2}\ddot{c}(t_0)(t - t_0)^2 + \mathrm{O}(|t - t_0|^3).$$

But $\langle \dot{c}(t_0), n(t_0)\rangle = 0$, so scalar multiplication by $n(t_0)$ gives

$$0 \leq \langle c(t) - c(t_0), n(t_0)\rangle = \tfrac{1}{2}\,\langle \ddot{c}(t_0), n(t_0)\rangle\,(t - t_0)^2 + \mathrm{O}(|t - t_0|^3).$$

We divide by the positive $(t - t_0)^2$ and obtain

$$0 \leq \tfrac{1}{2}\,\langle \ddot{c}(t_0), n(t_0)\rangle + \mathrm{O}(|t - t_0|) = \tfrac{1}{2}\kappa(t_0) + \mathrm{O}(|t - t_0|).$$

Letting $t \to t_0$ we see that

$$\kappa(t_0) \geq 0.$$

(b) Now let $\kappa \geq 0$. We show that the curve is convex. If the curve were not convex then there would be a t_0 such that the function

$$\varphi : \mathbb{R} \to \mathbb{R}, \qquad \varphi(t) = \langle c(t) - c(t_0), n(t_0) \rangle,$$

has both negative and positive values. The periodicity of c implies that φ attains its minimum at a point t_1 and its maximum at t_2. It follows that

$$\varphi(t_1) < 0 = \varphi(t_0) < \varphi(t_2). \tag{2.7}$$

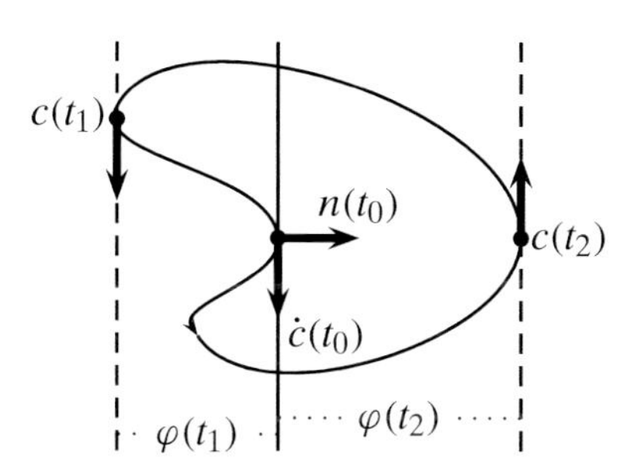

From $\dot{\varphi}(t_1) = 0$ it follows that $\langle \dot{c}(t_1), n(t_0) \rangle = 0$. Further $\dot{c}(t_1) = \pm \dot{c}(t_0)$. Analogously we obtain $\dot{c}(t_2) = \pm \dot{c}(t_0)$. At least two of the three unit vectors $\dot{c}(t_0)$, $\dot{c}(t_1)$ and $\dot{c}(t_2)$ must coincide. We choose $s_1, s_2 \in \{t_0, t_1, t_2\}$ with $s_1 < s_2$ such that

$$\dot{c}(s_1) = \dot{c}(s_2).$$

We therefore have $\vartheta(s_2) - \vartheta(s_1) = 2\pi k$ with $k \in \mathbb{Z}$. From $\dot{\vartheta} = \kappa \geq 0$ it follows that ϑ is monotonically increasing and thus $\vartheta(s_2) - \vartheta(s_1) \geq 0$. Hence $k \in \mathbb{N}_0$. Analogously $\vartheta(s_1 + L) - \vartheta(s_2) = 2\pi l$ with $l \in \mathbb{N}_0$. For the winding number it follows that $n_c = k + l$ (≥ 0). By Hopf's Umlaufsatz $n_c = 1$. Thus $k = 0$ or $l = 0$. Suppose that $k = 0$. Then $\kappa = \dot{\vartheta} = 0$ on $[s_1, s_2]$. Hence c parametrises a straight line on $[s_1, s_2]$. We thus have that

$$c(s) = c(s_1) + (s - s_1) \cdot \dot{c}(s_1) = c(s_1) \pm (s - s_1) \cdot \dot{c}(t_0)$$

for all $s \in [s_1, s_2]$. We can now compute the function φ for $s \in [s_1, s_2]$:

$$\begin{aligned}
\varphi(s) &= \langle c(s) - c(t_0), n(t_0) \rangle \\
&= \langle c(s_1) \pm (s - s_1) \cdot \dot{c}(t_0) - c(t_0), n(t_0) \rangle \\
&= \langle c(s_1) - c(t_0), n(t_0) \rangle,
\end{aligned}$$

i.e. φ is constant on $[s_1, s_2]$. But this contradicts (2.7) since at least two of the three values t_0, t_1 and t_2 lie in the interval $[s_1, s_2]$. □

Remark The following example shows that it is indeed important for the validity of this theorem that the curve is assumed to be *simple* closed.

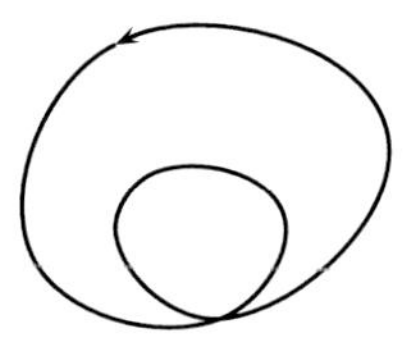

$\kappa > 0$, but not convex

However, the condition that the curve is simple closed was only used for one direction of the proof. We used Hopf's Umlaufsatz and $n_c = 1$ when we showed that convexity follows from $\kappa \geq 0$. We do not need to make this assumption for the other direction. We summarise:

For a parametrisation by arc-length of a convex (but not necessarily closed) curve we have $\kappa(t) \geq 0$ for all t or $\kappa(t) \leq 0$ for all t.

Definition 2.2.16 Let $c : I \to \mathbb{R}^2$ be a plane curve parametrised by arc-length. We say that c has a ***vertex*** at $t_0 \in I$ if $\dot{\kappa}(t_0) = 0$.

Example 2.2.17 Let us consider the ***ellipse***, parametrised by

$$c : \mathbb{R} \to \mathbb{R}^2, \qquad c(t) = \begin{pmatrix} a\cos(t) \\ b\sin(t) \end{pmatrix},$$

with $0 < a < b$.

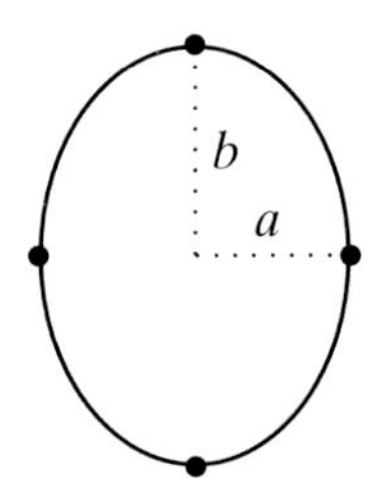

This is unfortunately not a parametrisation by arc-length. Instead of trying to reparametrise the ellipse by arc-length, we use the formula for curvature from exercise 2.10:

$$\begin{aligned}
\kappa(t) &= \frac{\det(\dot{c}(t), \ddot{c}(t))}{\|\dot{c}(t)\|^3} \\
&= \frac{\det\left(\begin{pmatrix} -a\sin(t) \\ b\cos(t) \end{pmatrix}, \begin{pmatrix} -a\cos(t) \\ -b\sin(t) \end{pmatrix}\right)}{\left\| \begin{pmatrix} -a\sin(t) \\ b\cos(t) \end{pmatrix} \right\|^3} \\
&= \frac{ab}{\left(a^2\sin(t)^2 + b^2\cos(t)^2\right)^{3/2}}.
\end{aligned}$$

Differentiating gives

$$\dot{\kappa}(t) = -\frac{3ab}{2}\left(a^2\sin(t)^2 + b^2\cos(t)^2\right)^{-5/2}\left(2a^2\sin(t)\cos(t) - 2b^2\cos(t)\sin(t)\right),$$

i.e. $\dot{\kappa}(t) = 0$ if and only if

$$a^2\sin(t)\cos(t) = b^2\cos(t)\sin(t).$$

But $a^2 \neq b^2$, so this is exactly the case when

$$\sin(t)\cos(t) = 0,$$

i.e. when $t \in \mathbb{Z}\cdot\pi/2$. In one traverse of $[0, 2\pi)$ the ellipse therefore has exactly four vertices, at points $t = 0, \pi/2, \pi$ and $3\pi/2$.

Indeed, a closed curve always has at least four vertices. More precisely,

Theorem 2.2.18 (Four-vertex theorem) *If $c : \mathbb{R} \to \mathbb{R}^2$ is a convex simple closed plane curve parametrised by arc-length and with period L, then c has at least four vertices in $[0, L)$.*

To prove the above theorem we need the following lemmas.

Lemma 2.2.19 *If a simple closed convex plane curve intersects a straight line in more than two points, then a whole segment of this straight line is part of the curve and there are thus infinitely many points of intersection.*

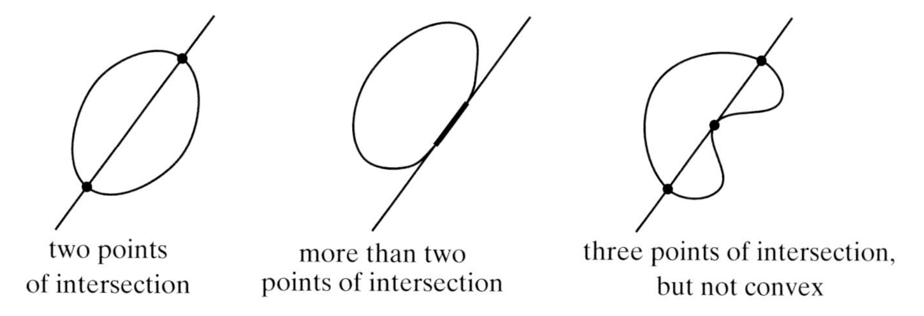

Proof of the lemma Let $c : \mathbb{R} \to \mathbb{R}^2$ be a parametrisation of the curve by arc-length with period L. With a parameter transformation of the form $t \mapsto t + t_0$ we can achieve that $c(0)$ is one of the three points of intersection with the straight line. According to theorem 2.2.15 we can, possibly after applying the parameter transformation $t \mapsto -t$, assume that the curvature satisfies $\kappa \geq 0$. The trigonometric function from lemma 2.2.5 also satisfies

$\dot{\vartheta} \geq 0$, i.e. it is monotonically increasing. According to Hopf's Umlaufsatz $\vartheta(L) - \vartheta(0) = 2\pi n_c = 2\pi$. Hence

$$\vartheta : [0, L] \to [\vartheta_0, \vartheta_0 + 2\pi]$$

is a smooth surjective monotonically increasing function, $\vartheta_0 = \vartheta(0)$.

Let the curve c intersect the straight line G at the points $t_0 = 0 < t_1 < t_2 < L$. Let G be parametrised by $t \mapsto p_0 + t \cdot v$. Let $n = \begin{pmatrix} 0 & -1 \\ 1 & 0 \end{pmatrix} \cdot v$ be the normal vector of G.

Now let I be one of the three intervals $[0, t_1]$, $[t_1, t_2]$ and $[t_2, L]$. The curve c lies on the straight line G at the end-points of I. If $c(t)$ lies on G for all $t \in I$, then a whole segment of G is part of c and the lemma is proved. Let us therefore suppose that there exist points $t \in I$ such that $c(t)$ does not lie on G. We now consider the straight lines G_s parallel to G that are parametrised by $t \mapsto p + s \cdot n + t \cdot v$. Set $s_1 := \sup\{s > 0 \mid G_s \text{ intersects } c|_I\}$. Should there not be any points of $c|_I$ on the side of G to which n points, then we consider $s_1 := \inf\{s < 0 \mid G_s \text{ intersects } c|_I\}$ instead.

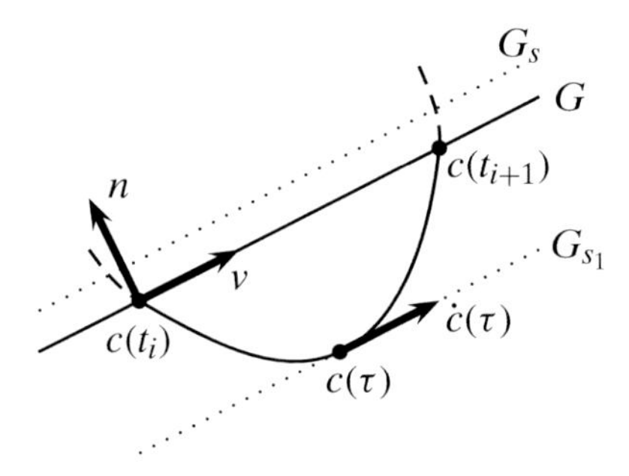

In any case G_{s_1} tangentially intersects the segment $c|_I$ at a point τ from the interior of I, i.e. $\dot{c}(\tau) = \pm v$. Applying this to all three intervals $I = [0, t_1]$, $I = [t_1, t_2]$ and $I = [t_2, t_3]$ we obtain the three points τ_1, τ_2 and τ_3, $0 < \tau_1 < t_1 < \tau_2 < t_2 < \tau_3 < L$, with $\dot{c}(\tau_j) = \pm v$.

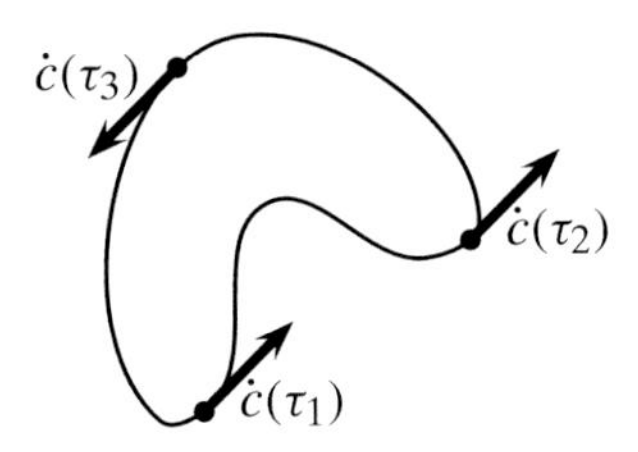

Let ϑ_1 denote the unique point from $[\vartheta_0, \vartheta_0 + 2\pi)$ for which $v = (\cos(\vartheta_1), \sin(\vartheta_1))^\top$ and $\vartheta_2 = \vartheta_1 \pm \pi$ the point satisfying $-v = (\cos(\vartheta_2), \sin(\vartheta_2))^\top$. We can assume without loss of generality that $\vartheta_2 = \vartheta_1 + \pi$, otherwise interchange the roles of ϑ_1 and ϑ_2.

First case Let ϑ_1 (and therefore also ϑ_2) be strictly greater than ϑ_0. Then ϑ must take one of the two values ϑ_1 or ϑ_2 at each of the three points τ_1, τ_2 and τ_3. In particular, ϑ must be the same at no less than two of the three points. Since ϑ is monotonically increasing, ϑ must be constant on one of the two intervals $[\tau_1, \tau_2]$ or $[\tau_2, \tau_3]$. But then $\dot{c} \equiv \pm v$ is in this interval, i.e. this part of the curve is a segment of a straight line parallel to G. As both of the two intervals in question contain a point in which c intersects the straight line G, namely t_1 and t_2 respectively, c must contain a segment of G.

Second case Now there is still the possibility that $\vartheta(\tau_1) = \vartheta_1 = \vartheta_0$, $\vartheta(\tau_2) = \vartheta_2$ and $\vartheta(\tau_3) = \vartheta_0 + 2\pi$. But then, because of monotonicity, ϑ must be constant on the interval $[0, \tau_1]$, and it follows as above that $c|_{[0,\tau_1]}$ coincides with a segment of the straight line G. □

Lemma 2.2.20 *If a simple closed convex plane curve intersects a straight line at more than one point tangentially, then the curve contains a segment of some straight line.*

Proof of the lemma If the curve has more than two points of intersection with the straight line G, then the claim follows from lemma 2.2.19. Hence we can assume that the curve intersects the straight line G at exactly two points. Because of the convexity of the curve it must lie entirely on one side of the straight line. We shift the straight line by a sufficiently small distance towards the curve and obtain a parallel straight line G'. Continuity implies that G' must intersect the curve in at least two points in the vicinity of each of the two points of intersection with G.

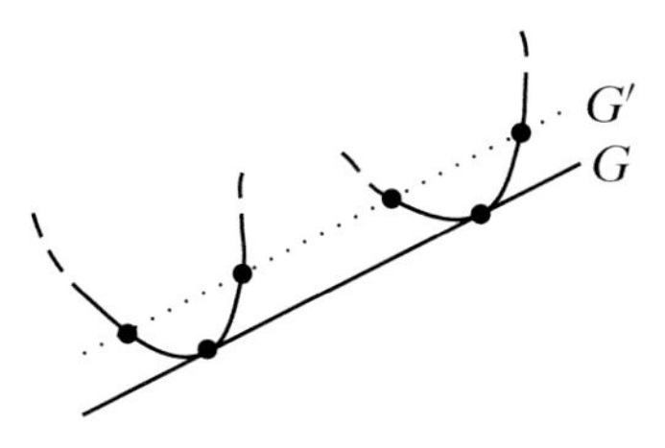

Hence G' has at least four points of intersection with the curve and because of lemma 2.2.19 the curve contains a segment of the straight line G'. □

In lemma 2.2.20 it is not claimed that the curve contains a segment of the tangentially intersecting straight line (although this is also true), but the segment contained in the curve may by all means be part of another straight line. However, for our application of the lemma in the proof of the four-vertex theorem this does not play a role.

Proof of the four-vertex theorem The curvature κ of c attains maxima and minima because of the periodicity of c, which already yields two vertices. Without loss of generality we can assume that the minimum is attained at $t = 0$ and the maximum at $t = t_0 \in (0, L)$. Let G be the straight line through the two points $c(0)$ and $c(t_0)$. If G has another point of intersection with the curve parametrised by c, then c contains by lemma 2.2.19 an entire straight line segment. But then the curvature is constantly 0 on an interval and we obtain infinitely many vertices.

Let us therefore consider the case that G does not have another point in common with the curve. After applying a Euclidean motion we can assume that G is the x-axis. Suppose that the curve does not have further vertices. Then $\dot{\kappa}$ does not vanish anywhere on the two intervals $(0, t_0)$ and (t_0, L). Since $\int_0^L \dot{\kappa}(t)dt = \kappa(L) - \kappa(0) = 0$, we have that $\dot{\kappa}$ is positive on one of the two intervals and negative on the other. For instance, let $\dot{\kappa}(t) > 0$ for $t \in (0, t_0)$ and $\dot{\kappa}(t) < 0$ for $t \in (t_0, L)$.

If c lay on one side of G, then c would intersect the straight line G tangentially in $t = 0$ and $t = t_0$. Then c would by lemma 2.2.20 contain a straight line segment and, in particular, would have infinitely many vertices. We can therefore suppose that, for instance, $c|_{(0,t_0)}$ lies above of G while $c|_{(t_0,L)}$ lies below. This means that we have for the y-component of c that $c^2(t) > 0$ for $t \in (0, t_0)$ and $c^2(t) < 0$ for $t \in (t_0, L)$.

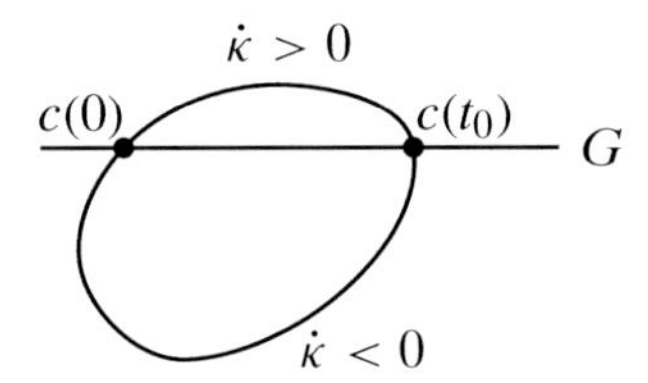

It follows that $\dot{\kappa}(t)c^2(t) > 0$ for all $t \in (0, t_0) \cup (t_0, L)$ and, in particular,

$$\int_0^L \dot{\kappa}(t)c^2(t)dt > 0. \tag{2.8}$$

Integration by parts and the Frenet formulae from proposition 2.2.4 give that

$$\int_0^L \dot{\kappa}(t)c(t)dt = -\int_0^L \kappa(t)\dot{c}(t)dt = \int_0^L \dot{n}(t)dt = n(L) - n(0) = \begin{pmatrix} 0 \\ 0 \end{pmatrix}.$$

We form the scalar product with the unit vector e_2 and obtain

$$\int_0^L \dot{\kappa}(t)c^2(t)dt = 0,$$

contradicting (2.8).

Hence there must be a third vertex, e.g. in $t_1 \in (t_0, L)$. Suppose that there were no fourth vertex. Then we have divided our curve in four arcs on which the derivative of the curvature does not vanish, corresponding to the intervals $(0, t_0)$, (t_0, t_1) and (t_1, L). We see as above that the sign of $\dot{\kappa}$ cannot be the same for all three arcs. We now consider the two arcs on which $\dot{\kappa}$ has the same sign as one arc. In this way we obtain a division of the closed curve into two arcs on which $\dot{\kappa}$ has opposite signs, apart from a single zero of $\dot{\kappa}$ in one of the two arcs. Despite this zero, the corresponding integral remains positive and the same argument as above provides a contradiction. □

To conclude this section about plane curves we consider the following optimization problem, whose agricultural formulation could read as follows: suppose a farmer has a given length of fence, e.g. 10 km. What would the largest paddock that the farmer could fence look like?

The following theorem gives us the answer: the paddock would be circular.

Theorem 2.2.21 (isoperimetric inequality) *Let $G \subset \mathbb{R}^2$ be a bounded region with the trace of the simple closed plane curve c as boundary. Let $A[G]$ be the area of the surface. Then*

$$4\pi A[G] \leq L[c]^2.$$

We have equality if and only if c is a circle.

To prove the isoperimetric inequality we first need a lemma which tells us how to find the surface area of G using the boundary curve c.

Lemma 2.2.22 *Let $G \subset \mathbb{R}^2$ be a bounded region, bounded by the simple closed plane curve c. Let $c(t) = (x(t), y(t))^\top$ be a periodic parametrisation of c with period L, which winds in a mathematically positive sense around the surface, i.e. with winding number $+1$. Then*

$$A[G] = -\int_0^L \dot{x}(t)y(t)dt = \int_0^L x(t)\dot{y}(t)dt = \frac{1}{2}\int_0^L (x(t)\dot{y}(t) - \dot{x}(t)y(t))\,dt.$$

Proof of the lemma We can assume without loss of generality that c is parametrised by arc-length, since the substitution rule shows that the values of the integrals do not change under an orientation-preserving reparametrisation.

To prove this lemma we will make an exception and refer to a theorem which is only proved later in this book, namely the divergence theorem, see theorem 5.1.7. It states in our case that for every smooth map $V = (v^1, v^2) : \bar{G} \to \mathbb{R}^2$

$$\int_G \left(\frac{\partial v^1}{\partial x} + \frac{\partial v^2}{\partial y} \right) dx\,dy = -\int_0^L \langle V(c(t)), n(t) \rangle\, dt. \tag{2.9}$$

The negative sign comes from the fact that the curve is traversed in a mathematically positive sense and hence $n(t)$ points into the surface. Applying this to the map $V = \mathrm{Id}$ we obtain on the one hand

$$\int_G \left(\frac{\partial v^1}{\partial x} + \frac{\partial v^2}{\partial y} \right) dx\, dy = \int_G \left(\frac{\partial x}{\partial x} + \frac{\partial y}{\partial y} \right) dx\, dy = 2 \int_G dx\, dy = 2A[G] \quad (2.10)$$

and on the other hand

$$\begin{aligned} \int_0^L \langle V(c(t)), n(t) \rangle\, dt &= \int_0^L \left\langle \begin{pmatrix} x(t) \\ y(t) \end{pmatrix}, \begin{pmatrix} -\dot{y}(t) \\ \dot{x}(t) \end{pmatrix} \right\rangle dt \\ &= \int_0^L (-x(t)\dot{y}(t) + \dot{x}(t)y(t))\, dt. \end{aligned} \quad (2.11)$$

From (2.9), (2.10) and (2.11) it follows that

$$2A[G] = \int_0^L (x(t)\dot{y}(t) - \dot{x}(t)y(t))\, dt.$$

The other two equations of the lemma follow by integration by parts. □

Proof of the isoperimetric inequality (a) Let $c(t) = (x(t), y(t))^\top$ be a parametrisation of c by arc-length. Then c has period $L = L[c]$. After possibly reversing the orientation of c we can assume that c traverses the surface in a mathematically positive sense. We consider the complex-valued function

$$z : \mathbb{R} \to \mathbb{C}, \qquad z(t) = x\left(\frac{L}{2\pi}t\right) + i \cdot y\left(\frac{L}{2\pi}t\right).$$

Apart from the parameter transformation $t \mapsto (L/2\pi)t$ the map z simply parametrises the curve c, if one identifies $\mathbb{R}^2$ with the complex plane $\mathbb{C}$. In any case z is a periodic function with period 2π. We expand z as a Fourier series, see [18, Ch. XII]:

$$z(t) = \sum_{k=-\infty}^{\infty} c_k e^{ikt},$$

with Fourier coefficients $c_k \in \mathbb{C}$. We can express the length of c in terms of the Fourier coefficients, since on the one hand

$$\int_0^{2\pi} |\dot{z}(t)|^2 dt = \int_0^{2\pi} \left(\frac{L}{2\pi}\right)^2 \underbrace{\left\| \dot{c}\left(\frac{Lt}{2\pi}\right) \right\|^2}_{=1} dt = \frac{L^2}{2\pi}.$$

On the other hand

$$\dot{z}(t) = \sum_{k=-\infty}^{\infty} c_k ik e^{ikt}$$

and hence

$$\int_0^{2\pi} |\dot{z}(t)|^2 dt = \int_0^{2\pi} \dot{z}(t)\bar{\dot{z}}(t) dt = \int_0^{2\pi} \sum_{k,\ell=-\infty}^{\infty} c_k \bar{c}_\ell k\ell e^{i(k-\ell)t} dt$$
$$= \sum_{k=-\infty}^{\infty} |c_k|^2 \cdot k^2 \cdot 2\pi.$$

It follows that

$$L[c]^2 = (2\pi)^2 \sum_{k=-\infty}^{\infty} k^2 |c_k|^2. \tag{2.12}$$

(b) We now express the area of G in terms of the Fourier coefficients. We denote the real part of a complex number w as $\Re(w)$. By lemma 2.2.22 we have that

$$\begin{aligned}
2A[G] &= \int_0^L (x(t)\dot{y}(t) - \dot{x}(t)y(t))\, dt \\
&= \int_0^L \Re\left(z(2\pi t/L) \cdot \frac{2\pi}{L} \cdot i \cdot \bar{\dot{z}}(2\pi t/L) \right) dt \\
&= \int_0^{2\pi} \Re\left(z(s) \cdot i \cdot \bar{\dot{z}}(s) \right) ds \\
&= \int_0^{2\pi} \Re\left(\sum_{k=-\infty}^{\infty} c_k e^{iks} \cdot i \cdot \sum_{\ell=-\infty}^{\infty} \bar{c}_\ell (-i)\ell e^{-i\ell s} \right) ds \\
&= \Re \int_0^{2\pi} \sum_{k,\ell=-\infty}^{\infty} \ell c_k \bar{c}_\ell e^{i(k-\ell)s} ds \\
&= \sum_{k=-\infty}^{\infty} k |c_k|^2 2\pi.
\end{aligned} \tag{2.13}$$

(c) From (2.12) and (2.13) we conclude that

$$\frac{A[G]}{\pi} = \sum_{k=-\infty}^{\infty} k|c_k|^2 \le \sum_{k=-\infty}^{\infty} k^2 |c_k|^2 = \frac{L[c]^2}{(2\pi)^2}$$

and thus

$$4\pi A[G] \le L[c]^2.$$

We have equality exactly when

$$\sum_{k=-\infty}^{\infty} k|c_k|^2 = \sum_{k=-\infty}^{\infty} k^2|c_k|^2,$$

i.e. exactly when all $c_k = 0$ for $k \neq 0, 1$. This means precisely that

$$z(t) = c_0 + c_1 \cdot e^{it},$$

i.e. c describes a circle. □

There are numerous other proofs of the isoperimetric inequality. The interested reader can find a collection in [4, §29].

Exercise 2.12 The set $P = \{(x, y)^\top \in \mathbb{R}^2 \mid y^2 = x^3, y > 0\}$ describes the upper branch of ***Neil's parabola***. Find a regular parametrisation and show that the length of the segment of the curve from the origin to the point $(x, y)^\top \in P$ is given by

$$\tfrac{1}{27}\left((9x + 4)^{3/2} - 8\right).$$

Further show that the curvature takes (for the right choice of orientation) all values from $(0, \infty)$.

Hint Use exercise 2.10.

Exercise 2.13 The ***clothoid*** is given by the regular parametrisation

$$c(t) = \begin{pmatrix} \sqrt{\pi} \displaystyle\int_0^t \cos\left(\frac{\pi\tau^2}{2}\right) d\tau \\ \sqrt{\pi} \displaystyle\int_0^t \sin\left(\frac{\pi\tau^2}{2}\right) d\tau \end{pmatrix}.$$

Show that the curvature at each point on the curve coincides, up to the sign of the curvature, with the length of the segment of curve from the point under consideration to the origin. Sketch the curve.

Exercise 2.14 Let $c : I \to \mathbb{R}^2$ be a plane curve parametrised by arc-length. Let $t_0 \in I$ with $\kappa(t_0) \neq 0$. The ***osculating circle*** to c at t_0 is the circle with centre $c(t_0) + (1/\kappa(t_0))n(t_0)$ and radius $1/|\kappa(t_0)|$.

Show that if the circle is parametrised by arc-length and the orientation is chosen correctly, then the osculating circle touches the curve c at the point $c(t_0)$ to second order, i.e. the first and second derivatives coincide.

Exercise 2.15 The set of all centres of osculating circles of a curve c is called the ***evolute*** of c. Show that the evolute of the parabola $\{(x, y)^\top \in \mathbb{R}^2 \mid y = x^2\}$ is given by

$$\left\{ \begin{pmatrix} X \\ Y \end{pmatrix} \in \mathbb{R}^2 \;\middle|\; \left(Y - \frac{1}{2}\right)^3 = \frac{27}{16} X^2 \right\}.$$

2.3 Space curves

We will now consider those curves that run through three-dimensional space, i.e. take values in $\mathbb{R}^3$.

Definition 2.3.1 A parametrised curve $c : I \to \mathbb{R}^3$ is called a ***parametrised space curve***. ***Regular parametrised space curves***, ***space curves*** and ***oriented space curves*** are defined analogously.

Unlike for plane curves it is now not straightforward to define a normal field. If $c : I \to \mathbb{R}^3$ is a space curve parametrised by arc-length, then the vectors perpendicular to the velocity vector $\dot{c}(t)$ form a plane, the *unit* vectors perpendicular to the velocity vector form a circle.

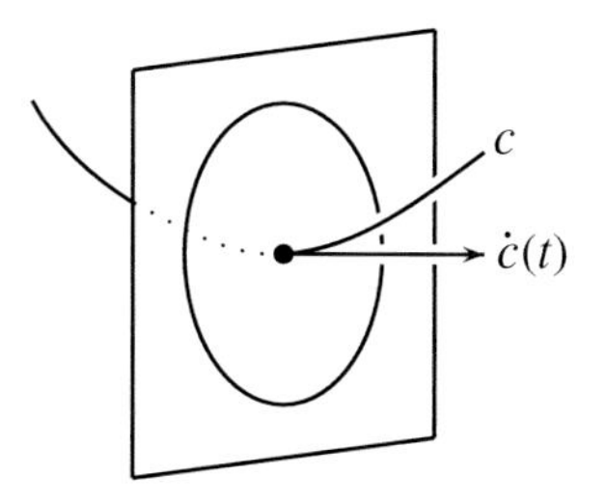

In the case of plane curves there were two perpendicular unit vectors. We used this fact to define the normal vector. Which normal vector should we choose in the case of space curves? The definition of curvature is a priori also a problem, since it required the normal vector. Let us recall that the sign of the curvature described whether a particular curve curves to the left or to the right. What should it mean in the context of space curves?

However, there is a way to avoid this problem. Let us recall that for plane curves

$$\ddot{c}(t) = \kappa(t) \cdot n(t). \tag{2.14}$$

It follows that

$$|\kappa(t)| = \|\ddot{c}(t)\|.$$

If we give up the sign of the curvature, then we can define it without reference to the normal field. We will therefore define the curvature of a space curve as follows.

Definition 2.3.2 Let $c : I \to \mathbb{R}^3$ be a space curve parametrised by arc-length. The function $\kappa : I \to \mathbb{R}$, $\kappa(t) := \|\ddot{c}(t)\|$, is called the ***curvature*** of c.

Again curvature is a measure of how much a curve deviates from a straight line. Concretely, if c is a space curve parametrised by arc-length, then c is a straight line if and only if $\ddot{c} \equiv 0$, i.e. if $\kappa \equiv 0$. But now the curvature is always ≥ 0. It no longer makes sense to talk about a curve curving to the left or to the right.

Attention We can also regard plane curves as space curves, since the plane is contained in three-dimensional space, e.g. as the x–y plane. We therefore now have two different definitions for the curvature of plane curves. If $\tilde{c} : I \to \mathbb{R}^2$ is a plane curve parametrised by arc-length and with curvature $\tilde{\kappa} : I \to \mathbb{R}$ and $c = (\tilde{c}, 0) : I \to \mathbb{R}^3$ is the same parametrised curve, considered as a space curve with curvature $\kappa : I \to \mathbb{R}$, then

$$\kappa(t) = \|\ddot{c}(t)\| = \|(\ddot{\tilde{c}}(t), 0)\| = \|\ddot{\tilde{c}}(t)\| = |\tilde{\kappa}(t)|.$$

Knowing what the curvature of a space curve is, we can now define the normal field using (2.14). However, this will only work if the curvature does not vanish.

Definition 2.3.3 Let $c : I \to \mathbb{R}^3$ be a space curve parametrised by arc-length. Let $t_0 \in I$ and $\kappa(t_0) \neq 0$. Then

$$n(t_0) := \frac{\ddot{c}(t_0)}{\kappa(t_0)} = \frac{\ddot{c}(t_0)}{\|\ddot{c}(t_0)\|}$$

is called the ***normal vector*** of c at t_0.

The same argument as for the planar case shows us that $n(t_0)$ is indeed always perpendicular to $\dot{c}(t_0)$:

$$0 = \frac{d}{dt} \langle \dot{c}, \dot{c} \rangle = 2 \langle \ddot{c}, \dot{c} \rangle .$$

As we now have the normal vector we can define the "binormal" vector to obtain a complete orthonormal basis of $\mathbb{R}^3$.

Definition 2.3.4 Let $c : I \to \mathbb{R}^3$ be a space curve parametrised by arc-length. Let $t_0 \in I$ and $\kappa(t_0) \neq 0$. Then

$$b(t_0) := \dot{c}(t_0) \times n(t_0)$$

is called the ***binormal vector*** of c at t_0.

Here $\times$ denotes the cross product in $\mathbb{R}^3$. The vector product $x \times y$ of two vectors $x, y \in \mathbb{R}^3$ has the property that $x \times y$ is perpendicular to x and y, and that $x, y, x \times y$ form a positively oriented basis of $\mathbb{R}^3$ if x and y are linearly independent. If x and y are orthogonal and of length 1, then $x, y, x \times y$ form a positively oriented orthonormal basis.

Definition 2.3.5 The orthonormal basis $(\dot{c}(t_0), n(t_0), b(t_0))$ is called the ***Frenet dreibein*** or ***Frenet trihedron*** of c at t_0.

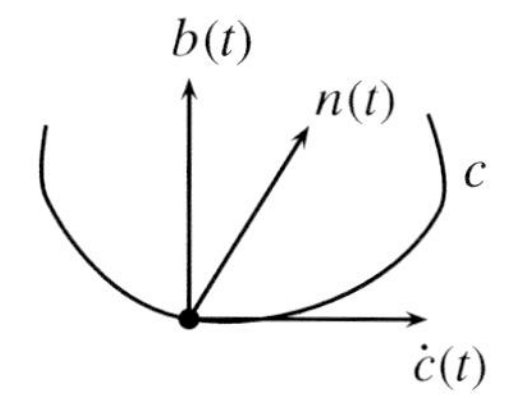

Attention The Frenet dreibein is only defined for those t for which $\kappa(t) \neq 0$. The curvature of a plane curve says how much the velocity vector turns in the direction of the normal vector. We can introduce a similar notion for a space curve. It will measure how much the normal vector turns out of the plane spanned by itself and the velocity vector, i.e. how much it moves in the direction of the binormal vector.

Definition 2.3.6 Let $c : I \to \mathbb{R}^3$ be a space curve parametrised by arc-length. Let $t_0 \in I$ with $\kappa(t_0) \neq 0$, let $(\dot{c}(t_0), n(t_0), b(t_0))$ be the Frenet dreibein of c at t_0. Then

$$\tau(t_0) := \langle \dot{n}(t_0), b(t_0) \rangle$$

is called the ***torsion*** of c at t_0.

Proposition 2.3.7 (Frenet formulae) Let $c : I \to \mathbb{R}^3$ be a space curve parametrised by arc-length with positive curvature, $\kappa(t) > 0$ for all $t \in I$. Let (v, n, b) be the Frenet dreibein of c, and let τ be the torsion. Then

$$(\dot{v}(t), \dot{n}(t), \dot{b}(t)) = (v(t), n(t), b(t)) \begin{pmatrix} 0 & -\kappa(t) & 0 \\ \kappa(t) & 0 & -\tau(t) \\ 0 & \tau(t) & 0 \end{pmatrix}.$$

Proof The equation $\dot{v} = \kappa \cdot n$ is exactly the definition of the normal vector. Thus the first column of the 3×3-matrix is correct.

The second column follows from $\langle \dot{n}, v \rangle = (d/dt) \langle n, v \rangle - \langle n, \dot{v} \rangle = 0 - \kappa = -\kappa$, from $\langle \dot{n}, n \rangle = \frac{1}{2}(d/dt) \langle n, n \rangle = 0$ and from the definition of τ.

The third column follows because $\langle \dot{b}, v \rangle = (d/dt) \langle b, v \rangle - \langle b, \dot{v} \rangle = 0 - \kappa \langle b, n \rangle = 0$, because $\langle \dot{b}, n \rangle = (d/dt) \langle b, n \rangle - \langle b, \dot{n} \rangle = 0 - \tau = -\tau$ and because $\langle \dot{b}, b \rangle = \frac{1}{2}(d/dt) \langle b, b \rangle = 0$. □

Example 2.3.8 Let us consider the ***helix*** as an example. We choose a parametrisation similar to the one from example 2.1.4:

$$c : \mathbb{R} \to \mathbb{R}^3, \qquad c(t) = \begin{pmatrix} \cos(t/\sqrt{2}) \\ \sin(t/\sqrt{2}) \\ t/\sqrt{2} \end{pmatrix}.$$

We calculate $\dot{c}(t) = \frac{1}{\sqrt{2}}\left(-\sin(t/\sqrt{2}), \cos(t/\sqrt{2}), 1\right)^\top$. In particular,

$$\|\dot{c}(t)\|^2 = \tfrac{1}{2}(1+1) = 1,$$

i.e. c is parametrised by arc-length. Further, we find

$$\ddot{c}(t) = \tfrac{1}{2}\left(-\cos(t/\sqrt{2}), -\sin(t/\sqrt{2}), 0\right)^\top$$

and thus

$$\kappa(t) = \|\ddot{c}(t)\| = \tfrac{1}{2}.$$

As the curvature does not vanish anywhere, we can find the normal vector for all t. We obtain

$$n(t) = \frac{\ddot{c}(t)}{\kappa(t)} = \left(-\cos(t/\sqrt{2}), -\sin(t/\sqrt{2}), 0\right)^\top.$$

For the binormal vector we find

$$\begin{aligned} b(t) &= \dot{c}(t) \times n(t) \\ &= \frac{1}{\sqrt{2}} \begin{pmatrix} -\sin(t/\sqrt{2}) \\ \cos(t/\sqrt{2}) \\ 1 \end{pmatrix} \times \begin{pmatrix} -\cos(t/\sqrt{2}) \\ -\sin(t/\sqrt{2}) \\ 0 \end{pmatrix} \\ &= \frac{1}{\sqrt{2}} \begin{pmatrix} \sin(t/\sqrt{2}) \\ -\cos(t/\sqrt{2}) \\ 1 \end{pmatrix}. \end{aligned}$$

We can now calculate the torsion as

$$\begin{aligned} \tau(t) &= \langle \dot{n}(t), b(t) \rangle \\ &= \left\langle \begin{pmatrix} \frac{1}{\sqrt{2}}\sin(t/\sqrt{2}) \\ -\frac{1}{\sqrt{2}}\cos(t/\sqrt{2}) \\ 0 \end{pmatrix}, \frac{1}{\sqrt{2}} \begin{pmatrix} \sin(t/\sqrt{2}) \\ -\cos(t/\sqrt{2}) \\ 1 \end{pmatrix} \right\rangle \\ &= \frac{1}{2}. \end{aligned}$$

Exercise 2.16 Show that the curvature and the torsion of a space curve are invariant under orientation-preserving Euclidean motions. More precisely, if $c : I \to \mathbb{R}^3$ is a space curve parametrised by arc-length with curvature $\kappa > 0$ and if $F \in E(3)$ is an orientation-preserving Euclidean motion, i.e. $F(x) = Ax + p$, $A \in \mathrm{SO}(3)$, $p \in \mathbb{R}^3$, then the curvature and the torsion of the curve $\tilde{c} := F \circ c$ satisfy

$$\tilde{\kappa} = \kappa, \ \tilde{\tau} = \tau.$$

What happens if the Euclidean motion is orientation-reversing?

We will now see that one can find a space curve if curvature and torsion are given. According to exercise 2.16 this space curve can at most be unique up to Euclidean motions. Indeed, this is the only ambiguity.

Theorem 2.3.9 (fundamental theorem of space curve theory) *Let $I \subset \mathbb{R}$ be an interval, let $\kappa, \tau : I \to \mathbb{R}$ be smooth functions, $\kappa > 0$. Then there exists a space curve $c : I \to \mathbb{R}^3$ parametrised by arc-length and with curvature κ and torsion τ. This space curve is unique up to post-composition with orientation-preserving Euclidean motions.*

Proof (a) We consider the system of first-order linear ordinary differential equations

$$\frac{d}{dt}(c, v, n, b) = (c, v, n, b) \cdot \begin{pmatrix} 0 & 0 & 0 & 0 \\ 1 & 0 & -\kappa & 0 \\ 0 & \kappa & 0 & -\tau \\ 0 & 0 & \tau & 0 \end{pmatrix}, \tag{2.15}$$

where $c, v, n, b : I \to \mathbb{R}^3$ are functions to be found.

Let $t_0 \in I$. The existence and uniqueness theorem for such systems of differential equations [18, Ch. XIX] states that we can find exactly one solution that satisfies the initial conditions

$$\begin{aligned} c(t_0) &= 0, \\ v(t_0) &= e_1, \\ n(t_0) &= e_2, \\ b(t_0) &= e_3. \end{aligned}$$

The linearity of the system ensures that the solution is defined on the whole interval I.

We have chosen the initial values in such a way that v, n and b at $t = t_0$ form an orthonormal basis of $\mathbb{R}^3$. Let us first show that this remains true for all $t \in I$. From the system of equations it follows that

$$\frac{d}{dt}\langle v,v\rangle = 2\langle\dot{v},v\rangle = 2\kappa\cdot\langle n,v\rangle$$

and

$$\frac{d}{dt}\langle n,v\rangle = \langle\dot{n},v\rangle + \langle n,\dot{v}\rangle = -\kappa\langle v,v\rangle + \tau\langle b,v\rangle + \kappa\langle n,n\rangle\,.$$

Using (2.15), one analogously calculates the derivatives of $\langle n,n\rangle$, $\langle b,b\rangle$, $\langle b,v\rangle$ and $\langle b,n\rangle$ and obtains the following system of differential equations

$$\frac{d}{dt}\begin{pmatrix}\langle v,v\rangle\\ \langle n,n\rangle\\ \langle b,b\rangle\\ \langle b,v\rangle\\ \langle b,n\rangle\\ \langle n,v\rangle\end{pmatrix} = \begin{pmatrix}0 & 0 & 0 & 0 & 0 & 2\kappa\\ 0 & 0 & 0 & 0 & 2\tau & -2\kappa\\ 0 & 0 & 0 & 0 & -2\tau & 0\\ 0 & 0 & 0 & 0 & \kappa & -\tau\\ 0 & -\tau & \tau & -\kappa & 0 & 0\\ -\kappa & \kappa & 0 & \tau & 0 & 0\end{pmatrix}\cdot\begin{pmatrix}\langle v,v\rangle\\ \langle n,n\rangle\\ \langle b,b\rangle\\ \langle b,v\rangle\\ \langle b,n\rangle\\ \langle n,v\rangle\end{pmatrix} \tag{2.16}$$

with the initial conditions

$$\left.\begin{pmatrix}\langle v,v\rangle\\ \langle n,n\rangle\\ \langle b,b\rangle\\ \langle b,v\rangle\\ \langle b,n\rangle\\ \langle n,v\rangle\end{pmatrix}\right|_{t=t_0} = \begin{pmatrix}1\\1\\1\\0\\0\\0\end{pmatrix}.$$

But it is obvious that the constant function

$$t\mapsto\begin{pmatrix}1\\1\\1\\0\\0\\0\end{pmatrix}$$

also satisfies the system of differential equations (2.16) with the same initial conditions. From the uniqueness theorem for ordinary differential equations we obtain that

$$\begin{pmatrix}\langle v,v\rangle\\ \langle n,n\rangle\\ \langle b,b\rangle\\ \langle b,v\rangle\\ \langle b,n\rangle\\ \langle n,v\rangle\end{pmatrix} \equiv \begin{pmatrix}1\\1\\1\\0\\0\\0\end{pmatrix}.$$

This means that $v(t)$, $n(t)$, $b(t)$ form an orthonormal basis of $\mathbb{R}^3$ for all $t\in I$, not only for $t=t_0$.

The orientation of this orthonormal basis remains positive for all $t \in I$, since for reasons of continuity the determinant of the matrix $\big(v(t), n(t), b(t)\big)$ cannot jump from $+1$ to -1.

Furthermore, it follows from the system of equations (2.15), that $\dot{c} = v$, i.e. c is a space curve parametrised by arc-length. The Frenet formulae state exactly that (v, n, b) is the Frenet dreibein with curvature κ and torsion τ. This proves the existence of a space curve as specified in the claim.

(b) For the uniqueness let $\tilde{c}$ be another space curve with curvature κ and torsion τ. We set $A := \big(\dot{\tilde{c}}(t_0), \tilde{n}(t_0), \tilde{b}(t_0)\big)^{-1}$ and $p := -A \cdot \tilde{c}(t_0)$, where $(\dot{\tilde{c}}, \tilde{n}, \tilde{b})$ is the Frenet dreibein of $\tilde{c}$. But the moving dreibein always forms a positively oriented orthonormal basis of $\mathbb{R}^3$, so $A \in \mathrm{SO}(3)$. We consider the orientation-preserving Euclidean motion F, defined by $F(x) := Ax + p$. By exercise 2.16, the space curve $\hat{c} := F \circ \tilde{c}$ also has curvature κ and torsion τ. Further, by the definition of F we have

$$\hat{c}(t_0) = 0, \quad \Big(\dot{\hat{c}}(t_0), \hat{n}(t_0), \hat{b}(t_0)\Big) = (e_1, e_2, e_3).$$

Thus (c, v, n, b) and $(\hat{c}, \dot{\hat{c}}, \hat{n}, \hat{b})$ both satisfy the system of differential equations (2.15) with the same initial conditions and are therefore the equal. In particular, $c = \hat{c} = F \circ \tilde{c}$. □

Exercise 2.17 Prove the analogous fundamental theorem of plane curve theory: *Let $I \subset \mathbb{R}$ be an interval, let $\kappa : I \to \mathbb{R}$ be a smooth function. Then there exists a plane curve $c : I \to \mathbb{R}^2$ parametrised by arc-length and with curvature κ. This plane curve is unique up to post-composition with orientation-preserving Euclidean motions.*

Let us recall definition 2.1.17, in which we called a tuple $P = (a_1, \ldots, a_m)$ of points in $\mathbb{R}^n$ a polygon, where the a_i are interpreted as the vertices of the polygon and imagine two successive vertices a_i and a_{i+1} to be connected by a line segment.

In the following we will say that a polygon $P = (a_1, \ldots, a_m)$ is a ***closed polygon*** if in addition to the condition $a_i \neq a_{i-1}$ for all $i = 2, \ldots, m$ it also satisfies that $a_1 \neq a_m$.

In this case we can also connect a_m with a_1 and use the convention $a_{i+m} = a_i$.

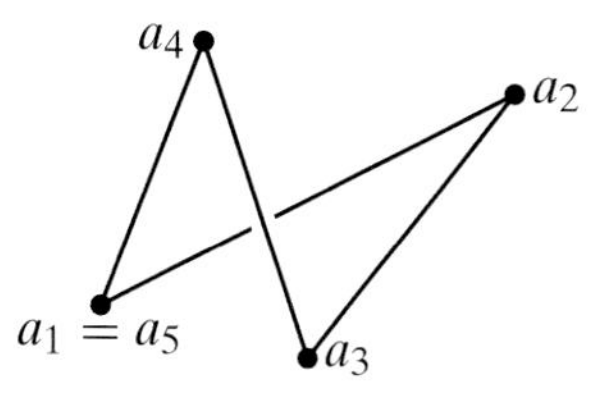

We use α_i to denote the unoriented angle at the vertex a_i, i.e. the number $\alpha_i \in [0, \pi]$, given by

$$\cos(\alpha_i) = \frac{\langle a_i - a_{i-1}, a_{i+1} - a_i \rangle}{\|a_i - a_{i-1}\| \cdot \|a_{i+1} - a_i\|}. \tag{2.17}$$

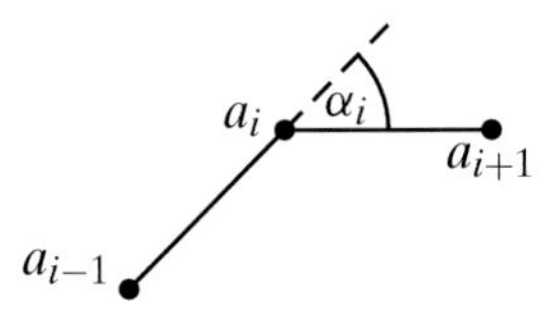

We see that the condition $a_i \neq a_{i-1}$ is now necessary to ensure that we do not divide by 0 in (2.17).

Obviously the angle at the vertex a_i vanishes exactly when a_{i-1}, a_i, a_{i+1} lie on a straight line in this order. This angle is therefore, like curvature for smooth curves, a measure for the deviation of the polygon from a straight line. We will now investigate the exact connection between curvature and angles.

Definition 2.3.10 Let $P = (a_1, \ldots, a_m)$ be a closed polygon with angles α_i. Then

$$\kappa(P) := \sum_{i=1}^{m} \alpha_i$$

is called the ***total curvature*** of P.

Example 2.3.11 Let us consider a triangle $P = (a_1, a_2, a_3)$. The exterior angle α_i and the interior angle β_i at a vertex a_i always add up to π, i.e. $\alpha_i + \beta_i = \pi$. According to theorem 1.2.7 the sum of the interior angles is $\beta_1 + \beta_2 + \beta_3 = \pi$. It follows that

$$\begin{aligned}\kappa(P) &= \alpha_1 + \alpha_2 + \alpha_3 \\ &= \alpha_1 + \beta_1 + \alpha_2 + \beta_2 + \alpha_3 + \beta_3 - (\beta_1 + \beta_2 + \beta_3) \\ &= 3\pi - \pi = 2\pi.\end{aligned}$$

Lemma 2.3.12 *Let P_1 and P_2 be closed polygons in $\mathbb{R}^3$. Suppose that the polygon P_2 is the result of adding a vertex to P_1. Then*

$$\kappa(P_1) \leq \kappa(P_2).$$

Suppose that the additional vertex a was added to P_1 between a_i and a_{i+1}. If we have equality $\kappa(P_1) = \kappa(P_2)$, then one of the following is the case:

(1) *the points* a_i, a, a_{i+1} *lie on a straight line, or*
(2) *the points* $a_{i-1}, a_i, a, a_{i+1}, a_{i+2}$ *lie in a plane.*

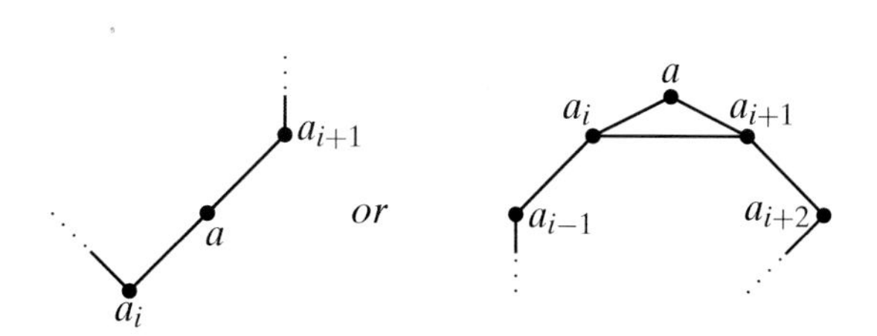

Proof (a) We denote the angles of P_1 as usual by α_k. For the angles of P_2 at the vertex a_k we write β_k, for the one at the vertex a we write β. Then $\alpha_k = \beta_k$, unless $k = i$ or $k = i + 1$.

Let us now consider the triangle (a_i, a, a_{i+1}) and denote its interior angles at the vertices a_k by γ_k, $k = i, i + 1$.

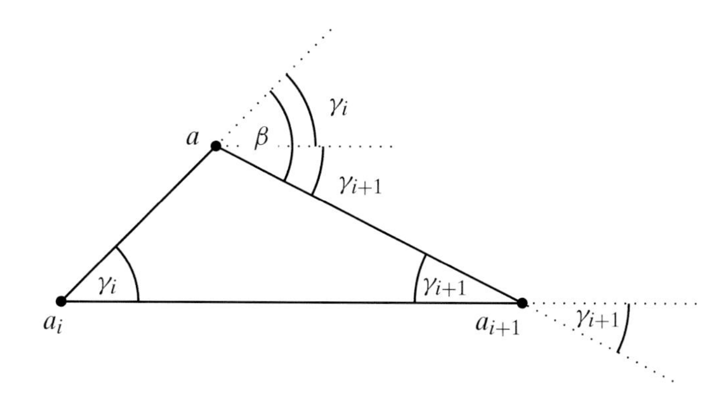

Then

$$\beta = \gamma_i + \gamma_{i+1}.$$

The angle between two unit vectors X and Y is precisely the spherical distance between X and Y, regarded as points on the two-dimensional sphere, i.e. the length of the shortest connecting great circle. Let us therefore consider the spherical triangle with vertices

$$\frac{a_i - a_{i-1}}{\|a_i - a_{i-1}\|}, \quad \frac{a - a_i}{\|a - a_i\|} \quad \text{and} \quad \frac{a_{i+1} - a_i}{\|a_{i+1} - a_i\|}.$$

The spherical side lengths are α_i, β_i and γ_i.

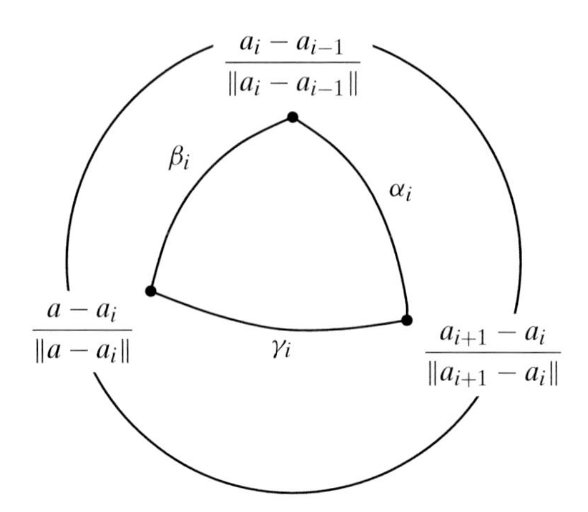

The triangle inequality gives that

$$\alpha_i \leq \beta_i + \gamma_i. \tag{2.18}$$

Using the spherical triangle with the vertices

$$\frac{a_{i+1} - a_i}{\|a_{i+1} - a_i\|}, \quad \frac{a_{i+1} - a}{\|a_{i+1} - a\|} \quad \text{and} \quad \frac{a_{i+2} - a_{i+1}}{\|a_{i+2} - a_{i+1}\|}$$

we obtain analogously:

$$\alpha_{i+1} \leq \beta_{i+1} + \gamma_{i+1}. \tag{2.19}$$

We summarise:

$$\begin{aligned} \kappa(P_2) - \kappa(P_1) &= \beta_i + \beta + \beta_{i+1} - \left(\alpha_i + \alpha_{i+1}\right) \\ &= (\beta_i - \alpha_i) + \beta + \left(\beta_{i+1} - \alpha_{i+1}\right) \\ &\geq -\gamma_i + \beta - \gamma_{i+1} \\ &= 0, \end{aligned}$$

and hence

$$\kappa(P_2) \geq \kappa(P_1).$$

(b) Let us now assume that we are dealing with the equality case:

$$\kappa(P_2) = \kappa(P_1).$$

Then we must have equality in the triangle inequality (2.18),

$$\alpha_i = \beta_i + \gamma_i.$$

Thus the vertices of the spherical triangle lie on a great circle, i.e. $a_i - a_{i-1}$, $a - a_i$ and $a_{i+1} - a_i$ lie in a two-dimensional subspace $V_1 \subset \mathbb{R}^3$. For suitable $\lambda_1, \lambda_2 \in \mathbb{R}$ we have that

$$a_i - a_{i-1} = \lambda_1 \cdot (a - a_i) + \lambda_2 \cdot \big(a_{i+1} - a_i\big)$$

and thus

$$a_{i-1} = -\lambda_1 \cdot a - \lambda_2 \cdot a_{i+1} + (1 + \lambda_1 + \lambda_2) \cdot a_i. \tag{2.20}$$

We now suppose that a_i, a and a_{i+1} are not collinear, else we would be dealing with case (1). Hence those three points span an affine plane $E \subset \mathbb{R}^3$. Because of (2.20) we also have $a_{i-1} \in E$. It can be shown analogously that $a_{i+2} \in E$. This proves (2). □

Definition 2.3.13 For a periodic space curve c parametrised by arc-length and with period L we can, analogously to the total curvature of a polygon, define the ***total curvature of a closed space curve*** by

$$\kappa(c) := \int_0^L \kappa(t)dt.$$

Remark The parametrisation by arc-length of a closed space curve is periodic and unique up to parameter transformations of the form $t \mapsto \pm t + t_0$. Such parameter transformations do not change the value of the total curvature. We can therefore not only talk about the total curvature of a parametrised closed space curve, but also about the total curvature of a closed space curve.

If c is a periodic space curve parametrised by arc-length and with period L, then we say that the polygon $P = (a_0, \ldots, a_{m-1})$ is ***inscribed in*** c, if there exists a partition $0 \leq t_1 < t_2 < \cdots < t_m < L$ with $a_i = c(t_i)$.

We have already met inscribed polygons in the context of lengths of curves in proposition 2.1.18. Like there we see that the total curvature of inscribed polygons approximates the total curvature of a closed space curve.

Proposition 2.3.14 (approximation of curvature by means of polygons) *Let c be a closed space curve. Then*

$$\kappa(c) = \sup_P \kappa(P),$$

where the supremum is taken over all polygons P inscribed in c.

Before proving the proposition we derive the following lemma.

Lemma 2.3.15 *Let $c : [t_0, t_1] \to \mathbb{R}^3$ be a segment of a curve parametrised by arc-length with $\|\ddot{c}(u) - \ddot{c}(v)\| < \varepsilon$ for all $u, v \in [t_0, t_1]$. Let $\tau := \frac{1}{2}(t_1 + t_0)$. Then*

$$\left\| \frac{c(t_1) - c(t_0)}{t_1 - t_0} - \dot{c}(\tau) \right\| < \frac{\varepsilon}{4}(t_1 - t_0).$$

Proof We compute

$$\int_{\tau}^{t_1}\left(\int_{\tau}^{v}(\ddot{c}(u)-\ddot{c}(\tau))du\right)dv$$
$$=\int_{\tau}^{t_1}(\dot{c}(v)-\dot{c}(\tau)-(v-\tau)\ddot{c}(\tau))\,dv$$
$$=(t_1-\tau)(-\dot{c}(\tau)+\tau\ddot{c}(\tau))+c(t_1)-c(\tau)-\tfrac{1}{2}\left(t_1^2-\tau^2\right)\ddot{c}(\tau)$$
$$=c(t_1)-c(\tau)-(t_1-\tau)\dot{c}(\tau)-\tfrac{1}{2}(t_1-\tau)^2\ddot{c}(\tau). \tag{2.21}$$

We obtain analogously that

$$\int_{t_0}^{\tau}\left(\int_{v}^{\tau}(\ddot{c}(\tau)-\ddot{c}(u))du\right)dv=c(\tau)-c(t_0)-(\tau-t_0)\dot{c}(\tau)+\tfrac{1}{2}(\tau-t_0)^2\ddot{c}(\tau). \tag{2.22}$$

As $t_1-\tau=\tau-t_0$, addition of (2.21) and (2.22) gives that

$$\int_{\tau}^{t_1}\left(\int_{\tau}^{v}(\ddot{c}(u)-\ddot{c}(\tau))du\right)dv+\int_{t_0}^{\tau}\left(\int_{v}^{\tau}(\ddot{c}(\tau)-\ddot{c}(u))du\right)dv=c(t_1)-c(t_0)-(t_1-t_0)\dot{c}(\tau). \tag{2.23}$$

Moreover,

$$\left\|\int_{\tau}^{t_1}\left(\int_{\tau}^{v}(\ddot{c}(u)-\ddot{c}(\tau))du\right)dv\right\|\le\int_{\tau}^{t_1}\int_{\tau}^{v}\varepsilon\,du\,dv=\frac{\varepsilon}{2}(t_1-\tau)^2 \tag{2.24}$$

and analogously

$$\left\|\int_{t_0}^{\tau}\left(\int_{v}^{\tau}(\ddot{c}(\tau)-\ddot{c}(u))du\right)dv\right\|\le\frac{\varepsilon}{2}(\tau-t_0)^2. \tag{2.25}$$

Substituting (2.24) and (2.25) into (2.23) yields

$$\|c(t_1)-c(t_0)-(t_1-t_0)\dot{c}(\tau)\|\le\varepsilon(t_1-\tau)^2=\frac{\varepsilon}{4}(t_1-t_0)^2.$$

Division by t_1-t_0 completes the proof. □

We now prove the proposition.

Proof of proposition 2.3.14 We parametrise c by arc-length and show that for every $\varepsilon>0$ there exists a $\delta>0$ such that for every partition $0\le t_1<$

$t_2 < \cdots < t_m < L$ with mesh $< \delta$ we have

$$|\kappa(c) - \kappa(P)| < \varepsilon,$$

where $P = (c(t_1), \ldots, c(t_m))$. The claim follows by lemma 2.3.12.

Let $\varepsilon > 0$. We compute the Taylor expansions

$$2\sin\left(\frac{\beta}{2}\right) = \beta + r_1(\beta)$$

and

$$\sin(\gamma) = \gamma + r_2(\gamma),$$

where the remainder terms satisfy $|r_j(x)| \leq K \cdot |x|^3$ for a suitable constant K. We choose $\varepsilon_1 > 0$ so small that

$$\frac{\varepsilon_1 + K \cdot \varepsilon_1^2 \cdot \kappa(c) + \varepsilon_1 \cdot L/2}{1 - K \cdot \varepsilon_1^2} < \varepsilon$$

and

$$\varepsilon_1^2 < \frac{1}{K}.$$

We observe that the total curvature of a space curve can be interpreted as a length, namely as the length of the spherical curve $\dot{c}$:

$$\kappa(c) = \int_0^L \kappa(t)dt = \int_0^L \|\ddot{c}(t)\|dt = L[\dot{c}].$$

We can already approximate lengths by means of inscribed polygons. According to proposition 2.1.18 there exists a $\delta_1 > 0$ such that for every partition $0 \leq t_0 < t_1 < \cdots < t_{m-1} < L$ with mesh $< \delta_1$ we have

$$\left|\kappa(c) - \sum_{j=1}^{m} \|\dot{c}(t_j) - \dot{c}(t_{j-1})\|\right| < \varepsilon_1.$$

We again used the convention $\dot{c}(t_0) = \dot{c}(t_m)$. We now set $\tau_j := \frac{1}{2}(t_{j+1} + t_j)$ and consider the angles

$$\alpha_j := \angle(c(t_{j+1} - c(t_j), c(t_j) - c(t_{j-1}),$$
$$\beta_j := \angle(\dot{c}(\tau_j), \dot{c}(\tau_{j-1})),$$
$$\gamma_j := \angle(\dot{c}(\tau_j), c(t_{j+1} - c(t_j)).$$

Then

$$\|\dot{c}(t_j) - \dot{c}(t_{j-1})\| = 2\sin\left(\frac{\beta_j}{2}\right)$$

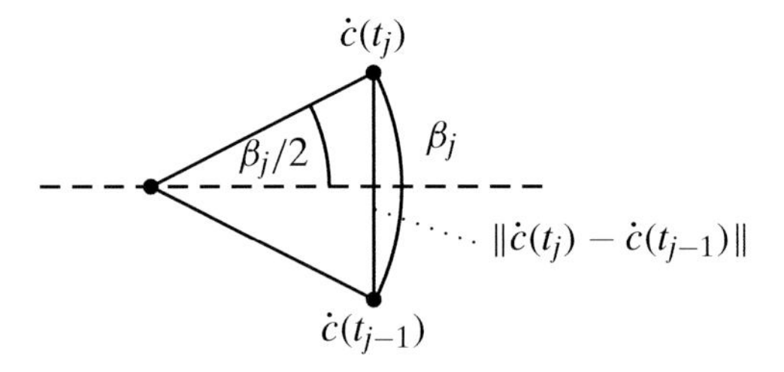

and

$$\sin(\gamma_j) \le \left\| \dot{c}(\tau_j) - \frac{c(t_{j+1}) - c(t_j)}{t_{j+1} - t_j} \right\|. \tag{2.26}$$

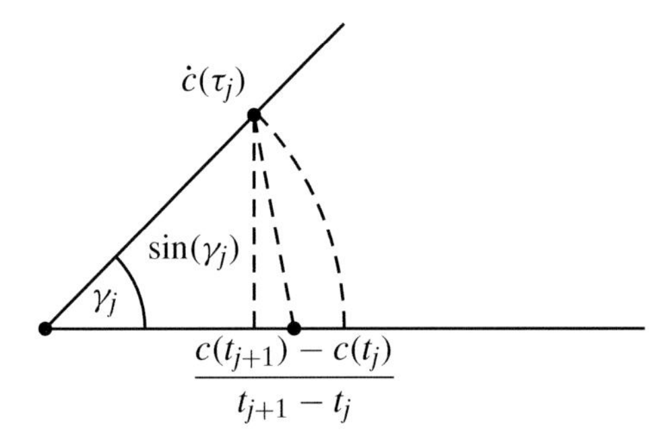

As $\dot{c}$ is uniformly continuous on the compact interval $[0, L]$, there exists a $\delta_2 > 0$ such that

$$\angle(\dot{c}(s), \dot{c}(t)) < \varepsilon_1,$$

whenever $|t-s| < \delta_2$. Analogously the uniform continuity of $\ddot{c}$ on $[0, L]$ implies that there exists a $\delta_3 > 0$ such that

$$\|\ddot{c}(s) - \ddot{c}(t)\| < \varepsilon_1,$$

whenever $|t - s| < \delta_3$. This allows us to apply lemma 2.3.15. Finally we find a $\delta_4 > 0$ such that $\gamma_j \le \varepsilon_1$ if the mesh of the partition is $< \delta_4$.

We set $\delta := \min\{\delta_1, \delta_2, \delta_3, \delta_4\}$. For partitions with mesh $< \delta$ we then have

$$\begin{aligned} |\kappa(c) - \kappa(P)| &= \left|\kappa(c) - \sum_j \alpha_j\right| \\ &\le \left|\kappa(c) - \sum_j \|\dot{c}(t_j) - \dot{c}(t_{j-1})\|\right| \\ &\quad + \left|\sum_j \|\dot{c}(t_j) - \dot{c}(t_{j-1})\| - \sum_j \beta_j\right| + \sum_j |\beta_j - \alpha_j|. \end{aligned} \tag{2.27}$$

We already know that the first term is smaller than ε_1. For the second we obtain

$$\begin{aligned}
&\Big|\sum_j \|\dot{c}(t_j) - \dot{c}(t_{j-1})\| - \sum_j \beta_j\Big| \\
&\quad \le \sum_j \big|r_1(\beta_j)\big| \\
&\quad \le K \cdot \sum_j \beta_j^3 \\
&\quad \overset{|\beta_j| \le \varepsilon_1}{\le} K \cdot \varepsilon_1^2 \sum_j \beta_j \\
&\quad \le K \cdot \varepsilon_1^2 \cdot \Big\{\kappa(c) + \Big|\kappa(c) - \sum_j \|\dot{c}(t_j) - \dot{c}(t_{j-1})\|\Big| \\
&\qquad\qquad + \Big|\sum_j \|\dot{c}(t_j) - \dot{c}(t_{j-1})\| - \sum_j \beta_j\Big|\Big\} \\
&\quad \le K \cdot \varepsilon_1^2 \cdot \kappa(c) + K \cdot \varepsilon_1^3 + K \cdot \varepsilon_1^2 \cdot \Big|\sum_j \|\dot{c}(t_j) - \dot{c}(t_{j-1})\| - \sum_j \beta_j\Big|.
\end{aligned}$$

It follows that

$$\left(1 - K \cdot \varepsilon_1^2\right) \cdot \Big|\sum_j \|\dot{c}(t_j) - \dot{c}(t_{j-1})\| - \sum_j \beta_j\Big| \le K \cdot \varepsilon_1^2 \cdot \kappa(c) + K \cdot \varepsilon_1^3$$

and thus

$$\Big|\sum_j \|\dot{c}(t_j) - \dot{c}(t_{j-1})\| - \sum_j \beta_j\Big| \le \frac{K \cdot \varepsilon_1^2 \cdot \kappa(c) + K \cdot \varepsilon_1^3}{1 - K \cdot \varepsilon_1^2}. \tag{2.28}$$

To estimate the third term of (2.27) we observe that the triangle inequality for the spherical distances gives the inequality

$$|\alpha_j - \beta_j| \le \gamma_{j-1} + \gamma_j. \tag{2.29}$$

By (2.26) and lemma 2.3.15 we have

$$\begin{aligned}\gamma_j &\leq \left\| \dot{c}(\tau_j) - \frac{c(t_{j+1}) - c(t_j)}{t_{j+1} - t_j} \right\| - r_2(\gamma_j) \\ &\leq \frac{\varepsilon_1}{4}(t_{j+1} - t_j) + K \cdot \gamma_j^3 \\ &\leq \frac{\varepsilon_1}{4}(t_{j+1} - t_j) + K\varepsilon_1^2 \cdot \gamma_j\end{aligned}$$

and thus

$$\gamma_j \leq \frac{\varepsilon_1}{4\left(1 - K\varepsilon_1^2\right)}(t_{j+1} - t_j). \tag{2.30}$$

From (2.29) and (2.30) it follows that

$$\sum_j |\alpha_j - \beta_j| \leq \frac{\varepsilon_1}{2\left(1 - K\varepsilon_1^2\right)} L. \tag{2.31}$$

We substitute (2.28) and (2.31) into (2.27) and obtain

$$\begin{aligned}|\kappa(c) - \kappa(P)| &\leq \varepsilon_1 + \frac{K \cdot \varepsilon_1^2 \cdot \kappa(c) + K \cdot \varepsilon_1^3}{1 - K \cdot \varepsilon_1^2} + \frac{\varepsilon_1}{2\left(1 - K\varepsilon_1^2\right)} L \\ &= \frac{\varepsilon_1 + K \cdot \varepsilon_1^2 \cdot \kappa(c) + \varepsilon_1 \cdot L/2}{1 - K \cdot \varepsilon_1^2} \\ &< \varepsilon.\end{aligned}$$

□

Let $c : \mathbb{R} \to \mathbb{R}^3$ be a periodic parametrised space curve with period L. Let $e \in \mathbb{R}^3$ be a unit vector, i.e. $e \in S^2$. We count the local maxima of c in direction e,

$$\begin{aligned}\mu(c, e) &:= |\{\text{local maxima of the function } \mathbb{R} \to \mathbb{R},\ t \mapsto \langle c(t), e\rangle \text{ in } [0, L)\}| \\ &\in \mathbb{N} \cup \{\infty\}.\end{aligned}$$

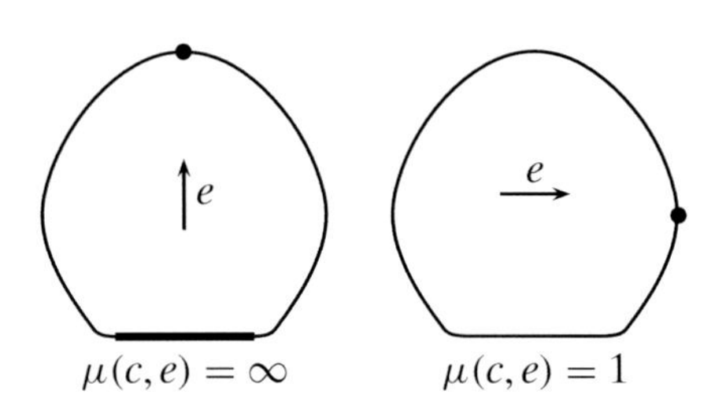

Exercise 2.18 Show that $\mu(c,e) = \mu(c,-e)$.

Definition 2.3.16 We call

$$\mu(c) := \min_{e \in S^2} \mu(c,e)$$

the ***bridge number***[3] of the curve c.

A reparametrisation of c does not change $\mu(c,e)$ and $\mu(c)$. Thus the bridge number of a closed space curve is also defined.

For the definition of $\mu(c,e)$ and $\mu(c)$ the parametrised curve c does not even need to be regular. In particular, we can talk about the bridge number of closed polygons. For this purpose we can, for example, consider a piecewise linear parametrisation c of the polygon $P = (a_0, \ldots, a_{m-1})$, which is given by

$$c(t) = \frac{t - t_j}{t_{j+1} - t_j} \cdot a_{j+1} + \frac{t_{j+1} - t}{t_{j+1} - t_j} \cdot a_j$$

for $t \in [t_j, t_{j+1}]$. This parametrisation satisfies $c(t_j) = a_j$ and connects the successive vertices a_j and a_{j+1} by the corresponding line segment. The $\cdots < t_{j-1} < t_j < t_{j+1} < \cdots$ may be given arbitrarily.

We now show that the mean of $\mu(c,e)$ over all directions e essentially gives the total curvature of c.

Proposition 2.3.17 *Let c be a closed space curve. Then*

$$\frac{1}{A[S^2]} \int_{S^2} \mu(c,e) dA(e) = \frac{\kappa(c)}{2\pi}.$$

Here $A[S^2] = 4\pi$ denotes the surface area of the two-dimensional sphere. We will at this point naively integrate the function $e \mapsto \mu(c,e)$ over the two-dimensional sphere and use the usual properties of the Lebesgue integral, as we know them for integration over subsets of $\mathbb{R}^n$. The formal justification will follow in section 3.7, where an introduction to integration of functions over surfaces, such as S^2, can be found.

Proof (a) We prove the claim first for polygons. Let $P = (a_1, \ldots, a_m)$ be a closed polygon. We set $b_j := (a_j - a_{j-1})/\|a_j - a_{j-1}\| \in S^2$. For $e \in S^2$ we define a circular line $S^1_e := \{x \in S^2 \mid x \perp e\}$.

[3] Called *crookedness* in [21].

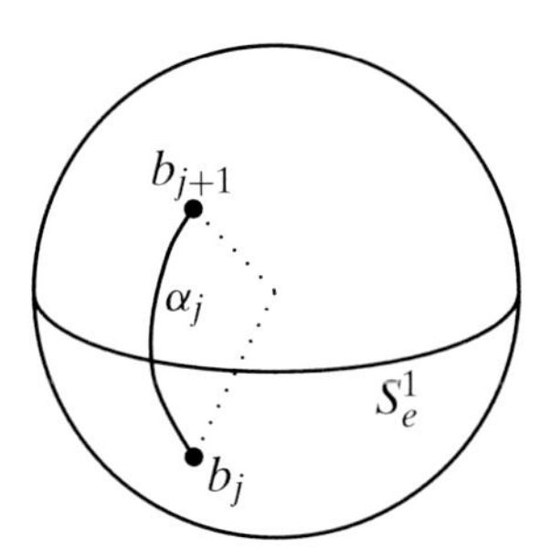

The great circle from b_j to b_{j+1} intersects S_e^1 if and only if $\langle b_j, e\rangle$ and $\langle b_{j+1}, e\rangle$ have different signs, i.e. when $\langle a_j - a_{j-1}, e\rangle$ and $\langle a_{j+1} - a_j, e\rangle$ have different signs. This is the case exactly when for the piecewise linear parametrisation c of P with $c(t_j) = a_j$ the function $t \mapsto \langle c(t), e\rangle$ has a local maximum or minimum in t_j .

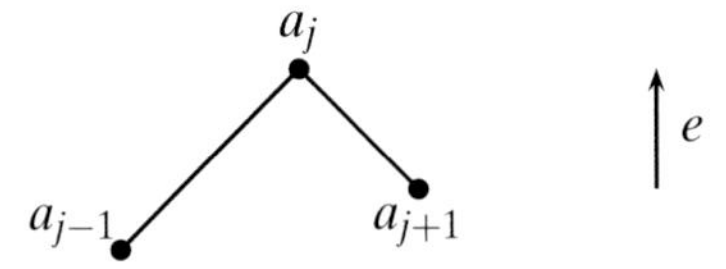

Consider the set of all $e \in S^2$ whose circular line S_e^1 intersects a part of length α of a given great circle. This set consists of two spherical segments of width α and therefore makes up the $(2\alpha/2\pi = \alpha/\pi)$th fraction of the total surface area of the sphere.

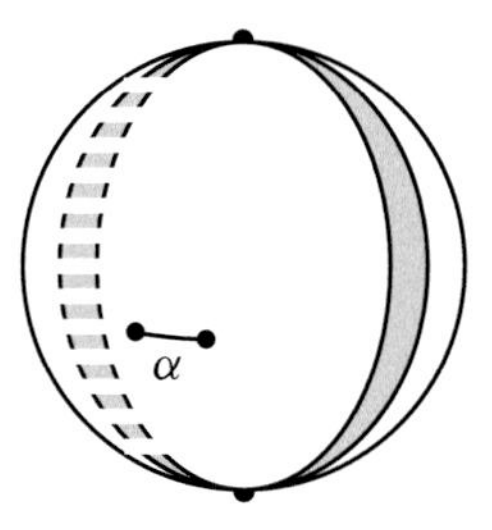

It follows that

$$
\begin{aligned}
&A\Big[\{e \in S^2 \mid t \mapsto \langle c(t), e\rangle \text{ has a local maximum or minimum at } t_j\}\Big] \\
&\quad = A\Big[\{e \in S^2 \mid S_e^1 \text{ intersects the great circle from } b_{j+1} \text{ to } b_j\}\Big] \\
&\quad = \frac{\alpha_j}{\pi} \cdot A[S^2].
\end{aligned}
$$

Summation over j gives

$$
\int_{S^2} (\mu(P, e) + \mu(P, -e))\, dA(e) = \frac{A[S^2]}{\pi} \sum_j \alpha_j = \frac{A[S^2]}{\pi} \kappa(P)
$$

and thus as desired

$$\int_{S^2} \mu(P,e)dA(e) = \frac{A[S^2]}{2\pi}\kappa(P).$$

(b) Now let c be a periodic space curve parametrised by arc-length and with period L. We choose a sequence of partitions $\mathcal{U}_k$ of the interval $[0,L)$ such that their mesh converges to 0. Further, we suppose that the partition $\mathcal{U}_{k+1}$ is always obtained by a refinement of $\mathcal{U}_k$. Let P_k be the corresponding polygons inscribed in c. By proposition 2.3.14 we have

$$\lim_{k\to\infty} \kappa(P_k) = \kappa(c).$$

It remains to be shown that

$$\lim_{k\to\infty} \int_{S^2} \mu(P_k,e)dA(e) = \int_{S^2} \mu(c,e)dA(e). \tag{2.32}$$

For this purpose we show that the sequence of functions $\mu(P_k,\cdot)$ is monotonically increasing and converges to $\mu(c,\cdot)$ as $k\to\infty$ everywhere but on a null set. Equation (2.32) follows by the theorem about monotone convergence.

Let c_k be a piecewise linear parametrisation of P_k. The function $t \mapsto \langle c_k(t),e\rangle$ has an infinite number of local maxima if it is constant along one of the line segments $\overline{a_j a_{j+1}}$, i.e. if $a_{j+1} - a_j \perp e$. The a_j are the vertices of P_k. Apart from that, local maxima can only be found at vertices and are thus automatically isolated. The set $\mathcal{N}_k := \{e \in S^2 \mid e \perp a_{j+1} - a_j \text{ for some } j\}$ is a union of finitely many circular lines and thus a null set in S^2. Countable unions of null sets are null sets themselves, thus

$$\mathcal{N} := \bigcup_k \mathcal{N}_k$$

is again a null set.

We have shown that for $e \in S^2 - \mathcal{N}$ all functions $t \mapsto \langle c_k(t),e\rangle$ only have finitely many local maxima, i.e. $\mu(P_k,e) < \infty$ for all k. Further, every subsequent partition $\mathcal{U}_{k+1}$ is a partition of $\mathcal{U}_k$ and the addition of vertices can at most increase the number of local maxima. Thus

$$\mu(P_{k+1},e) \geq \mu(P_k,e).$$

We have established that the sequence of functions $\mu(P_k,\cdot)$ is monotonically increasing on $S^2 - \mathcal{N}$. If the function $t \mapsto \langle c_k(t),e\rangle$ in $t = t_j$ has a local maximum, then the function $t \mapsto \langle c(t),e\rangle$ in (t_{j-1},t_{j+1}) must thus also have a local maximum.

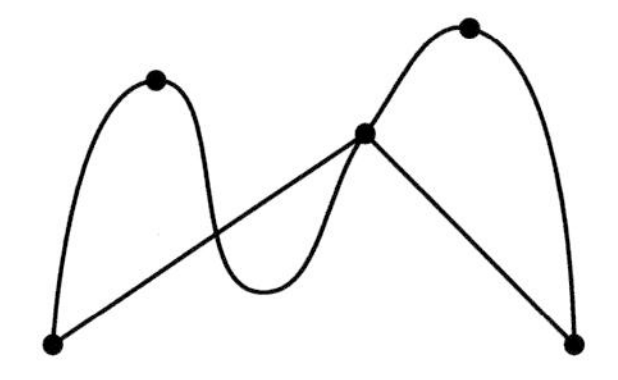

It follows that for all $e \in S^2 - \mathcal{N}$

$$\mu(P_k, e) \leq \mu(c, e).$$

Finally, let $\nu \in \mathbb{N}$, $\nu \leq \mu(c,e)$. Then the function $t \mapsto \langle c(t), e\rangle$ has at least ν local maxima $s_1, \ldots, s_\nu$. We choose $\delta > 0$ so small that every one of those local maxima is a global maximum in the δ-neighbourhood $(s_j - \delta, s_j + \delta)$.

First case The function $t \mapsto \langle c(t), e\rangle$ is constant on one of the intervals $(s_j - \delta, s_j]$, $[s_j, s_j + \delta)$, $j = 1, \ldots, \nu$.

For a sufficiently large k the partition $\mathcal{U}_k$ has mesh $<\delta/2$. Then there are at least two of the points partitioning the interval in each of those subintervals. It follows that $e \in \mathcal{N}_k$. Thus this cannot be the case for $e \in S^2 - \mathcal{N}$.

Second case The function $t \mapsto \langle c(t), e\rangle$ is not constant on any of the intervals $(s_j - \delta, s_j]$, $[s_j, s_j + \delta)$, $j = 1, \ldots, \nu$.

Then there exist $u_j^- \in (s_j - \delta, s_j)$ and $u_j^+ \in (s_j, s_j + \delta)$ with $\langle c(u_j^\pm), e\rangle < \langle c(s_j), e\rangle$. We choose $\eta > 0$ so small that $(u_j^\pm - \eta, u_j^\pm + \eta) \subset (s_j - \delta, s_j + \delta)$ and $\langle c(u), e\rangle < \langle c(s), e\rangle$ for all $u \in (u_j^\pm - \eta, u_j^\pm + \eta)$ and $s \in (s_j - \eta, s_j + \eta)$.

If k is sufficiently large, then the mesh of $\mathcal{U}_k$ is smaller than η. Hence there is one of the points partitioning the interval in each of the intervals $(u_j^\pm - \eta, u_j^\pm + \eta)$ and $(s_j - \eta, s_j + \eta)$. Then the function $t \mapsto \langle c_k(t), e\rangle$ has a local maximum in $(s_j - \delta, s_j + \delta)$ as well. Thus $\mu(P_k, e) \geq \nu$.

We have shown that for $e \in S^2 - \mathcal{N}$:

$$\mu(P_k, e) \underset{k\to\infty}{\nearrow} \mu(c, e). \qquad \square$$

Corollary 2.3.18 *Let c be a closed space curve. Then*

$$\kappa(c) \geq 2\pi\,\mu(c).$$

Proof $\kappa(c)/2\pi$ is the mean of the function $\mu(c, \cdot)$ by proposition 2.3.17, while $\mu(c)$ is the minimum. $\square$

As a very pleasing application of the concept of the bridge number of space curves we obtain Fenchel's theorem, which states how much a space curve needs to curve in order to become a closed curve.

Theorem 2.3.19 (Fenchel's theorem) *Let c be a simple closed space curve. Then*

$$\kappa(c) \geq 2\pi .$$

We have the equality $\kappa(c) = 2\pi$ exactly if c is a convex plane curve.

Proof (a) For every $e \in S^2$ the function $t \mapsto \langle c(t), e\rangle$ has at least one maximum and one minimum. Thus $\mu(c) \geq 1$. By corollary 2.3.18 it follows that

$$\kappa(c) \geq 2\pi\,\mu(c) \geq 2\pi .$$

(b) Now let c be a convex plane curve. We denote by $\tilde{\kappa}$ the curvature of c considered as a plane curve. $\tilde{\kappa}$ does not change sign according to theorem 2.2.15. It therefore follows that $\kappa(t) = \tilde{\kappa}(t)$ for all t or $\kappa(t) = -\tilde{\kappa}(t)$ for all t. By Hopf's Umlaufsatz the winding number is $n_c = \pm 1$. Using theorem 2.2.9 we find that

$$\pm 2\pi = 2\pi n_c = \int_0^L \tilde{\kappa}(t)dt = \pm \int_0^L \kappa(t)dt = \pm\kappa(c).$$

As $\kappa(t) \geq 0$, it follows that $\kappa(c) = 2\pi$.

(c) Now let $\kappa(c) = 2\pi$. We want to show that c lies in a plane and is convex. Let P_1 be a polygon that is inscribed in c and consists of three non-collinear points of c. We denote the plane spanned by the triangle P_1 by E. We have to show that c is fully contained in E. For this purpose let p be another point on c. We add the vertex p to P_1 and obtain the polygon P_2. Triangles have total curvature 2π. Lemma 2.3.12 and proposition 2.3.14 imply that

$$2\pi = \kappa(P_1) \leq \kappa(P_2) \leq \kappa(c) = 2\pi .$$

In particular, it follows that $\kappa(P_1) = \kappa(P_2)$ and, again by lemma 2.3.12, p must also lie on E.

It remains to prove the convexity of c. Again let $\tilde{\kappa}$ be the curvature (with sign) of c as a plane curve. By Hopf's Umlaufsatz $n_c = \pm 1$ and hence

$$\begin{aligned} 2\pi = |2\pi n_c| &= \left|\int_0^L \tilde{\kappa}(t)dt\right| \\ &\leq \int_0^L |\tilde{\kappa}(t)|\,dt = \int_0^L \kappa(t)dt = 2\pi . \end{aligned}$$

Thus $\tilde{\kappa}$ has a constant sign and by theorem 2.2.15 curve c is convex. □

We will conclude this chapter with a treatment of the concept of knotted curves and an investigation into how much a space curve needs to curve so that it can be knotted.

Definition 2.3.20 An ***isotopy*** of $\mathbb{R}^3$ is a continuous map $\Phi : [0,1] \times \mathbb{R}^3 \to \mathbb{R}^3$ such that for every fixed $t \in [0,1]$ the map $\Phi(t,\cdot) : \mathbb{R}^3 \to \mathbb{R}^3$ is a homeomorphism. Two simple closed space curves c_0 and c_1 are called ***ambient isotopic***, if there exists an isotopy Φ of $\mathbb{R}^3$ with $\Phi(0,x) = x$ for all $x \in \mathbb{R}^3$ and $\Phi(1, \mathrm{Trace}(c_0)) = \mathrm{Trace}(c_1)$.

We graphically interpret $t \in [0,1]$ as a deformation parameter. Φ bends the curve c_0 into the curve c_1. The word "ambient" in the definition suggests that not only the curves are deformed: the homeomorphism $\Phi(t,\cdot)$ always deforms the entire surrounding $\mathbb{R}^3$.

Exercise 2.19 Show that ambient isotopy defines an equivalence relation on the set of simple closed space curves.

Definition 2.3.21 An ambient isotopy class of simple closed space curves is a ***knot***. A simple closed space curve is said to be ***unknotted*** if it is ambient isotopic to a simple closed plane curve. Otherwise it is called ***knotted***.

So a simple closed *space* curve is unknotted if it can be bent into a simple closed *plane* curve without the curve intersecting itself in the process of deformation. The following curve for instance is unknotted:

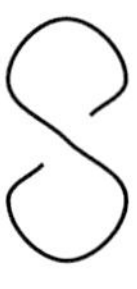  

The so-called ***trefoil*** is knotted. Readers can convince themselves of this using a piece of rope (at this point the mathematical proof for this would be too involved).

For a curve to be knotted it probably needs to curve even more than in Fenchel's theorem. Indeed the following holds:

Theorem 2.3.22 (Fáry–Milnor theorem) *Let c be a knotted, simple closed space curve. Then*

$$\kappa(c) \geq 4\pi .$$

Proof Let c be a simple closed space curve with $\kappa(c) < 4\pi$. We need to unknot c, i.e. find an isotopy of $\mathbb{R}^3$ which maps the trace of c to a simple closed plane curve. By the assumption and by corollary 2.3.18 it follows that

$$2\pi\,\mu(c) \leq \kappa(c) < 4\pi,$$

hence $\mu(c) < 2$ and therefore

$$\mu(c) = 1.$$

There therefore exists a unit vector $e \in \mathbb{R}^3$ with $\mu(c, e) = 1$. If $t \mapsto \langle c(t), e\rangle$ is parametrised by arc-length, then it will have exactly one local maximum (and therefore global maximum) h^+ and exactly one local minimum h^- in the one-period interval $[0, L)$, since one would otherwise find another local maximum between two distinct local minima. The corresponding points $p_{\max}$ and $p_{\min}$ from the trace of c divide the closed space curve into two parts c_L and c_R. As there are no additional local maxima or minima, the functions $t \mapsto \langle c_L(t), e\rangle$ and $t \mapsto \langle c_R(t), e\rangle$ are strictly monotonic.

Let E_0 be the orthogonal complement of e in $\mathbb{R}^3$. For $h \in \mathbb{R}$ let E_h be the affine plane parallel to E_0, translated in direction e by an amount h: i.e. $E_h = \{x \in \mathbb{R}^3 \mid \langle x - he, e\rangle = 0\}$. Thus the trace of c intersects the plane E_h

- not at all, if $h < h^-$ or $h > h^+$,
- at exactly one point, the maximal point $p_{\max}$, if $h = h^+$,
- at exactly one point, the minimal point $p_{\min}$, if $h = h^-$,
- at exactly two points $p_L(h)$ on c_L and $p_R(h)$ on c_R, if $h^- < h < h^+$.

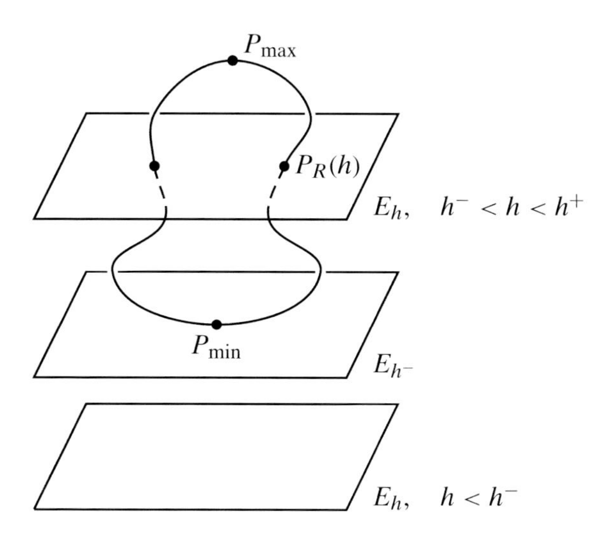

Using a first isotopy we centre the line segments $\overline{p_L(h)p_R(h)}$ on the e-axis. For this purpose, let $\Pi : \mathbb{R}^3 \to E_0$ be the orthogonal projection. We set

$$\Phi_1(t,x) := \begin{cases} x - t \cdot \Pi\left(p_{\min}\right), & \langle x, e\rangle \le h^-, \\ x - t \cdot \Pi\left(\dfrac{p_L(\langle x,e\rangle) + p_R(\langle x,e\rangle)}{2}\right), & h^- < \langle x,e\rangle < h^+, \\ x - t \cdot \Pi\left(p_{\max}\right), & \langle x,e\rangle \ge h^+. \end{cases}$$

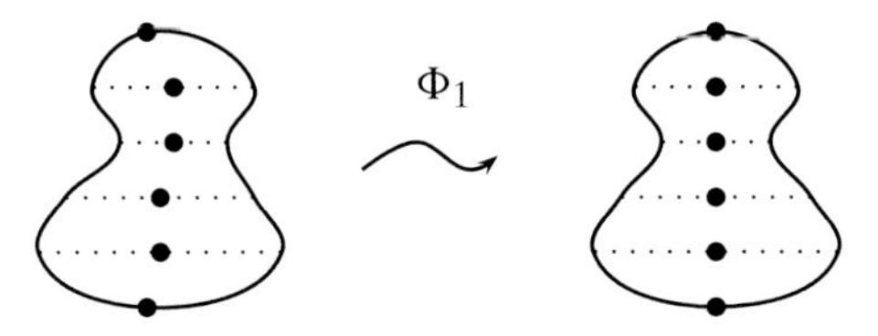

Generally, c may also be twisted around the e-axis. We avoid this by means of a second isotopy. We define the continuous map $\zeta : \mathbb{R} \to E_0$ by

$$\zeta(h) := \begin{cases} \Pi\left(\dfrac{p_R(h) - p_L(h)}{\|p_R(h) - p_L(h)\|}\right), & h^- < h < h^+, \\ \Pi\left(\dot{c}(t_{\max})\right), & h \ge h^+, \\ \Pi\left(-\dot{c}(t_{\min})\right), & h \le h^-. \end{cases}$$

We have that $c(t_{\max}) = p_{\max}$ and $c(t_{\min}) = p_{\min}$. We choose an orthonormal basis e_1, e_2 of E_0. By the lifting lemma 2.2.12 we can find a continuous map $\vartheta : \mathbb{R} \to \mathbb{R}$ that satisfies

$$\zeta(h) = \cos(\vartheta(h)) \cdot e_1 + \sin(\vartheta(h)) \cdot e_2.$$

We now undrill c using the isotopy

$$\Phi_3(t,x) := \langle x,e\rangle\, e + \begin{pmatrix} \cos(t\vartheta(\langle x,e\rangle)) & \sin(t\vartheta(\langle x,e\rangle)) \\ -\sin(t\vartheta(\langle x,e\rangle)) & \cos(t\vartheta(\langle x,e\rangle)) \end{pmatrix} \Pi(x).$$

The matrix represents a rotation in the plane E_0 by an angle of $-t\vartheta(\langle x,e\rangle)$, with respect to the basis e_1, e_2.

We have $\Phi_3(0,x) = \langle x,e\rangle\, e + \Pi(x) = x$, and $\Phi_3(1,\cdot)$ rotates in the plane E_h by the angle $-\vartheta(h)$. The result is a simple closed curve in the $(e$–$e_1)$-plane. We have therefore unknotted c. □

John Milnor and István Fáry discovered this theorem at roughly the same time independently of each other, see [11] and [21]. As a young student John Milnor attended a course on differential geometry taught by Al Tucker in 1948 in Princeton. Milnor managed to solve a conjecture of Karol Borsuk from a collection of open problems presented in class. This problem was nothing other than our theorem 2.3.22. According to an anecdote Milnor came late to a lecture and mistook the problems for (unusually difficult) homework exercises.

3 Classical surface theory

We introduce regular surfaces and their tangent planes. We investigate what normal fields have to do with orientability of surfaces. We discover that the geometry of regular surfaces is largely determined by the first and second fundamental forms, which also give rise to different types of curvature. We learn how to integrate over regular surfaces and, in particular, how their surface area is defined. Finally we examine some special classes of surfaces more closely: ruled surfaces, minimal surfaces, surfaces of revolution and tubular surfaces.

3.1 Regular surfaces

Surfaces in three-dimensional space are two-dimensional objects, i.e. the points on a surface can be described by two independent parameters. The definition which follows is *local*. Unlike with curves, which we always parametrised as a whole, we only require that small parts of the surface can be described by a parametrisation.

Definition 3.1.1 Let $S \subset \mathbb{R}^3$ be a subset. We call S a ***regular surface*** if there exists for every point $p \in S$ an open neighbourhood V of p in $\mathbb{R}^3$, and if, in addition, there exists an open subset $U \subset \mathbb{R}^2$ and a smooth map $F : U \to \mathbb{R}^3$ such that

(i) $F(U) = S \cap V$ and $F : U \to S \cap V$ is a homeomorphism and
(ii) the Jacobian $D_u F$ has rank 2 for every point $u \in U$.

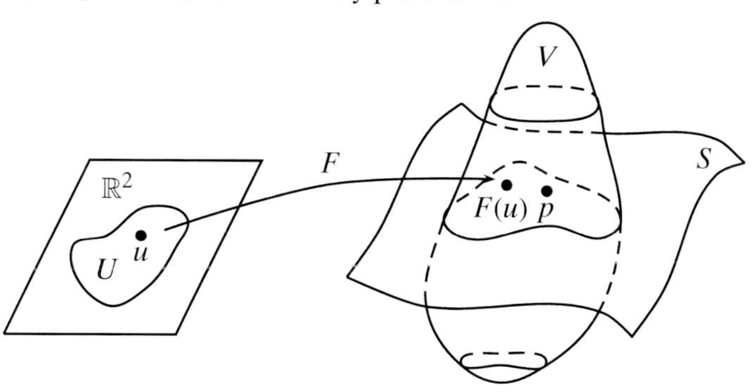

Condition (i) says that the points on the surface S that are close to p, namely those that are also in V, are through the map F given by two parameters, namely the coordinates of the points of $U \subset \mathbb{R}^2$. Condition (ii) ensures that the two parameters are really independent of each other.

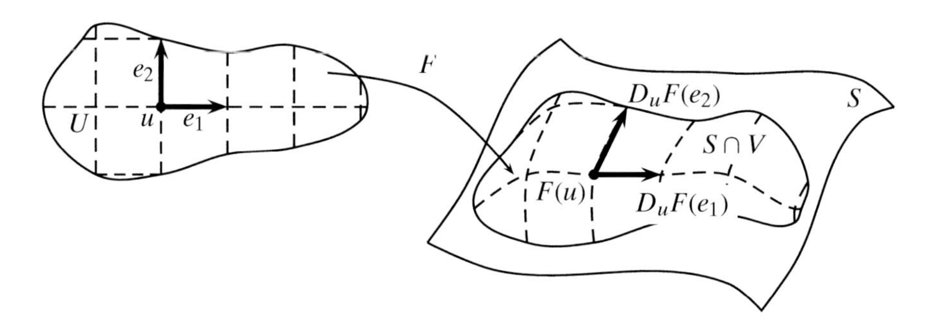

$D_uF(e_1)$ and $D_uF(e_2)$ are linearly independent

Definition 3.1.2 The map $F: U \to S \cap V$ from definition 3.1.1 and also the triple (U, F, V) is called a ***local parametrisation*** of S at p. The set $S \cap V$ is called the ***coordinate neighbourhood*** of p. The components u^1 and u^2 of $u = (u^1, u^2)^\top$ are also called ***coordinates*** of the point $F(u) \in S$ (w.r.t. the parametrisation F).

Example 3.1.3 (affine planes) The simplest examples of regular surfaces are affine planes. The affine plane through a point $p \in \mathbb{R}^3$, spanned by the linearly independent vectors $X, Y \in \mathbb{R}^3$, is the set

$$S = \left\{ p + u^1 \cdot X + u^2 \cdot Y \,\middle|\, u^1, u^2 \in \mathbb{R} \right\}.$$

We can use a single parametrisation. We set $V := \mathbb{R}^3$, $U := \mathbb{R}^2$ and $F : U \to \mathbb{R}^3$, $F(u^1, u^2) := p + u^1 \cdot X + u^2 \cdot Y$.

Example 3.1.4 (graph of a function) Let $U \subset \mathbb{R}^2$ be open, $f : U \to \mathbb{R}$ a smooth function. We consider the graph of f,

$$S = \left\{ (x, y, z)^\top \in \mathbb{R}^3 \,\middle|\, (x, y)^\top \in U,\ z = f(x, y) \right\}.$$

In this case we can also use a single coordinate neighbourhood. Again we set $V := \mathbb{R}^3$ and

$$F : U \to \mathbb{R}^3, \qquad F(x, y) := (x, y, f(x, y))^\top.$$

Then obviously $F(U) = S = S \cap V$. Further, F is smooth and the inverse map $G : S \to U$, $G(x, y, z) = (x, y)^\top$, is continuous as well. In particular, $F : U \to S$

is a homeomorphism. Condition (i) from the definition of a surface is therefore satisfied.

We check condition (ii): the matrix

$$D_{(x,y)}F = \begin{pmatrix} 1 & 0 \\ 0 & 1 \\ \frac{\partial f}{\partial x}(x,y) & \frac{\partial f}{\partial y}(x,y) \end{pmatrix}$$

has full rank for every $(x,y)^\top \in U$.

Example 3.1.5 (the sphere) We consider

$$S = S^2 = \left\{ (x,y,z)^\top \in \mathbb{R}^3 \,\middle|\, x^2 + y^2 + z^2 = 1 \right\}.$$

Let us first set $V := V_3^+ := \{(x,y,z)^\top \in \mathbb{R}^3 \mid z > 0\}$. Then $S^2 \cap V_3^+$ is the graph of the function $f(x,y) = \sqrt{1 - (x^2 + y^2)}$ for $x^2 + y^2 < 1$. Then, according to the discussion in example 3.1.4

$$U := \left\{ (x,y)^\top \in \mathbb{R}^2 \,\middle|\, x^2 + y^2 < 1 \right\}$$

and

$$F_3^+ : U \to \mathbb{R}^3,$$

$$F_3^+(x,y) = \left(x, y, \sqrt{1 - (x^2 + y^2)} \right)^\top,$$

is a local parametrisation. The points from $S^2 \cap V_3^-$, $V_3^- := \{(x,y,z)^\top \in \mathbb{R}^3 \mid z < 0\}$ are obtained analogously through the parametrisation

$$F_3^- : U \to \mathbb{R}^3,$$

$$F_3^-(x,y) = \left(x, y, -\sqrt{1 - (x^2 + y^2)} \right)^\top.$$

The points $p \in S^2$ with z-coordinate 0 remain to be considered. These points can be obtained by interchanging the role of the z-coordinate with that of the x-coordinate or the y-coordinate, depending on the position of p. For this purpose we set

$$V_1^+ := \left\{(x, y, z)^\top \in \mathbb{R}^3 \,\middle|\, x > 0\right\},$$

$$V_1^- := \left\{(x, y, z)^\top \in \mathbb{R}^3 \,\middle|\, x < 0\right\},$$

$$V_2^+ := \left\{(x, y, z)^\top \in \mathbb{R}^3 \,\middle|\, y > 0\right\},$$

$$V_2^- := \left\{(x, y, z)^\top \in \mathbb{R}^3 \,\middle|\, y < 0\right\},$$

$$F_1^\pm : U \to \mathbb{R}^3, \quad F_1^\pm(y, z) = \left(\pm\sqrt{1 - (y^2 + z^2)}, y, z\right)^\top,$$

$$F_2^\pm : U \to \mathbb{R}^3, \quad F_2^\pm(x, z) = \left(x, \pm\sqrt{1 - (x^2 + z^2)}, z\right)^\top.$$

Then $F_i^\pm(U) = S^2 \cap V_i^\pm$, $i = 1, 2, 3$, and all $F_i^\pm$ are local parametrisations. Every point $p \in S^2$ appears in at least one of the sets $V_i^\pm$, hence S^2 is a regular surface.

Exercise 3.1 We have covered the sphere S^2 with six coordinate neighbourhoods. In fact, two coordinate neighbourhoods would have sufficed using different local parametrisations. Find such local parametrisations.

Finding local parametrisations for a given surface can at times be tedious. It is thus often useful to also have other criteria to decide whether a subset $S \subset \mathbb{R}^3$ is a regular surface or not.

The set S is in many cases given by an equation of the form, $S := \left\{(x, y, z)^\top \in \mathbb{R}^3 \mid f(x, y, z) = 0\right\}$. The following criterion says that when the gradient of f does not vanish anywhere along S, then the set S is a regular surface.

Proposition 3.1.6 *Let $V_0 \subset \mathbb{R}^3$ be open, let $f : V_0 \to \mathbb{R}$ be a smooth function. We set $S := \left\{(x, y, z)^\top \in V_0 \mid f(x, y, z) = 0\right\}$. If*

$$\operatorname{grad} f(p) \neq (0, 0, 0)^\top$$

for all $p \in S$, then S is a regular surface.

Proof Let $p := (x_0, y_0, z_0)^\top \in S$. Since

$$\operatorname{grad} f(p) = \left(\frac{\partial f}{\partial x}(p), \frac{\partial f}{\partial y}(p), \frac{\partial f}{\partial z}(p)\right)^\top \neq (0, 0, 0)^\top$$

we can assume without loss of generality that

$$\frac{\partial f}{\partial z}(p) \neq 0.$$

By the implicit mapping theorem [18, p. 529, theorem 5.4] there exist an open neighbourhood $V \subset V_0$ of p, an open neighbourhood $U \subset \mathbb{R}^2$ of (x_0, y_0) and a smooth function $g : U \to \mathbb{R}$, such that

$$S \cap V = \left\{ (x, y, g(x, y))^\top \,\middle|\, (x, y)^\top \in U \right\}.$$

If we set $F : U \to V$, $F(x, y) := (x, y, g(x, y))^\top$, then it follows by the same argument as in example 3.1.4 that F is a local parametrisation. □

Example 3.1.7 (Ellipsoid) Let us consider

$$S := \left\{ (x, y, z)^\top \in \mathbb{R}^3 \,\middle|\, \frac{x^2}{a^2} + \frac{y^2}{b^2} + \frac{z^2}{c^2} = 1 \right\}$$

for non-vanishing constants $a, b, c \in \mathbb{R}$.

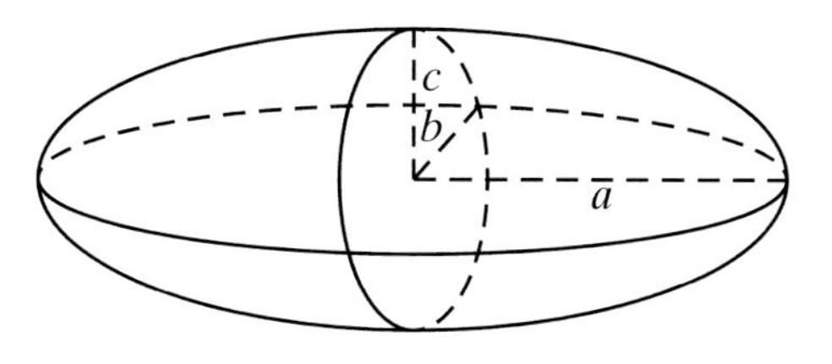

ellipsoid

If we set

$$V_0 := \mathbb{R}^3, \quad f : \mathbb{R}^3 \to \mathbb{R}, \quad f(x, y, z) := \frac{x^2}{a^2} + \frac{y^2}{b^2} + \frac{z^2}{c^2} - 1,$$

then

$$S = \left\{ (x, y, z)^\top \in \mathbb{R}^3 \,\middle|\, f(x, y, z) = 0 \right\}.$$

In order to apply proposition 3.1.6, we need to check that the gradient of f does not vanish for any $p \in S$. The expression

$$\operatorname{grad} f(x,y,z) = \left(\frac{2x}{a^2}, \frac{2y}{b^2}, \frac{2z}{c^2}\right)^\top$$

vanishes only for $p_0 = (0,0,0)^\top$. But $p_0 \notin S$, hence S is a regular surface.

Attention If S is given in the form

$$S = \left\{(x,y,z)^\top \in \mathbb{R}^3 \,\middle|\, f(x,y,z) = 0\right\},$$

then the fact that $\operatorname{grad} f$ does not vanish along S is *sufficient* for S to be a regular surface, but not *necessary*. We can, for example, describe the sphere $S = S^2$ as follows:

$$S^2 = \left\{(x,y,z)^\top \in \mathbb{R}^3 \,\middle|\, (x^2+y^2+z^2-1)^2 = 0\right\},$$

i.e. the set of zeros of the function

$$f(x,y,z) = (x^2+y^2+z^2-1)^2.$$

For the gradient of f we obtain

$$\operatorname{grad} f(x,y,z) = 2(x^2+y^2+z^2-1)\cdot(2x,2y,2z)^\top.$$

We observe that $\operatorname{grad} f(x,y,z)$ even vanishes for all $p \in S^2$. Nevertheless $S = S^2$ is a regular surface, the describing function f was simply chosen very clumsily.

Exercise 3.2 Let $S \subset \mathbb{R}^3$ be a regular surface, $W \subset \mathbb{R}^3$ open. Show that $W \cap S$ is also a regular surface.

Exercise 3.3 Show that the property of being a regular surface is a local property. More precisely, let $S \subset \mathbb{R}^3$ be a subset. For every point $p \in S$ there exists an open neighbourhood V of p in $\mathbb{R}^3$, such that $V \cap S$ is a regular surface. Then S itself is a regular surface as well.

Example 3.1.8 We now want to think about whether the ***double cone***

$$S = \left\{(x,y,z)^\top \in \mathbb{R}^3 \,\middle|\, x^2+y^2 = z^2\right\}$$

is a regular surface or not. S is defined as the set of zeros of the function $f : \mathbb{R}^3 \to \mathbb{R}$, $f(x,y,z) = x^2+y^2-z^2$. The gradient

$$\operatorname{grad} f(x,y,z) = (2x,2y,-2z)^\top$$

vanishes only in $(x,y,z)^\top = (0,0,0)^\top$. If we restrict f to the open subset $V_0 := \mathbb{R}^3 - \{(0,0,0)^\top\}$, then we can apply proposition 3.1.6 and observe that

$$S \cap V_0 = S - \left\{(0,0,0)^\top\right\}$$

is a regular surface. The question of whether S is also a regular surface at $(0,0,0)^\top$ remains. Proposition 3.1.6 does not make any statement about this; function f may have been chosen in a clumsy way for the point $(0,0,0)^\top$.

Suppose that S were a regular surface. Then a local parametrisation around $p = (0,0,0)^\top$ would exist, i.e. there would exist open subsets $V \subset \mathbb{R}^3$, $U \subset \mathbb{R}^2$ and a smooth map $F : U \to V$, such that $F(U) = S \cap V$ and $F : U \to S \cap V$ would be a homeomorphism.

Set $u_0 := F^{-1}\left((0,0,0)^\top\right) \in U$. Since U is an open neighbourhood of u_0, we can find an open disc $U' \subset U$ with centre u_0. As $F : U \to S \cap V$ is a homeomorphism it follows that $F(U')$ is open in $S \cap V$. This means that there exists a V' open in V (and hence also open in $\mathbb{R}^3$), such that $F(U') = S \cap V'$. As $V' \subset \mathbb{R}^3$ is an open neighbourhood of $(0,0,0)^\top$, all vectors $(x,y,z)^\top \in \mathbb{R}^3$ of sufficiently small length are contained in V'. In particular, there are points $p_1 = (x_1,y_1,z_1)^\top$ with $z_1 > 0$ and $p_2 = (x_2,y_2,z_2)^\top$ with $z_2 < 0$ in $S \cap V'$.

Let $u_i := F^{-1}(p_i)$, $i = 1,2$. In the disc U' we join u_1 with u_2 by a continuous path c that does not run through the centre u_0. The path given by the image $F \circ c$, which joins p_1 and p_2, must run through $(0,0,0)^\top = F(u_0)$ because of the intermediate value theorem. This is a contradiction.

S is therefore not a regular surface, but has a so-called *singularity* at $(0,0,0)^\top$.

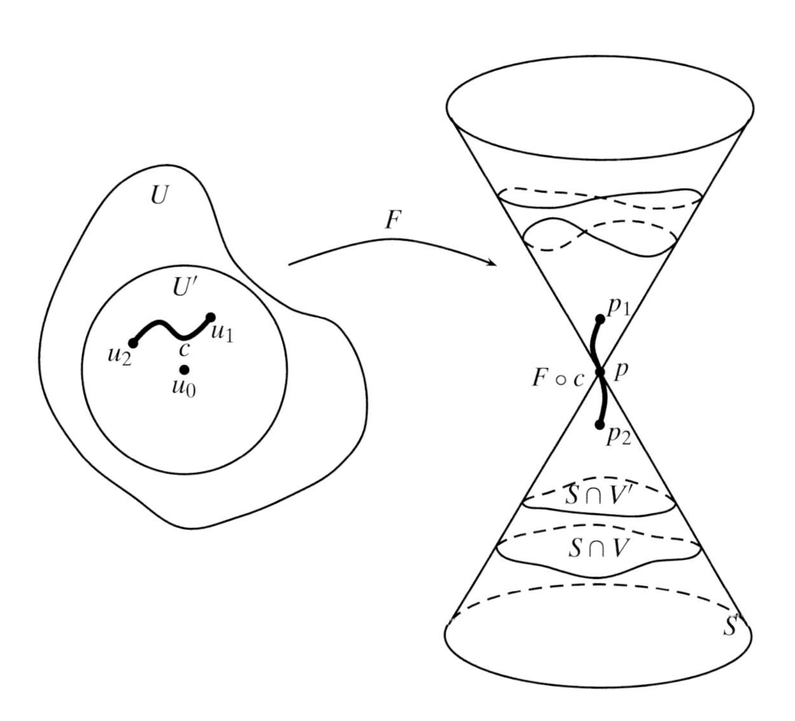

In the following we investigate the differentiability of maps whose domain or target is a regular surface. In this discussion "smooth" always means "infinitely often differentiable".

Proposition 3.1.9 *Let $S \subset \mathbb{R}^3$ be a regular surface. Let (U, F, V) be a local parametrisation of S. Let $W \subset \mathbb{R}^n$ be an open set, and $\varphi : W \to \mathbb{R}^3$ a map with $\varphi(W) \subset S \cap V$. Then φ, considered as a map from W to $\mathbb{R}^3$, is smooth if and only if $F^{-1} \circ \varphi : W \to U \subset \mathbb{R}^2$ is smooth.*

If we investigate differentiability of a map that takes values in a regular surface S, then it is therefore irrelevant if we consider this map as a map with values in $\mathbb{R}^3$ or, using coordinates, as a map with values $\mathbb{R}^2$.

Proof One direction is trivial, since if $\psi := F^{-1} \circ \varphi$ is a smooth map, then $\varphi = F \circ \psi$ is also smooth, being the composition of two smooth maps.

Now suppose that $\varphi : W \to \mathbb{R}^3$ is smooth. Let $p \in W$. We set $q := \varphi(p) \in S \cap V$ and $u_0 := F^{-1}(q) \in U$.

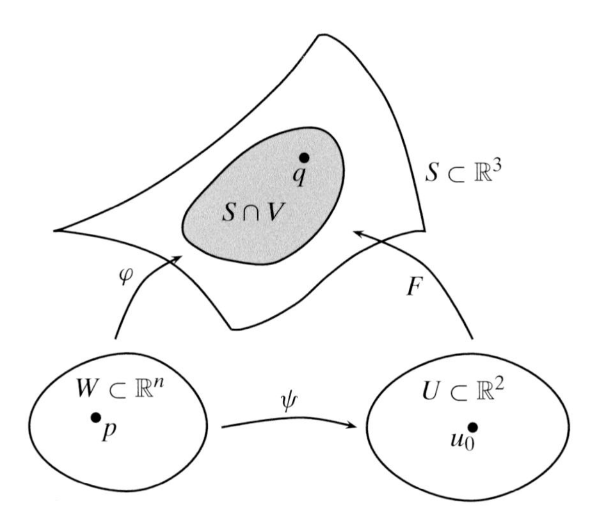

We write $F(u^1, u^2) := \left(x(u^1, u^2), y(u^1, u^2), z(u^1, u^2)\right)^\top$. Since the differential $D_{u_0}F$ has full rank we can assume without loss of generality that the 2×2 matrix $\left(\left(\partial(x, y)/\partial(u^1, u^2)\right)(u_0)\right)$ is invertible. If it is not, then we can simply exchange the x-coordinate or y-coordinate by the z-coordinate.

We define the map

$$G : U \times \mathbb{R} \to \mathbb{R}^3,$$

$$G(u^1, u^2, t) = \left(x(u^1, u^2), y(u^1, u^2), z(u^1, u^2) + t\right)^\top$$

and find its differential at the point $(u^1, u^2, t) = (u_0^1, u_0^2, 0)$:

$$D_{(u_0^1, u_0^2, 0)} G = \begin{pmatrix} \frac{\partial x}{\partial u^1}(u_0) & \frac{\partial x}{\partial u^2}(u_0) & 0 \\ \frac{\partial y}{\partial u^1}(u_0) & \frac{\partial y}{\partial u^2}(u_0) & 0 \\ \frac{\partial z}{\partial u^1}(u_0) & \frac{\partial z}{\partial u^2}(u_0) & 1 \end{pmatrix}.$$

An expansion of the determinant about the last column gives:

$$\det D_{(u_0^1, u_0^2, 0)} G = \det\left(\frac{\partial(x, y)}{\partial(u^1, u^2)}(u_0)\right) \neq 0.$$

Hence $D_{(u_0^1, u_0^2, 0)} G$ is invertible and we can, according to the inverse mapping theorem [18, p. 515, theorem 3.1], find an open neighbourhood $U_1 \subset U \times \mathbb{R}$ of $(u_0^1, u_0^2, 0)$ and an open neighbourhood $V_1 \subset V$ of q, such that

$$G|_{U_1} : U_1 \to V_1$$

is a diffeomorphism. We set $W_1 := \varphi^{-1}(V_1)$. Then W_1 is an open neighbourhood of p. For $p' \in W_1$ we have

$$G^{-1} \circ \varphi(p') = (F^{-1} \circ \varphi(p'), 0),$$

because $F(u^1, u^2) = G(u^1, u^2, 0)$. As $G^{-1} \circ \varphi$ is smooth, being the composition of two smooth maps, we have smoothness for $F^{-1} \circ \varphi$ on W_1. Now W_1 is an open neighbourhood of an arbitrary given point p, and the claim is proved. □

We easily obtain that parameter transformations are diffeomorphisms. More precisely, we obtain:

Corollary 3.1.10 *Let S be a regular surface with local parametrisations (U_1, F_1, V_1) and (U_2, F_2, V_2). Then*

$$F_2^{-1} \circ F_1 : F_1^{-1}(V_1 \cap V_2) \to F_2^{-1}(V_1 \cap V_2)$$

is smooth.

Proof This follows from the application of proposition 3.1.9 to

$$W = F_1^{-1}(V_1 \cap V_2), \quad \varphi = F_1 \text{ and } (U, F, V) = (U_2, F_2, V_2).$$

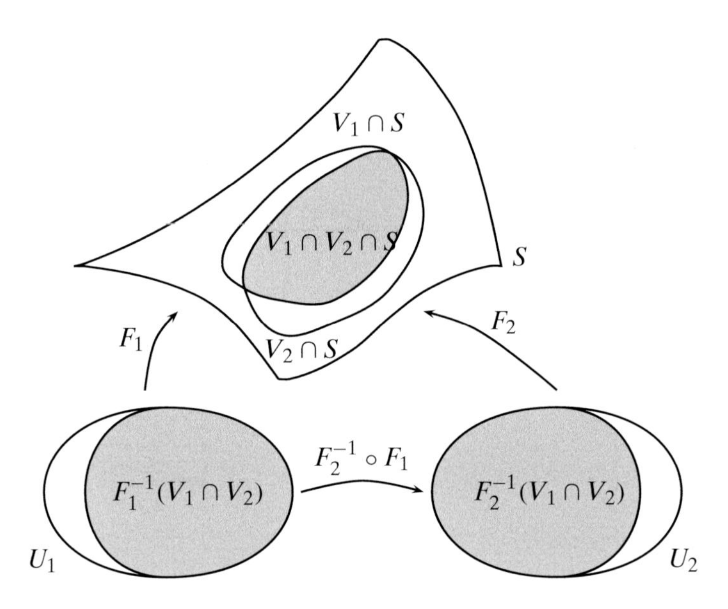

To illustrate the situation from the above proposition we calculate a parameter transformation for the sphere $S = S^2$. For this purpose let $F_1 = F_1^+$ and $F_2 = F_2^-$ be as in example 3.1.5. Then

$$V_1 \cap V_2 = V_1^+ \cap V_2^- = \left\{(x, y, z)^\top \in \mathbb{R}^3 \,\middle|\, x > 0 \text{ and } y < 0\right\}$$

and hence

$$F_1^{-1}(V_1 \cap V_2) = \left\{(y, z)^\top \in \mathbb{R}^2 \,\middle|\, y^2 + z^2 < 1 \text{ and } y < 0\right\}$$

as well as

$$F_2^{-1}(V_1 \cap V_2) = \left\{(x, z)^\top \in \mathbb{R}^2 \,\middle|\, x^2 + z^2 < 1 \text{ and } x > 0\right\}.$$

For $F_2^{-1} \circ F_1$ this gives

$$\begin{aligned} F_2^{-1}\left(F_1(y, z)\right) &= F_2^{-1}\left(\sqrt{1 - y^2 - z^2}, y, z\right) \\ &= \begin{pmatrix} \sqrt{1 - y^2 - z^2} \\ z \end{pmatrix}. \end{aligned}$$

Indeed, $F_2^{-1} \circ F_1$ is a smooth map. □

We now investigate the differentiability of maps whose domain is on a surface, and obtain several equivalent statements:

Proposition 3.1.11 *Let $S \subset \mathbb{R}^3$ be a regular surface, $p \in S$, and $f : S \to \mathbb{R}^n$ a continuous map. Then the following are equivalent:*

(1) *There is an open neighbourhood V of p in $\mathbb{R}^3$ and an extension $\tilde{f}$ of $f|_{S \cap V}$ to V, which is smooth around p.*
(2) *There exists a local parametrisation (U, F, V) with $p \in V$, such that $f \circ F : U \to \mathbb{R}^n$ is smooth around $F^{-1}(p)$.*
(3) *For all local parametrisations (U, F, V) with $p \in V$ the map $f \circ F : U \to \mathbb{R}^n$ is smooth around $F^{-1}(p)$.*

Proof

(a) (1) implies (3): F is smooth and $\tilde{f}$ is smooth around p, hence

$$f \circ F = \tilde{f} \circ F$$

is also a smooth map on a neighbourhood of $F^{-1}(p)$.
(b) (3) implies (2) trivially.
(c) (2) implies (1): we consider the local diffeomorphism

$$G(u^1, u^2, t) = \left(x(u^1, u^2), y(u^1, u^2,), z(u^1, u^2) + t \right)$$

from the proof of proposition 3.1.9 once again. We set

$$g(u^1, u^2, t) := f \circ F(u^1, u^2) = f \circ G\left(u^1, u^2, 0\right).$$

Now g is smooth near $(F^{-1}(p), 0)$ and we can take

$$\tilde{f} := g \circ G^{-1}$$

as an extension. □

Definition 3.1.12 If the equivalent conditions (1) to (3) from proposition 3.1.11 hold, then we call f ***smooth near*** p.

To conclude this section we consider maps where both the domain and the target are surfaces.

Definition 3.1.13 Let $S_1, S_2 \subset \mathbb{R}^3$ be regular surfaces. Let $p \in S_1$ and f: $S_1 \to S_2$ be a continuous map. We call f ***smooth near*** p, if there exists a local parametrisation (U_1, F_1, V_1) of S_1 around p and a local parametrisation (U_2, F_2, V_2) of S_2 around $f(p)$ in such a way that $F_2^{-1} \circ f \circ F_1$: $F_1^{-1}(f^{-1}(V_2) \cap V_1) \to U_2$ is smooth near p.

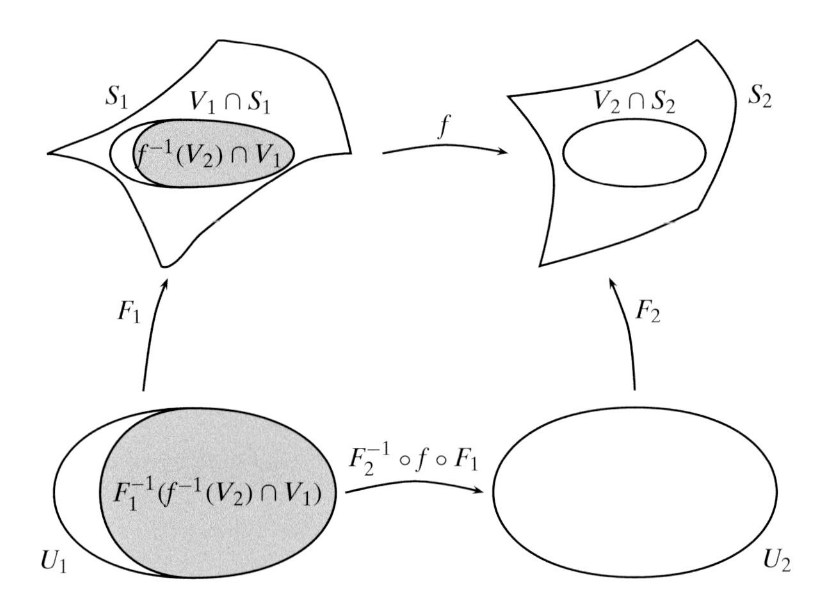

A map f between two surfaces is therefore called smooth if it is smooth when expressed in suitable coordinates. Could it now happen that a such map, expressed in other coordinates, is not smooth? This cannot happen since, as we have seen as an implication of proposition 3.1.9, parameter transformations are always C^∞: if, in addition to (U_i, F_i, V_i), $(\tilde{U}_i, \tilde{F}_i, \tilde{V}_i,)$ is also a local parametrisation of S_i, then smoothness of $F_2^{-1} \circ f \circ F_1$ implies that

$$\tilde{F}_2^{-1} \circ f \circ \tilde{F}_1 = \underbrace{\tilde{F}_2^{-1} \circ F_2}_{C^\infty} \circ \underbrace{F_2^{-1} \circ f \circ F_1}_{C^\infty} \circ \underbrace{F_1^{-1} \circ \tilde{F}_1}_{C^\infty}$$

is a smooth map as well. This is a very useful remark, since it means that one can check the differentiability of a map f in cleverly chosen coordinates.

Exercise 3.4 Let $V \subset \mathbb{R}^3$ be open, let $S_1, S_2 \subset \mathbb{R}^3$ be regular surfaces with $S_1 \subset V$. Let $f : V \to \mathbb{R}^3$ be a smooth map with $f(S_1) \subset S_2$. Show that

$$f|_{S_1} : S_1 \to S_2$$

is smooth in the sense defined above.

Example 3.1.14 Let A be an orthogonal 3×3 matrix. Being a linear map $A : \mathbb{R}^3 \to \mathbb{R}^3$ is certainly C^∞. Because of the orthogonality, A maps the unit sphere to itself. By exercise 3.4,

$$f = A|_{S^2} : S^2 \to S^2$$

is a smooth map.

Suggestion Check this in the coordinates for S^2 from the beginning of this section.

Definition 3.1.15 Let $S_1, S_2 \subset \mathbb{R}^3$ be regular surfaces. A map $f: S_1 \to S_2$ is called a ***diffeomorphism***, if f is bijective and both f and f^{-1} are smooth. If such a diffeomorphism $f: S_1 \to S_2$ exists, then the surfaces S_1 and S_2 are ***diffeomorphic***.

Example 3.1.16 Let

$$S_1 = \left\{ (x,y,z)^\top \in \mathbb{R}^3 \,\middle|\, \frac{x^2}{a^2} + \frac{y^2}{b^2} + \frac{z^2}{c^2} = 1 \right\}, a, b, c > 0,$$

an ellipsoid. Let $S_2 = S^2$ be the sphere. Then S_1 and S_2 are diffeomorphic. We may take

$$f: S_1 \to S_2,$$
$$f(x,y,z) = \left(\frac{x}{a}, \frac{y}{b}, \frac{z}{c} \right)^\top,$$

as the diffeomorphism, for example.

Example 3.1.17 Let $S_1 = \{(x,y,\varphi(x,y))^\top \mid (x,y)^\top \in U\}$ be the graph of a C^∞-function $\varphi: U \to \mathbb{R}$. Let $S_2 = U \times \{0\} \subset \mathbb{R}^3$ be U interpreted as a surface in $\mathbb{R}^3$. Then S_1 and S_2 are diffeomorphic by virtue of the following diffeomorphism:

$$f: S_1 \to S_2, \quad f(x,y,z) = (x,y,0)^\top,$$
$$f^{-1}: S_2 \to S_1, \quad f^{-1}(x,y,0) = (x,y,\varphi(x,y))^\top.$$

3.2 The tangent plane

The simplest regular surfaces are certainly planes, just as straight lines are the simplest curves. We would therefore now like to approximate possibly very complicated surfaces by planes. This concept is very similar to the differential of a smooth map. For a smooth map $F: U \to \mathbb{R}^m$, $U \subset \mathbb{R}^n$ open and $p \in U$, the map $\mathbb{R}^n \to \mathbb{R}^m$, $x \mapsto F(p) + D_pF(x-p)$ is the first (the affine linear) approximation at the point p. What is now the geometric equivalent to D_pF for a regular surface?

Definition 3.2.1 Let $S \subset \mathbb{R}^3$ be a regular surface, let $p \in S$. Then

$$T_pS = \{X \in \mathbb{R}^3 \mid \text{there exists an } \varepsilon > 0 \text{ and a smooth parametrised curve } c: (-\varepsilon, \varepsilon) \to S \text{ with } c(0) = p \text{ and } \dot{c}(0) = X\}$$

is called the ***tangent plane*** of S at p. The elements of the tangent plane are called ***tangent vectors***.

To illustrate the tangent plane one often draws the affine plane $T_pS + p$ obtained by a translation of T_pS to the root point p.

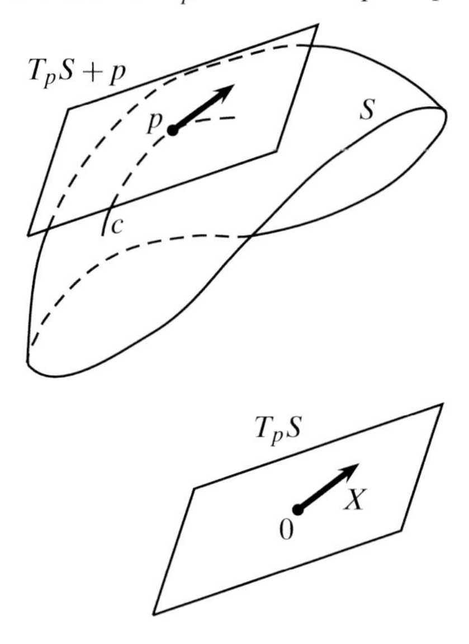

It is not immediately clear from the definition itself that T_pS is indeed a plane. However, we can also describe the tangent plane using local parametrisations as follows.

Proposition 3.2.2 *Let $S \subset \mathbb{R}^3$ be a regular surface, let $p \in S$. Further, let (U, F, V) be a local parametrisation of S around p. We set $u_0 := F^{-1}(p) \in U$.*

Then

$$T_pS = \text{Image}\ (D_{u_0}F) = D_{u_0}F(\mathbb{R}^2).$$

Proof (a) We show the inclusion "$\supset$". Let $X \in \text{Image}\ (D_{u_0}F)$, i.e. there exists a $Y \in \mathbb{R}^2$ with

$$X = D_{u_0}F(Y).$$

We set $c(t) := F(u_0 + tY)$. For a sufficiently small $\varepsilon > 0$ we have $u_0 + tY \in U$ if $|t| < \varepsilon$. Hence c is well defined on $(-\varepsilon, \varepsilon)$. It follows that

$$c(0) = F(u_0) = p$$

and

$$\dot{c}(0) = \frac{d}{dt} F(u_0 + tY)\Big|_{t=0} = D_{u_0}F(Y) = X.$$

Thus $X \in T_pS$.

(b) Now we show the inclusion "$\subset$". Let $X \in T_pS$, i.e. there exists a smooth curve $c : (-\varepsilon, \varepsilon) \to S$ with $c(0) = p$ and $\dot{c}(0) = X$. After possibly reducing the size of ε we can assume that c is fully contained in V. According to proposition 3.1.9

$$u := F^{-1} \circ c : (-\varepsilon, \varepsilon) \to U$$

is a smooth (plane) parametrised curve. We set $Y := \dot{u}(0) \in \mathbb{R}^2$. Then

$$D_{u_0} F(Y) = \frac{d}{dt}(F \circ u)\Big|_{t=0} = \frac{d}{dt} c\Big|_{t=0} = X.$$

Hence $X \in$ Image $(D_{u_0} F)$. □

Corollary 3.2.3 *$T_p S \subset \mathbb{R}^3$ is a two-dimensional vector subspace since $D_{u_0} F$ has full rank 2.*

We are often given a regular surface as the set of zeros of a function as in proposition 3.1.6. Then we can find the tangent plane using this function as well.

Proposition 3.2.4 *Let $V \subset \mathbb{R}^3$ be open, let $f : V \to \mathbb{R}$ be a smooth function and let $S = f^{-1}(0) \subset \mathbb{R}^3$. Suppose that $\operatorname{grad} f(p) \neq 0$ for all $p \in S$. Then for $p \in S$ the gradient of f is perpendicular to the tangent plane:*

$$T_p S = \operatorname{grad} f(p)^{\perp}.$$

Proof Let $X \in T_p S$. We choose a smooth parametrised curve $c : (-\varepsilon, \varepsilon) \to S$ with $c(0) = p$ and $\dot{c}(0) = X$. As c is completely contained in S we have $(f \circ c)(t) = 0$ for all $t \in (-\varepsilon, \varepsilon)$. Differentiating gives

$$0 = \frac{d}{dt} f \circ c\Big|_{t=0} = \langle \operatorname{grad} f(c(0)), \dot{c}(0) \rangle = \langle \operatorname{grad} f(p), X \rangle.$$

Hence X is perpendicular to $\operatorname{grad} f(p)$. We have thus shown that $T_p S \subset \operatorname{grad} f(p)^{\perp}$. As both subspaces $T_p S$ and $\operatorname{grad} f(p)^{\perp}$ of $\mathbb{R}^3$ have dimension 2 it follows that $T_p S = \operatorname{grad} f(p)^{\perp}$. □

Example 3.2.5 The sphere is described by

$$S^2 = f^{-1}(0),$$

where $f(x, y, z) = x^2 + y^2 + z^2 - 1$. We calculate

$$\operatorname{grad} f(x, y, z) = 2(x, y, z).$$

The tangent plane $T_p S^2$ is hence exactly the orthogonal complement of the root point vector p.

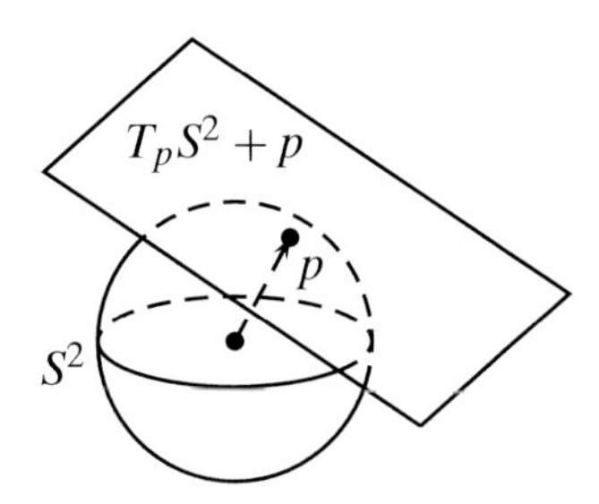

We know the concept of linear approximations from smooth maps (differential) which are defined on open subsets of $\mathbb{R}^n$, and the concept of linear approximations of regular surfaces (tangent plane). Combined these lead to the concept of linear approximations of smooth maps defined on regular surfaces, again called differentials.

Definition 3.2.6 Let $S_1, S_2 \subset \mathbb{R}^3$ be regular surfaces, let $f : S_1 \to S_2$ be a smooth map and let $p \in S_1$. The ***differential*** of f at p is the map

$$d_pf : T_pS_1 \to T_{f(p)}S_2,$$

which is given by the rule: for $X \in T_pS_1$ choose a smooth parametrised curve $c : (-\varepsilon, \varepsilon) \to S_1$ with $c(0) = p$ and $\dot{c}(0) = X$ and set

$$d_pf(X) := \frac{d}{dt}(f \circ c)\Big|_{t=0} \in T_{f(p)}S_2.$$

Proposition 3.2.7 *This definition makes d_pf well-defined, i.e. $d_pf(X)$ depends only on X, not on the particular choice of the curve c. Further d_pf is linear.*

Proof We express d_pf using local parametrisations. Let (U_1, F_1, V_1) be a local parametrisation of S_1 around p and let (U_2, F_2, V_2) be a local parametrisation of S_2 around $f(p)$. After possibly reducing the size of U_1 and V_1 we can assume that $f(S \cap V_1) \subset V_2$. We set

$$\tilde{f} := F_2^{-1} \circ f \circ F_1 : U_1 \to U_2$$

and $u_0 := F_1^{-1}(p) \in U_1$. For the curve $c : (-\varepsilon, \varepsilon) \to S_1$ with $c(0) = p$ and $\dot{c}(0) = X$ we set

$$u := F_1^{-1} \circ c : (-\varepsilon, \varepsilon) \to U_1.$$

Again we may need to choose a smaller ε to ensure that c is fully contained in V_1. Then we have, as in the proof of proposition 3.2.2, that $D_{u_0}F_1(\dot{u}(0)) = X$ and we find

$$\begin{aligned}
d_p f(X) &= \frac{d}{dt}(f \circ c)\Big|_{t=0} \\
&= \frac{d}{dt}(f \circ F_1 \circ u)\Big|_{t=0} \\
&= \frac{d}{dt}(F_2 \circ \tilde{f} \circ u)\Big|_{t=0} \\
&= D_{u_0}(F_2 \circ \tilde{f})(\dot{u}(0)) \\
&= D_{u_0}(F_2 \circ \tilde{f}) \circ \left(D_{u_0} F_1\right)^{-1}(X).
\end{aligned}$$

The last expression is no longer a function of c, but only of X. Hence the definition is independent of the particular choice of c.

Further the map $d_p f$ can be written as the composition of the linear maps $D_{u_0}(F_2 \circ \tilde{f})$ and $\left(D_{u_0} F_1\right)^{-1}$ and is thus linear itself. □

Remark The proof showed that by virtue of the local parametrisation (U_1, F_1, V_1) and (U_2, F_2, V_2) the differential $d_p f$ is given by the Jacobian matrix $D_{u_0}\tilde{f}$. More precisely, the following diagram commutes:

$$\begin{array}{ccc}
T_p S_1 & \xrightarrow{d_p f} & T_{f(p)} S_2 \\
{\scriptstyle D_{u_0} F_1} \uparrow \cong & & \cong \uparrow {\scriptstyle D_{\tilde{f}(u_0)} F_2} \\
\mathbb{R}^2 & \xrightarrow{D_{u_0} \tilde{f}} & \mathbb{R}^2
\end{array}$$

Example 3.2.8 Let $A : \mathbb{R}^3 \to \mathbb{R}^3$ be an orthogonal map, i.e. $A \in \mathrm{O}(3)$. We set $f : S^2 \to S^2$, $f := A|_{S^2}$. Let $p \in S^2$. We find the differential of f at p. For this purpose let $c : (-\varepsilon, \varepsilon) \to S^2$ be a smooth parametrised curve with $c(0) = p$ and $\dot{c}(0) = X \in T_p S^2$. Because of the linearity of A we have

$$\begin{aligned}
\frac{d}{dt}(f \circ c)\Big|_{t=0} &= \frac{d}{dt}(A \circ c)\Big|_{t=0} \\
&= A \circ \frac{d}{dt} c\Big|_{t=0} \\
&= A(X).
\end{aligned}$$

Thus $d_p f = A|_{T_p S^2} : T_p S^2 \to T_{Ap} S^2$.

Analogously one can define the differential

$$d_p f : T_p S \to \mathbb{R}^n$$

for smooth maps $f : S \to \mathbb{R}^n$ that are defined on a regular surface. Suppose that $c : (-\varepsilon, \varepsilon) \to S$ is a smooth parametrised curve with $c(0) = p$ and $\dot{c}(0) = X$. Then we set $d_pf(X) := (d/dt)(f \circ c)|_{t=0}$.

If the domain is an open subset $U \subset \mathbb{R}^n$ and $f : U \to S$ takes the values of a regular surface S, then we simply set

$$d_pf : \mathbb{R}^n \to T_{f(p)}S \subset \mathbb{R}^3,$$
$$d_pf(X) := D_pf(X),$$

for $p \in U$. It is easy to see that $D_pf(X)$ indeed lies in the subspace $T_{f(p)}S$. In any case one obtains a well-defined linear map d_pf.

3.3 The first fundamental form

In order to do geometry on a regular surface $S \subset \mathbb{R}^3$, we need to be able to measure, for example, lengths of curves that lie in S or the angle between two tangent vectors to the surface. Since the tangent plane at each point $p \in S$ is a two-dimensional subspace of $\mathbb{R}^3$, we can restrict the conventional scalar product $\langle \cdot, \cdot \rangle$ of $\mathbb{R}^3$ to T_pS, and obtain a Euclidean scalar product on T_pS. The map, which assigns each point $p \in S$ this restriction $g_p := \langle \cdot, \cdot \rangle|_{T_pS \times T_pS}$, is called ***first fundamental form*** of S. We often write the first fundamental form as

$$I_p(X, Y) = g_p(X, Y) = \langle X, Y \rangle,$$

where $X, Y \in T_pS$.

As learnt in linear algebra, every Euclidean scalar product on a vector space, here g_p on T_pS, can be represented as a positive definite symmetric matrix after a choice of basis. A basis of T_pS is usually obtained by a local parametrisation (U, F, V) of S at p. If e_1, e_2 are the standard basis vectors of $\mathbb{R}^2$, then $D_uF(e_1) = (\partial F/\partial u^1)(u)$ and $D_uF(e_2) = (\partial F/\partial u^2)(u)$, $u = F^{-1}(p)$, form a basis of T_pS. With respect to this basis, the matrix representation of g_p is then given by

$$g_{ij}(u) := g_p(D_uF(e_i), D_uF(e_j)) = \left\langle \frac{\partial F}{\partial u^i}(u), \frac{\partial F}{\partial u^j}(u) \right\rangle.$$

The 2×2 matrix $(g_{ij}(u))_{i,j=1,2}$ is therefore symmetric and positive definite. Further, the above formula shows immediately that the matrix entries g_{ij} depend smoothly on u, i.e. $g_{ij} : U \to \mathbb{R}$ is a smooth function for every i and j.

Example 3.3.1 Let $S \subset \mathbb{R}^3$ be a plane. Then S can be described by an affine-linear parametrisation as in example 3.1.3,

$$F : \mathbb{R}^2 \to \mathbb{R}^3,$$

$$F(u^1, u^2) = p_0 + u^1 \cdot X + u^2 \cdot Y, \qquad p_0,\ X,\ Y \in \mathbb{R}^3.$$

Hence S is here the plane spanned by the vectors X and Y through the point p_0. We find the first fundamental form w.r.t. the parametrisation

$$g_{11}(u^1, u^2) = \left\langle \frac{\partial F}{\partial u^1}(u^1, u^2), \frac{\partial F}{\partial u^1}(u^1, u^2) \right\rangle = \langle X, X \rangle,$$

$$g_{12}(u^1, u^2) = g_{21}(u^1, u^2) = \left\langle \frac{\partial F}{\partial u^1}(u^1, u^2), \frac{\partial F}{\partial u^2}(u^1, u^2) \right\rangle = \langle X, Y \rangle = \langle Y, X \rangle,$$

$$g_{22}(u^1, u^2) = \left\langle \frac{\partial F}{\partial u^2}(u^1, u^2), \frac{\partial F}{\partial u^2}(u^1, u^2) \right\rangle = \langle Y, Y \rangle.$$

For example, if S is the x–y plane and if (u^1, u^2) are Cartesian coordinates, i.e. $p_0 = 0$, $X = e_1$ and $Y = e_2$, then the first fundamental form is given by the matrix

$$\left(g_{ij}(u)\right)_{ij} = \begin{pmatrix} 1 & 0 \\ 0 & 1 \end{pmatrix}.$$

The functions $g_{ij} : \mathbb{R}^2 \to \mathbb{R}$ from our example are therefore constant. If one uses another local parametrisation for the same surface, then this will in general no longer be the case.

Let us consider polar coordinates as an example. We still assume that S is the x–y plane in $\mathbb{R}^3$. Polar coordinates $(\tilde{u}^1, \tilde{u}^2) = (r, \varphi)$ give the local parametrisation

$$\tilde{F} : (0, \infty) \times (0, 2\pi) \to \mathbb{R}^3,$$
$$\tilde{F}(r, \varphi) = (r \cdot \cos\varphi, r \cdot \sin\varphi, 0)^\top.$$

The first fundamental form is now

$$\begin{aligned} \tilde{g}_{11}(r, \varphi) &= \left\langle \frac{\partial \tilde{F}}{\partial r}(r, \varphi), \frac{\partial \tilde{F}}{\partial r}(r, \varphi) \right\rangle \\ &= \left\langle \begin{pmatrix} \cos(\varphi) \\ \sin(\varphi) \\ 0 \end{pmatrix}, \begin{pmatrix} \cos(\varphi) \\ \sin(\varphi) \\ 0 \end{pmatrix} \right\rangle \\ &= \cos^2(\varphi) + \sin^2(\varphi) \\ &= 1, \end{aligned}$$

$$\begin{aligned}
\tilde{g}_{12}(r,\varphi) = \tilde{g}_{21}(r,\varphi) &= \left\langle \frac{\partial \tilde{F}}{\partial \varphi}(r,\varphi), \frac{\partial \tilde{F}}{\partial r}(r,\varphi) \right\rangle \\
&= \left\langle \begin{pmatrix} -r\sin(\varphi) \\ r\cos(\varphi) \\ 0 \end{pmatrix}, \begin{pmatrix} \cos(\varphi) \\ \sin(\varphi) \\ 0 \end{pmatrix} \right\rangle \\
&= r \cdot (-\sin(\varphi)\cdot\cos(\varphi) + \cos(\varphi)\cdot\sin(\varphi)) \\
&= 0, \\
\tilde{g}_{22}(r,\varphi) &= \left\langle \begin{pmatrix} -r\sin(\varphi) \\ r\cos(\varphi) \\ 0 \end{pmatrix}, \begin{pmatrix} -r\sin(\varphi) \\ r\cos(\varphi) \\ 0 \end{pmatrix} \right\rangle \\
&= r^2\sin^2(\varphi) + r^2\cos^2(\varphi) \\
&= r^2.
\end{aligned}$$

The first fundamental form of the x–y plane with respect to polar coordinates is therefore given by the matrix

$$\left(\tilde{g}_{ij}(r,\varphi)\right)_{ij} = \begin{pmatrix} 1 & 0 \\ 0 & r^2 \end{pmatrix}. \tag{3.1}$$

At least the $\tilde{g}_{22}$-component is not constant this time.

This example already shows that the formulae for the first fundamental form depend strongly on the choice of the local parametrisation. The more clumsily the parametrisation is chosen, the more complicated the formulae will be.

Example 3.3.2 Let us consider the cylindrical surface

$$S = \{(x,y,z)^\top \in \mathbb{R}^3 \mid x^2 + y^2 = 1\}.$$

We use the local parametrisation

$$F : (0, 2\pi) \times \mathbb{R} \to \mathbb{R}^3,$$

$$F(\varphi, h) = \begin{pmatrix} \cos(\varphi) \\ \sin(\varphi) \\ h \end{pmatrix}.$$

For the first fundamental form we obtain, with respect to the coordinates $(u^1, u^2) = (\varphi, h)$,

$$\begin{aligned}
g_{11}(\varphi,h) &= \left\langle \frac{\partial F}{\partial \varphi}(\varphi,h),\ \frac{\partial F}{\partial \varphi}(\varphi,h) \right\rangle \\
&= \left\langle \begin{pmatrix} -\sin(\varphi) \\ \cos(\varphi) \\ 0 \end{pmatrix}, \begin{pmatrix} -\sin(\varphi) \\ \cos(\varphi) \\ 0 \end{pmatrix} \right\rangle \\
&= \sin^2(\varphi) + \cos^2(\varphi) \\
&= 1, \\
g_{12}(\varphi,h) &= g_{21}(\varphi,h) = \left\langle \frac{\partial F}{\partial \varphi}(\varphi,h),\ \frac{\partial F}{\partial h}(\varphi,h) \right\rangle \\
&= \left\langle \begin{pmatrix} -\sin(\varphi) \\ \cos(\varphi) \\ 0 \end{pmatrix}, \begin{pmatrix} 0 \\ 0 \\ 1 \end{pmatrix} \right\rangle \\
&= 0, \\
g_{22}(\varphi,h) &= \left\langle \frac{\partial F}{\partial h}(\varphi,h),\ \frac{\partial F}{\partial h}(\varphi,h) \right\rangle \\
&= \left\langle \begin{pmatrix} 0 \\ 0 \\ 1 \end{pmatrix}, \begin{pmatrix} 0 \\ 0 \\ 1 \end{pmatrix} \right\rangle \\
&= 1.
\end{aligned}$$

We (astonishingly?) discover that the first fundamental form of the cylindrical surface with respect to the chosen coordinates has the same form as that of the plane with respect to Cartesian coordinates, i.e. $\left(g_{ij}\right)_{ij} = \left(\begin{smallmatrix} 1 & 0 \\ 0 & 1 \end{smallmatrix}\right)$. This suggests that the two surfaces have certain things in common. We will investigate these later.

Example 3.3.3 We find the first fundamental form of the sphere

$$S^2 = \left\{ (x,y,z)^\top \in \mathbb{R}^3 \,\middle|\, x^2 + y^2 + z^2 = 1 \right\}$$

in polar coordinates $(u^1, u^2) = (\theta, \varphi)$:

$$F : \left(-\frac{\pi}{2}, \frac{\pi}{2}\right) \times (0, 2\pi) \to \mathbb{R}^3,$$

$$F(\theta,\varphi) = \begin{pmatrix} \cos(\theta)\cdot\cos(\varphi) \\ \cos(\theta)\cdot\sin(\varphi) \\ \sin(\theta) \end{pmatrix}$$

From

$$\frac{\partial F}{\partial \theta}(\theta,\varphi) = \begin{pmatrix} -\sin(\theta)\cdot\cos(\varphi) \\ -\sin(\theta)\cdot\sin(\varphi) \\ \cos(\theta) \end{pmatrix}, \quad \frac{\partial F}{\partial \varphi}(\theta,\varphi) = \begin{pmatrix} -\cos(\theta)\cdot\sin(\varphi) \\ \cos(\theta)\cdot\cos(\varphi) \\ 0 \end{pmatrix}$$

we obtain

$$\big(g_{ij}(\theta,\varphi)\big)_{ij} = \begin{pmatrix} 1 & 0 \\ 0 & \cos^2(\theta) \end{pmatrix}.$$

We finally investigate what happens if we change the local parametrisation. Suppose that (U, F, V) and $(\tilde{U}, \tilde{F}, \tilde{V})$ are two local parametrisations. Let $(\tilde{g}_{ij})_{ij}$ be the corresponding matrix that describes the first fundamental form. Let us now denote the parameter transformation by $\varphi := \tilde{F}^{-1} \circ F$, then using the chain rule we obtain

$$\begin{aligned}
g_{ij}(u) &= I\Big(\frac{\partial F}{\partial u^i}(u), \frac{\partial F}{\partial u^j}(u)\Big) \\
&= I\Big(\frac{\partial(\tilde{F}\circ\varphi)}{\partial u^i}(u), \frac{\partial(\tilde{F}\circ\varphi)}{\partial u^j}(u)\Big) \\
&= I\Big(\sum_k \frac{\partial \tilde{F}}{\partial \tilde{u}^k}(\varphi(u)) \cdot \frac{\partial \varphi^k}{\partial u^i}(u), \sum_\ell \frac{\partial \tilde{F}}{\partial \tilde{u}^\ell}(\varphi(u)) \cdot \frac{\partial \varphi^\ell}{\partial u^j}(u)\Big) \\
&= \sum_{k\ell} \frac{\partial \varphi^k}{\partial u^i}(u) \frac{\partial \varphi^\ell}{\partial u^j}(u) \cdot I\Big(\frac{\partial \tilde{F}}{\partial \tilde{u}^k}(\varphi(u)), \frac{\partial \tilde{F}}{\partial \tilde{u}^\ell}(\varphi(u))\Big) \\
&= \sum_{k\ell} \frac{\partial \varphi^k}{\partial u^i}(u) \frac{\partial \varphi^\ell}{\partial u^j}(u) \tilde{g}_{k\ell}(\varphi(u)).
\end{aligned}$$

In matrix notation this equation is

$$\big(g_{ij}(u)\big)_{ij} = (D_u\varphi)^\top \cdot \big(\tilde{g}_{k\ell}(\varphi(u))\big)_{k\ell} \cdot D_u\varphi. \tag{3.2}$$

Exercise 3.5 Calculate the first fundamental form of the sphere w.r.t. the local parametrisation (U, F_3^+, V_3^+) from example 3.1.5.

Exercise 3.6 Find the first fundamental form of a graph of a function w.r.t. the parametrisation from example 3.1.4.

Exercise 3.7 Calculate the first fundamental form of the ***cone***

$$S = \{(x,y,z)^\top \in \mathbb{R}^3 \mid x^2 + y^2 = z^2,\ z > 0\}$$

w.r.t. the parametrisation

$$F : (0, 2\pi) \times (0, \infty) \to \mathbb{R}^3,$$
$$F(\varphi, r) = r(\cos(\varphi), \sin(\varphi), 1)^\top.$$

3.4 Normal fields and orientability

Definition 3.4.1 Let $S \subset \mathbb{R}^3$ be a regular surface. A ***normal field*** on S is a map

$$N : S \to \mathbb{R}^3,$$

such that $N(p) \perp T_pS$ for all $p \in S$. A normal field on S is said to be a ***unit normal field***, if in addition $\|N(p)\| = 1$ for all $p \in S$.

Remark Note that if N is a (unit) normal field on S, then the same is true for $-N$. Continuous unit normal fields may, but do not necessarily, exist on regular surfaces.

Example 3.4.2 Let $S = \{(x, y, 0)^\top \mid x, y \in \mathbb{R}\}$ be the x–y plane in $\mathbb{R}^3$. Then $N(x, y, 0) = (0, 0, 1)^\top$ is a constant unit normal field on S.

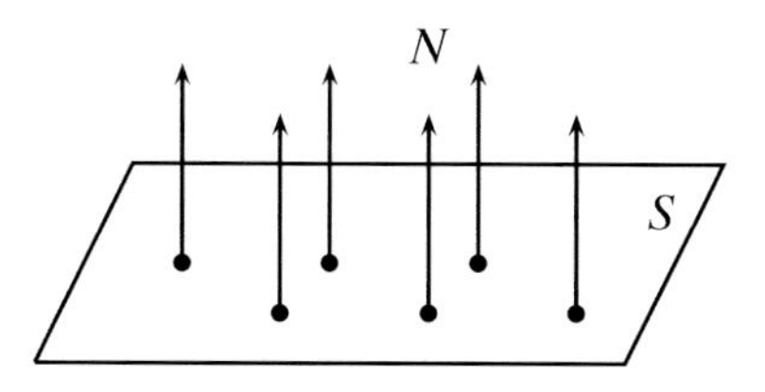

Example 3.4.3 Let $S = S^2$. Then we obtain a unit normal field by setting $N = \mathrm{Id}$.

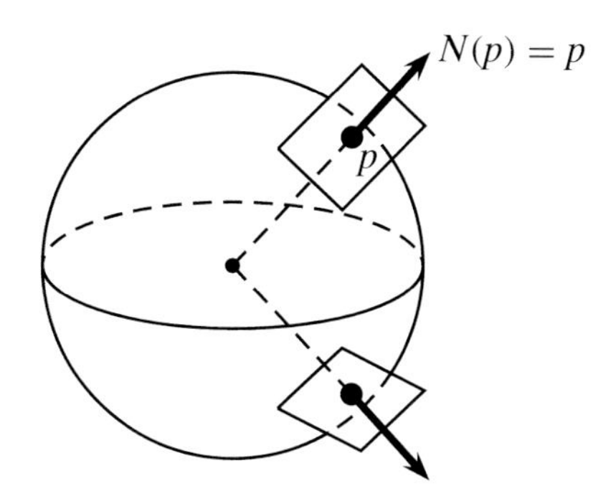

Example 3.4.4 Let $S = S^1 \times \mathbb{R}$ be the cylindrical surface. Then $N(x, y, z) = (x, y, 0)^\top$ defines a unit normal field.

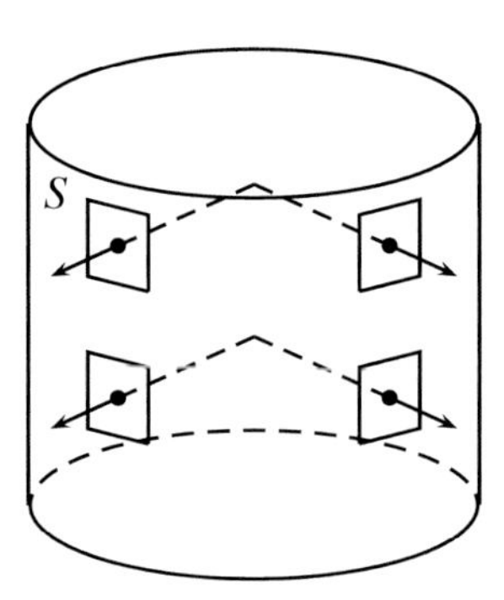

Example 3.4.5 Let S be the ***Möbius strip***. It is obtained by gluing together a strip of paper at its left and its right end, but, in order not to obtain the cylindrical surface, twisting it once.

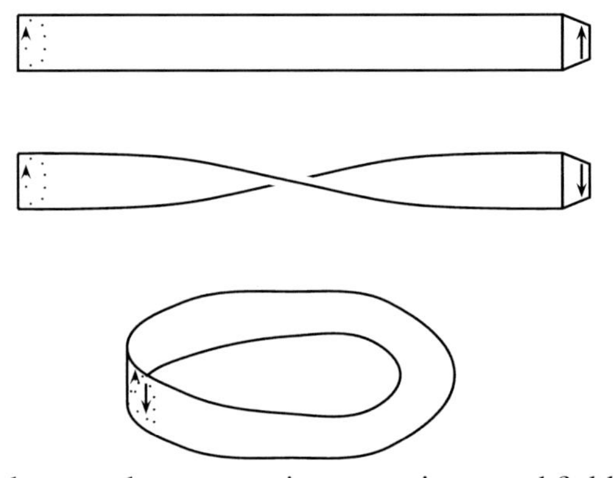

The Möbius strip does not have a continuous unit normal field.

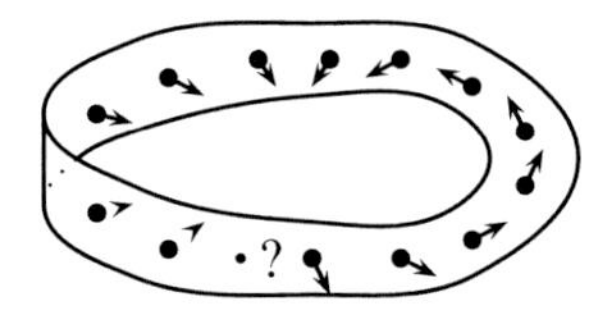

This also means that the Möbius strip only has one side, there is no "inside" or "outside". If one starts colouring the Möbius strip at one point, then one will find after a while that the strip is coloured everywhere on both sides.

Definition 3.4.6 A regular surface $S \subset \mathbb{R}^3$ is ***orientable*** if there exists a smooth unit normal field on S.

Remark Hence the plane, the sphere and the cylinder are orientable, while the Möbius strip is not. The main condition in the definition is that the unit normal field is *smooth*. It is always possible to find some unit normal field by choosing for each point $p \in S$ one of the two unit normal vectors to $T_pS \subset \mathbb{R}^3$ and calling it $N(p)$. But this N will usually not be continuous, let alone smooth. Indeed, one could change *smooth* to *continuous* in the definition without changing the concept of orientability at all, as the following exercise shows.

Exercise 3.8 Let $S \subset \mathbb{R}^3$ be a regular surface, $N : S \to S^2 \subset \mathbb{R}^3$ a unit normal field. Show that N is continuous if and only if N is smooth.

We now want to observe that regular surfaces S are always locally orientable. For this purpose let (U, F, V) be a local parametrisation of S. Using the vector product we obtain a normal field $\tilde{N}$ on $S \cap V$ via

$$\tilde{N}(p) = D_{F^{-1}(p)}F(e_1) \times D_{F^{-1}(p)}F(e_2).$$

As $D_{F^{-1}(p)}F$ has full rank, the vectors $D_{F^{-1}(p)}F(e_1)$ and $D_{F^{-1}(p)}F(e_2)$ are linearly independent and hence $\tilde{N}(p) \neq (0,0,0)^\top$.

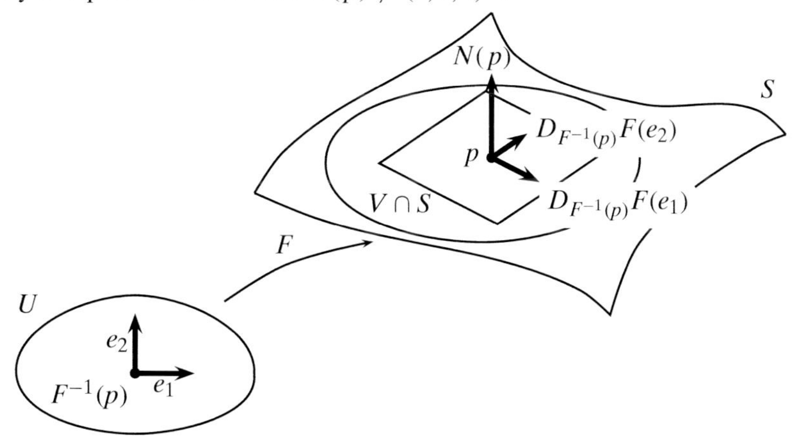

A smooth unit normal field on $S \cap V$ is then obtained via

$$N(p) := \frac{\tilde{N}(p)}{\|\tilde{N}(p)\|}.$$

Of course there is a certain arbitrariness in this choice. We could have taken $-N : S \cap V \to S^2 \subset \mathbb{R}^3$ instead of $N : S \cap V \to S^2 \subset \mathbb{R}^3$. If this construction is carried out for two local parametrisations (U_1, F_1, V_1) and (U_2, F_2, V_2), then the two corresponding unit normal vectors at points from $S \cap V_1 \cap V_2$ can either agree or be negatives of each other. We express this by a condition on the parameter transformation $\varphi := F_2^{-1} \circ F_1$.

For this purpose let $p \in S \cap V_1 \cap V_2$, $u_i := F_i^{-1}(p)$. Let $N_i(p)$ the unit normal vector of (U_i, F_i, V_i) at p, $i = 1, 2$. By construction $(D_{u_i}F_i(e_1), D_{u_i}F_i(e_2), N_i(p))$ form a positively oriented basis of $\mathbb{R}^3$. Hence $N_1(p)$ and $N_2(p)$ agree if and only if $(D_{u_1}F_1(e_1), D_{u_1}F_1(e_2))$ and $(D_{u_2}F_2(e_1), D_{u_2}F_2(e_2))$ have the same orientation in T_pS, otherwise $N_1(p) = -N_2(p)$. Hence $N_1(p) = N_2(p)$ if and only if φ is orientation-preserving at u_1, i.e. if $\det(D_{u_1}\varphi) > 0$.

Conclusion

$$N_1(p) = N_2(p) \quad \Leftrightarrow \quad \det(D_{u_1}\varphi) > 0$$
$$N_1(p) = -N_2(p) \quad \Leftrightarrow \quad \det(D_{u_1}\varphi) < 0.$$

Thc following theorem follows immediately.

Theorem 3.4.7 *A regular surface $S \subset \mathbb{R}^3$ is orientable if and only if S can be covered by local parametrisations such that for all parameter transformations φ*

$$\det(D\varphi) > 0.$$

3.5 The second fundamental form

Let $S \subset \mathbb{R}^3$ be an orientable regular surface with smooth unit normal field N. Regarded as a map between surfaces, i.e. $N : S \to S^2$, the map N is also called the ***Gauss map***.

Let $p \in S$. We consider the differential of N at p:

$$d_pN : T_pS \to T_{N(p)}S^2.$$

Now $T_{N(p)}S^2 = N(p)^\perp = T_pS$. Hence d_pN is an endomorphism on T_pS.

Definition 3.5.1 Let $S \subset \mathbb{R}^3$ be a regular surface with orientation given by the unit normal field N. The endomorphism

$$W_p : T_pS \to T_pS,$$

$$W_p(X) = -d_pN(X),$$

is called the ***Weingarten map***.

The negative sign appears for historic reasons. If orientation is reversed, i.e. if $-N$ is substituted for N, then W also changes its sign.

Example 3.5.2 Let $S = S^2$, and let N be the outer unit normal field, $N(p) = p$. Then

$$W_p = -\mathrm{Id} : T_pS^2 \to T_pS^2.$$

Example 3.5.3 Let $S = \{(x, y, 0)^\top \mid x, y \in \mathbb{R}\}$ be the x–y plane, $N(x, y, z) = (0, 0, 1)^\top$. Then N is constant and thus $W_p = 0$ for all $p \in S$.

Example 3.5.4 Let $S = S^1 \times \mathbb{R}$ be the cylinder, $N(x, y, z) = (x, y, 0)^\top$.

At a point $p = (x, y, z)^\top \in S$ the tangent plane T_pS is spanned by the basis vectors $(-y, x, 0)^\top$ and $(0, 0, 1)^\top$. We calculate

$$W_p\begin{pmatrix}0\\0\\1\end{pmatrix}=-d_pN\begin{pmatrix}0\\0\\1\end{pmatrix}=-\frac{d}{dt}N\left.\begin{pmatrix}x\\y\\z+t\end{pmatrix}\right|_{t=0}$$

$$=-\frac{d}{dt}\left.\begin{pmatrix}x\\y\\0\end{pmatrix}\right|_{t=0}=\begin{pmatrix}0\\0\\0\end{pmatrix}.$$

To find the image of $(-y,x,0)$ under W_p, we choose $t_0\in\mathbb{R}$, such that $(\cos(t_0),\sin(t_0))=(x,y)$. Then for $c(t):=(\cos(t+t_0),\sin(t+t_0),z)^\top$ we have that $c(0)=(x,y,z)^\top=p$ and $\dot{c}(0)=(-\sin(t_0),\cos(t_0),0)^\top=(-y,x,0)^\top$. It follows that

$$W_p\begin{pmatrix}-y\\x\\0\end{pmatrix}=-d_pN\begin{pmatrix}-y\\x\\0\end{pmatrix}=-\frac{d}{dt}N\left.\begin{pmatrix}\cos(t+t_0)\\\sin(t+t_0)\\z\end{pmatrix}\right|_{t=0}$$

$$=-\frac{d}{dt}\left.\begin{pmatrix}\cos(t+t_0)\\\sin(t+t_0)\\0\end{pmatrix}\right|_{t=0}=\begin{pmatrix}\sin(t_0)\\-\cos(t_0)\\0\end{pmatrix}=-\begin{pmatrix}-y\\x\\0\end{pmatrix}.$$

With respect to the basis vectors $(-y,x,0)^\top$ and $(0,0,1)^\top$, the map W_p therefore has the matrix representation

$$\begin{pmatrix}-1&0\\0&0\end{pmatrix}.$$

Proposition 3.5.5 *Let $S\subset\mathbb{R}^3$ be an orientable regular surface with Weingarten map $W_p:T_pS\to T_pS$, $p\in S$. Then W_p is self-adjoint with respect to the first fundamental form.*

Proof Let N be the unit normal field of S that induces the Weingarten map, $W_p=-d_pN$. We choose a local parametrisation (U,F,V) at p and set $u:=F^{-1}(p)$.

Let

$$X_1:=D_uF(e_1)=\frac{\partial F}{\partial u^1}(u)$$

and

$$X_2:=D_uF(e_2)=\frac{\partial F}{\partial u^2}(u)$$

be the corresponding basis vectors of T_pS. As N is perpendicular to S everywhere, we have

$$\left\langle \frac{\partial F}{\partial u^i}(u + te_j),\ N(F(u + te_j)) \right\rangle \equiv 0.$$

Differentiating this equation with respect to t gives

$$\begin{aligned}
0 &= \frac{d}{dt} \left\langle \frac{\partial F}{\partial u^i}(u + te_j), N(F(u + te_j)) \right\rangle \Bigg|_{t=0} \\
&= \left\langle \frac{d}{dt} \frac{\partial F}{\partial u^i}(u + te_j)\Bigg|_{t=0}, N(p) \right\rangle + \left\langle \frac{\partial F}{\partial u^i}(u), d_pN \circ D_uF(e_j) \right\rangle \\
&= \left\langle \frac{\partial^2 F}{\partial u^j \partial u^i}(u), N(p) \right\rangle + \langle X_i, -W_p(X_j) \rangle .
\end{aligned}$$

Thus

$$I_p(X_i, W_p(X_j)) = \langle X_i, W_p(X_j) \rangle = \left\langle \frac{\partial^2 F}{\partial u^j \partial u^i}(u), N(p) \right\rangle . \tag{3.3}$$

By the theorem of Schwarz [18, p. 372, theorem 1.1] the two partial derivatives of F can be exchanged and we obtain

$$\begin{aligned}
I_p(X_i, W_p(X_j)) &= \left\langle \frac{\partial^2 F}{\partial u^j \partial u^i}(u), N(p) \right\rangle \\
&= \left\langle \frac{\partial^2 F}{\partial u^i \partial u^j}(u), N(p) \right\rangle \\
&= I_p(X_j, W_p(X_i)).
\end{aligned}$$

We now know for our basis vectors X_1 and X_2 of T_pS that

$$I_p(X_i, W_p(X_j)) = I_p(X_j, W_p(X_i)) = I_p(W_p(X_i), X_j).$$

Since any two vectors $X, Y \in T_pS$ can be written as a linear combination of X_1 and X_2, it immediately follows from the bilinearity of I and the linearity of W_p that

$$I_p(X, W_p(Y)) = I_p(W_p(X), Y),$$

i.e. W_p is self-adjoint with respect to I. □

Let us recall from linear algebra that if V is a finite-dimensional real vector space with Euclidean scalar product $\langle \cdot, \cdot \rangle$, then the self-adjoint endomorphisms

W on V are uniquely associated with the symmetric bilinear forms β on V. The relation between W and β is

$$\beta(X,Y) = \langle W(X), Y\rangle, \qquad X, Y \in V.$$

Definition 3.5.6 The bilinear form that corresponds to the Weingarten map W_p is called the ***second fundamental form*** (of the surface S at the point p):

$$II_p(X,Y) = I_p(W_p(X), Y), \qquad X, Y \in T_pS.$$

For simplicity the root point vector p is often omitted in the notation, and we write II instead of II_p, and W instead of W_p.

Expression in local coordinates Let $S \subset \mathbb{R}^3$ be a regular surface, $p \in S$. Let (U, F, V) be a local parametrisation of S at p. We set $u := F^{-1}(p)$. We have already learnt how to express the first fundamental form locally. For the basis $D_uF(e_1) = (\partial F/\partial u^1)(u)$ and $D_uF(e_2) = (\partial F/\partial u^2)(u)$ the first fundamental form is given by the symmetric matrix $(g_{ij}(u))_{i,j=1,2}$, where

$$\begin{aligned} g_{ij}(u) &= \left\langle \frac{\partial F}{\partial u^i}(u), \frac{\partial F}{\partial u^j}(u) \right\rangle \\ &= I_p\left(D_uF(e_i), D_uF(e_j)\right). \end{aligned}$$

We now define

$$\begin{aligned} h_{ij}(u) &:= II_p(D_uF(e_i), D_uF(e_j)) \\ &= I_p\left(W_p(D_uF(e_i)), D_uF(e_j)\right) \\ &\overset{(3.3)}{=} \left\langle \frac{\partial^2 F}{\partial u^j \partial u^i}(u), N(p) \right\rangle, \ i,j = 1,2. \end{aligned} \tag{3.4}$$

Then $(h_{ij}(u))_{i,j=1,2}$ is the symmetric matrix which gives the second fundamental form w.r.t. the basis given above. The entries of the matrix representing the Weingarten map are denoted as w_i^j, i.e. we define

$$W_p(D_uF(e_i)) =: \sum_{j=1}^{2} w_i^j(u) D_uF(e_j).$$

As the Weingarten map is closely related to the second fundamental form, we expect that the matrices $(h_{ij}(u))_{i,j}$ and $\left(w_i^j(u)\right)_{i,j}$ can be determined from each other. This works as follows:

$$\begin{aligned}
h_{ij}(u) &= II(D_uF(e_i), D_uF(e_j)) \\
&= I(W(D_uF(e_i)), D_uF(e_j)) \\
&= I\Big(\sum_{k=1}^{2} w_i^k(u) D_uF(e_k), D_uF(e_j)\Big) \\
&= \sum_{k=1}^{2} w_i^k(u) I\left(D_uF(e_k), D_uF(e_j)\right) \\
&= \sum_{k=1}^{2} w_i^k(u) g_{kj}(u).
\end{aligned}$$

The matrix $(h_{ij}(u))_{i,j}$ therefore results from the matrix multiplication of the matrices $\left(w_i^k(u)\right)_{i,k}$ and $(g_{kj}(u))_{k,j}$. As the matrix $(g_{kj}(u))_{k,j}$ is positive definite, and thus in particular is invertible, the equation can be solved for $\left(w_i^k(u)\right)_{i,k}$. For this purpose let $(g^{ij}(u))_{i,j}$ be the inverse matrix of $(g_{ij}(u))_{i,j}$, i.e.

$$\left(g^{ij}(u)\right)_{ij} = \frac{1}{g_{11}(u)g_{22}(u) - g_{12}(u)^2} \begin{pmatrix} g_{22}(u) & -g_{12}(u) \\ -g_{21}(u) & g_{11}(u) \end{pmatrix}.$$

It then follows that

$$\sum_{k=1}^{2} h_{ik}(u) g^{kj}(u) = w_i^j(u). \tag{3.5}$$

3.6 Curvature

We now come to a central notion of surface theory and of differential geometry: ***curvature***. We will meet several concepts of curvature, and we begin with normal curvature.

For this purpose let $S \subset \mathbb{R}^3$ be an orientable regular surface with smooth unit normal field N, $p \in S$. Let $c : (-\varepsilon, \varepsilon) \to S$ be a curve parametrised by arc-length with $c(0) = p$. Regarded as a space curve in $\mathbb{R}^3$, the curve c has at point 0 curvature $\kappa(0)$, which in the case $\kappa(0) \neq 0$ is given by

$$\ddot{c}(0) = \kappa(0) \cdot n(0),$$

where n is the normal vector of c. We now want to split this curvature into a part which results from the fact that c curves *within* S, and a part which reflects the curvature of S in $\mathbb{R}^3$. For this purpose we decompose $n(0)$ into a part that is tangential to S and one that is perpendicular to S:

$$n(0) = n(0)^{\text{tang}} + n(0)^{\text{perp}},$$

where $n(0)^{\text{perp}} = \langle n(0), N(p) \rangle \, N(p)$. We therefore have

$$\ddot{c}(0) = \kappa(0) \cdot n(0) = \kappa(0) \cdot n(0)^{\text{tang}} + \kappa(0) \cdot \langle n(0), N(p) \rangle \, N(p).$$

The tangential part, which gives the extent to which c curves within S, leads to the ***geodesic curvature*** of c in S, to which we will come back in section 4.5. For now we are interested in the curvature of S in $\mathbb{R}^3$, and for this reason make the definition

$$\kappa_{\text{nor}} := \langle \ddot{c}(0), N(p) \rangle = \begin{cases} \kappa(0) \cdot \langle n(0), N(p) \rangle, & \text{if } \kappa(0) \neq 0, \\ 0, & \text{if } \kappa(0) = 0. \end{cases}$$

We call κ_{nor} the ***normal curvature*** of S at the point p in direction $\dot{c}(0)$. If, in the case of $\kappa(0) \neq 0$, θ denotes the angle between $N(p)$ and $n(0)$, then we therefore have

$$\kappa_{\text{nor}} = \kappa(0) \cdot \cos(\theta).$$

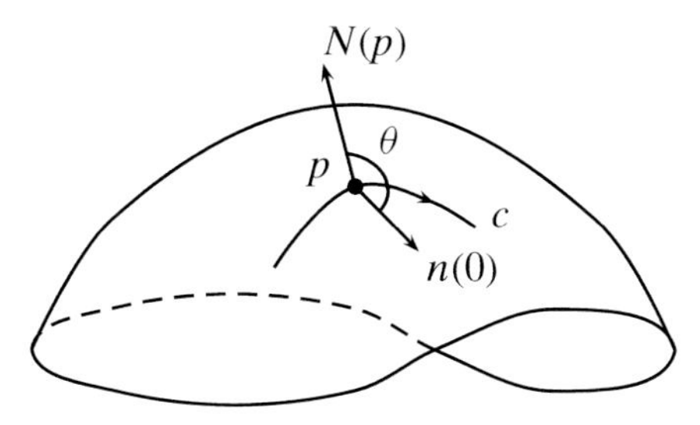

In particular, we always have

$$|\kappa_{\text{nor}}| \leq \kappa(0).$$

The following theorem tells us how to find the normal curvature from the second fundamental form.

Theorem 3.6.1 (Meusnier's theorem) *Let $S \subset \mathbb{R}^3$ be an orientable regular surface with unit normal field N and second fundamental form II. Let $p \in S$. Let $c : (-\varepsilon, \varepsilon) \to S$ be a curve parametrised by arc-length with $c(0) = p$. Then we have for the normal curvature κ_{nor} of c:*

$$\kappa_{\text{nor}} = II(\dot{c}(0), \dot{c}(0)).$$

In particular, all curves parametrised by arc-length in S through p with the same tangent vector have the same normal curvature.

This also justifies that we refer to κ_{nor} as the normal curvature of S at the point p in direction $\dot{c}(0)$, since κ_{nor} depends, apart from on S and p, only on $\dot{c}(0)$, but not on the particular choice of the curve c.

Proof As c lies on S, we have

$$\langle N(c(t)), \dot{c}(t)\rangle = 0$$

for all $t \in (-\varepsilon, \varepsilon)$. Differentiating this equation gives

$$\begin{aligned}
0 &= \frac{d}{dt}\langle N(c(t)), \dot{c}(t)\rangle \Big|_{t=0} \\
&= \left\langle \frac{d}{dt} N(c(t))\Big|_{t=0}, \dot{c}(0)\right\rangle + \langle N(p), \ddot{c}(0)\rangle \\
&= \langle d_p N(\dot{c}(0)), \dot{c}(0)\rangle + \kappa_{\mathrm{nor}} \\
&= \langle -W_p(\dot{c}(0)), \dot{c}(0)\rangle + \kappa_{\mathrm{nor}} \\
&= -II(\dot{c}(0), \dot{c}(0)) + \kappa_{\mathrm{nor}}.
\end{aligned}$$

□

Remark A change of orientation of the curve c does not change the value of the normal curvature κ_{nor}, since

$$II(-\dot{c}(0), -\dot{c}(0)) = II(\dot{c}(0), \dot{c}(0)).$$

However, if the orientation of the surface S is reversed with $N(p)$ being replaced by $-N(p)$, curvature κ_{nor} changes sign:

$$\langle \ddot{c}(0), -N(p)\rangle = -\langle \ddot{c}(0), N(p)\rangle.$$

Let $S \subset \mathbb{R}^3$ be an orientable regular surface with unit normal field N, let $p \in S$. Let $X \in T_pS$ be a tangent vector of length 1.

Exercise 3.9 Let E be the plane spanned by $N(p)$ and X. Use the implicit mapping theorem to show that for a neighbourhood V of p in $\mathbb{R}^3$ the set

$S \cap (E+p) \cap V$ can be parametrised by a regular curve. Here $E+p$ denotes the plane E translated to the point p.

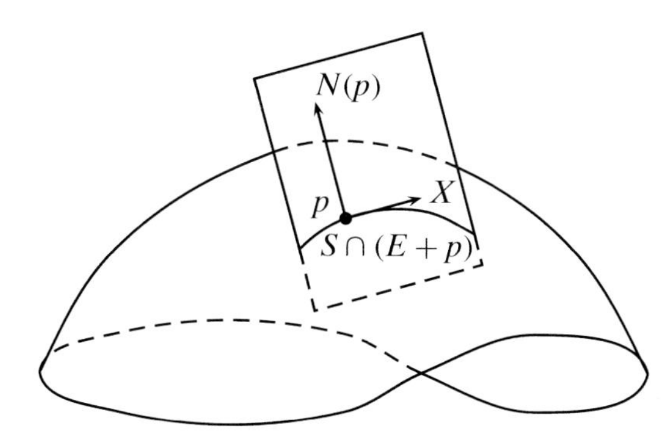

According to theorem 3.6.1 the normal curvature $II(X, X)$ can be found from the unit speed curve c which describes $S \cap (E+p) \cap V$. If we consider c as a plane curve in the plane $E+p \cong \mathbb{R}^2$, then the normal vector of the curve $n(0) = \pm N(p)$ and hence the normal curvature is

$$II(X, X) = \pm\kappa(0),$$

where κ is the curvature of c considered as a plane curve. This explains the term *normal* curvature. The normal curvature is the curvature of the plane curve $S \cap (E+p) \cap V$ in the plane E spanned by X and the normal vector.

Example 3.6.2 Let $S = S^1 \times \mathbb{R}$ be the cylinder. Let $p \in S$. The intersection of S with the normal plane at the point p is a circle, an ellipse or a straight line. The normal curvature therefore varies between 1 and 0, depending on the direction.

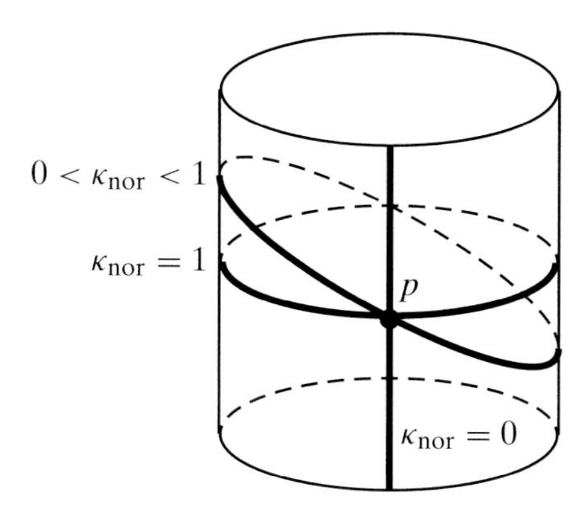

By proposition 3.5.5 the Weingarten map $W_p : T_pS \to T_pS$ is always self-adjoint. We can therefore find an orthonormal basis X_1, X_2 of T_pS, which consists of eigenvectors of W_p only,

$$W_p(X_i) = \kappa_i \cdot X_i, \qquad i = 1, 2.$$

Definition 3.6.3 The eigenvalues κ_1 and κ_2 are called the ***principal curvatures*** of S at the point p. The corresponding eigenvectors $\pm X_1$ and $\pm X_2$ are called ***principal curvature directions***.

Unless stated otherwise, we use the convention $\kappa_1 \leq \kappa_2$. An arbitrary unit vector $X \in T_pS$ can be expressed in terms of the basis X_1, X_2 by

$$X = \cos(\varphi) \cdot X_1 + \sin(\varphi) \cdot X_2$$

for a suitable $\varphi \in \mathbb{R}$. Substitution into the second fundamental form gives the *Euler formula* for the normal curvature in direction X:

$$II(X, X) = \cos^2(\varphi) \cdot \kappa_1 + \sin^2(\varphi) \cdot \kappa_2.$$

In particular, we observe that κ_1 and κ_2 are the minimum and maximum of all normal curvatures of S at p, if X runs through all directions, i.e. for all unit vectors $X \in T_pS$.

Example 3.6.4 Let $S = \mathbb{R}^2 \times \{0\}$ be the x–y plane in $\mathbb{R}^3$. As $W = 0$, we have that $\kappa_1 = \kappa_2 = 0$ and thus every direction is a principal curvature direction.

Example 3.6.5 Let $S = S^2$ be the sphere. Then for the inner unit normal field the Weingarten map is $W = \mathrm{Id}$. Hence $\kappa_1 = \kappa_2 = 1$ and every direction is a principal curvature direction.

Example 3.6.6 Let $S = S^1 \times \mathbb{R}$ be the cylinder, $p = (x, y, z)^\top$. As we have seen, the Weingarten map W_p with respect to the inner unit normal field and the basis $X_1 = (-y, x, 0)^\top$ and $X_2 = (0, 0, 1)^\top$ has the matrix representation

$$\begin{pmatrix} 1 & 0 \\ 0 & 0 \end{pmatrix}.$$

This means precisely that X_1 and X_2 are principal curvature directions for the principal curvatures $\kappa_1 = 1$ and $\kappa_2 = 0$.

Definition 3.6.7 Let $S \subset \mathbb{R}^3$ be a regular surface, let $c : I \to S$ be a curve parametrised by arc-length. If $\dot{c}(t)$ is a principal curvature direction for all $t \in I$, then c is called the ***curvature line***.

Example 3.6.8 On the cylinder $S = S^1 \times \mathbb{R}$ the curvature lines are horizontal circular lines or vertical straight lines (or parts of them).

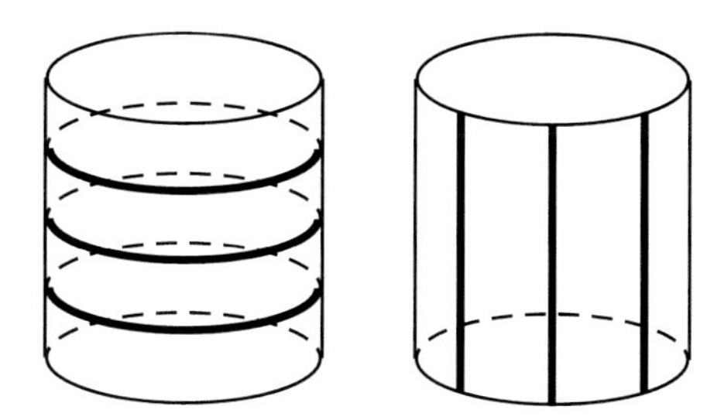

On the plane or the sphere all curves parametrised by arc-length are curvature lines.

Remark If one reverses orientation on an orientable regular surface, i.e. substitutes $-N(p)$ for $N(p)$, then W_p becomes $-W_p$. Hence κ_1 and κ_2 become $-\kappa_1$ and $-\kappa_2$. The eigenvectors of W_p and hence the principal curvature directions on the other hand remain unchanged. For this reason principal curvature directions and curvature lines are defined on non-orientable surfaces as well. However, the principal curvatures are only defined up to the sign on non-orientable surfaces.

Exercise 3.10 Prove Rodriguez's theorem:

Let $S \subset \mathbb{R}^3$ be an orientable regular surface, let $N : S \to S^2$ be the Gauss map. Let $c : I \to S$ be a curve parametrised by arc-length. Then c is a curvature line on S if and only if there exists a function $\lambda : I \to \mathbb{R}$ with

$$\frac{d}{dt}N(c(t)) = \lambda(t) \cdot \dot{c}(t), \qquad t \in I.$$

In this case $-\lambda(t)$ is the corresponding principal curvature.

Definition 3.6.9 Let $S \subset \mathbb{R}^3$ be an oriented regular surface, let $p \in S$ be a point. Let κ_1 and κ_2 be the principal curvatures of S at p. Then

$$K(p) := \kappa_1 \cdot \kappa_2 = \det(W_p)$$

is the ***Gauss curvature*** of S at p. Further,

$$H(p) := \frac{\kappa_1 + \kappa_2}{2} = \frac{1}{2}\mathrm{trace}(W_p)$$

is called the ***mean curvature*** of S at p.

Both concepts of curvature represent an average of the two principal curvatures: the mean curvature is the arithmetic mean, the Gauss curvature is the square of the geometric mean. We will investigate the geometric meaning of these notions of curvature later.

Reversing the orientation of an oriented regular surface changes the signs of the principal curvatures. Hence $H(p)$ changes to $-H(p)$, while $K(p)$ remains unchanged. Hence the Gauss curvature is also defined for non-orientable surfaces, while the mean curvature is only defined up to the sign. In order to obtain a version of the mean curvature that does not vary with orientation and that is defined even on non-orientable surfaces, we often consider not the real-valued function H but the ***mean curvature field*** $\mathscr{H}$, which is defined by

$$\mathscr{H} := H \cdot N. \tag{3.6}$$

Unlike the mean curvature H, $\mathscr{H}$ is not a function, but a normal field on the surface.

Now some more terminology.

Definition 3.6.10 Let $S \subset \mathbb{R}^3$ be an orientable regular surface, let $p \in S$. We call the point p

(i) ***elliptic***, if $K(p) > 0$,
(ii) ***hyperbolic***, if $K(p) < 0$,
(iii) ***parabolic***, if $K(p) = 0$, but $W_p \neq 0$, i.e. if one of the two principal curvatures vanishes while the other does not,
(iv) ***planar***, if $W_p = 0$, i.e. $\kappa_1 = \kappa_2 = 0$.

Remark The notions of elliptic, hyperbolic, parabolic and planar points also make sense on non-orientable surfaces, although in this case W_p is only defined up to a change of sign.

Let us look at some examples.

Example 3.6.11 For the plane $S = \mathbb{R}^2 \times \{0\}$ we have $W_p = 0$ for all $p \in S$. Hence all points are planar. We have $K \equiv 0$ and $H \equiv 0$.

Example 3.6.12 For the sphere $S = S^2$ with the orientation given by the inner unit normal field we have $W_p = \mathrm{Id}$ for all $p \in S$, thus $K \equiv 1$. It follows that all points are elliptic. For the mean curvature we have $H \equiv 1$ and for the mean curvature field $\mathscr{H}(p) = -p$.

Example 3.6.13 For the cylinder $S := S^1 \times \mathbb{R}$ with the orientation given by the inner unit normal field we have calculated: $\kappa_1 = 0$ and $\kappa_2 = 1$. It follows that $K \equiv 0$ (as for the plane!) and $H \equiv \frac{1}{2}$. All points are parabolic.

Example 3.6.14 The ***hyperbolic paraboloid***

$$S = \left\{ (x, y, z)^\top \in \mathbb{R}^3 \,\middle|\, z = y^2 - x^2 \right\}$$

is the graph of the function $\varphi : \mathbb{R}^2 \to \mathbb{R}$, $\varphi(x, y) = y^2 - x^2$, and hence a regular surface.

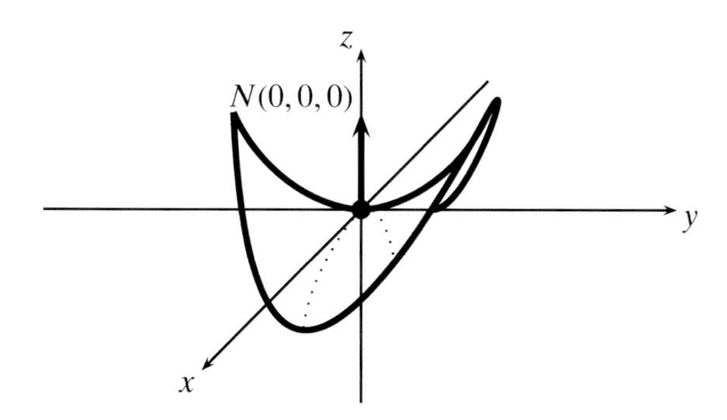

To find a normal field, we write S as a set of zeros $S = f^{-1}(0)$, where $f : \mathbb{R}^3 \to \mathbb{R}$, $f(x, y, z) = z - y^2 + x^2$. The gradient is then

$$\operatorname{grad} f(x, y, z) = \begin{pmatrix} 2x \\ -2y \\ 1 \end{pmatrix}.$$

As the gradient of f is everywhere perpendicular to S, we obtain a unit normal field, and hence an orientation, via

$$N(x, y, z) = \frac{\operatorname{grad} f(x, y, z)}{\| \operatorname{grad} f(x, y, z)\|} = \frac{(2x, -2y, 1)^\top}{\sqrt{4x^2 + 4y^2 + 1}}.$$

At the point $(x, y, z)^\top = (0, 0, 0)^\top =: p$ we get $N(p) = (0, 0, 1)^\top$. We now find the Weingarten map W_p for this point p. The tangent plane T_pS has the orthonormal basis $X_1 = (1, 0, 0)^\top$ and $X_2 = (0, 1, 0)^\top$. The curve $c : \mathbb{R} \to S$, $c(t) = (t, 0, -t^2)^\top$, satisfies $c(0) = p$ and $\dot{c}(0) = X_1$. Hence

$$\begin{aligned} d_pN(X_1) &= \frac{d}{dt} N(c(t)) \Big|_{t=0} \\ &= \frac{d}{dt} \frac{(2t, 0, 1)^\top}{\sqrt{4t^2 + 1}} \Bigg|_{t=0} \\ &= (2, 0, 0)^\top \\ &= 2 \cdot X_1. \end{aligned}$$

Thus $W_p(X_1) = -2 \cdot X_1$ and $\kappa_1 = -2$. Analogously the curve $\tilde{c}(t) = (0, t, t^2)^\top$ is used to find that

$$W_p(X_2) = 2X_2 \quad \text{and} \quad \kappa_2 = 2.$$

It follows that $p = (0, 0, 0)^\top$ is a hyperbolic point and $K(p) = -4$, $H(p) = 0$.

We now want to look at special local parametrisations of regular surfaces that show us that the second fundamental form approximately tells us how the surface moves in the proximity of a point away from the tangent plane. In particular, we will see that every surface can be given locally as a graph on the tangent plane. We will then investigate what curvatures tell us about the local geometric behaviour of the surface.

Theorem 3.6.15 *Let $S \subset \mathbb{R}^3$ be a regular surface, let $p \in S$ and let X_1, X_2 be an orthonormal basis of T_pS. Let N be a smooth unit normal field on S, defined on a neighbourhood of the point p, such that $(X_1, X_2, N(p))$ form a positively oriented orthonormal basis of $\mathbb{R}^3$.*

There then exists a local parametrisation (U, F, V) of S around p, such that:

(i) $(0,0)^\top \in U$ *and* $F(0,0) = p$.
(ii) $g_{ij}(0,0) = \delta_{ij}$, $i,j = 1,2$.
(iii) $(\partial g_{ij}/\partial u^k)(0,0) = 0$, $i,j,k = 1,2$.
(iv) $F(u) - p = u^1 \cdot X_1 + u^2 \cdot X_2 + \frac{1}{2}\sum_{i,j=1}^{2} h_{ij}(0,0)u^i u^j \cdot N(p) + \mathrm{O}(\|u\|^3)$.

Here (g_{ij}) and (h_{ij}) denote the local representations of the first and the second fundamental form with respect to the local parametrisation (U, F, V). The symbol $\mathrm{O}(\|u\|^k)$ denotes (as is conventional in analysis) a function φ with the property that $\varphi(u)/\|u\|^k$ is bounded in a neighbourhood of $(0,0)^\top$.

Proof Let us for now begin with an arbitrary local parametrisation (U_1, F_1, V_1) of S around p. We will repeatedly transform this parametrisation, and step by step establish the desired properties.

(a) Let $x_0 \in U_1$ be the point for which $F_1(x_0) = p$. Set $V_2 := V_1$, $U_2 := U_1 - x_0$ and

$$F_2 : U_2 \to V_2 \cap S,$$
$$F_2(x) = F_1(x + x_0).$$

Then (U_2, F_2, V_2) is a local parametrisation of S around p with $F_2(0,0) = p$. Satisfying condition (i) was easy.

(b) Now let $Y_1, Y_2 \in \mathbb{R}^2$ such that

$$D_{(0,0)}F_2(Y_j) = X_j, \qquad j = 1,2.$$

Let A be the 2×2 matrix with the columns Y_1 and Y_2, $A = (Y_1, Y_2)$. Considering A as a linear map $A : \mathbb{R}^2 \to \mathbb{R}^2$, we see that A is exactly the isomorphism that maps the standard basis to the basis Y_1, Y_2, $A(e_j) = Y_j$.

Set $V_3 := V_2$, $U_3 := A^{-1}(U_2)$ and $F_3 := F_2 \circ A$. Then (U_3, F_3, V_3) is a local parametrisation of S around p with $F_3(0) = p$ and

$$\begin{aligned}\frac{\partial F_3}{\partial u^i}(0,0) &= D_{(0,0)}F_3(e_i) = D_{(0,0)}(F_2 \circ A)(e_i) \\ &= (D_{(0,0)}F_2 \circ A)(e_i) = D_{(0,0)}F_2(Y_i) = X_i.\end{aligned}$$

It follows that with condition (i) condition (ii) is also already satisfied, since

$$g_{ij}^{(F_3)}(0,0) = \left\langle \frac{\partial F_3}{\partial u^i}(0,0), \frac{\partial F_3}{\partial u^j}(0,0) \right\rangle = \langle X_i, X_j \rangle = \delta_{ij}.$$

(c) We consider the Taylor expansion of F_3 around $(0,0)$ with a third-order error term, denote the coordinates as (v^1, v^2) and obtain

$$\begin{aligned}F_3(v^1, v^2) - p &= \frac{\partial F_3}{\partial v^1}(0,0) \cdot v^1 + \frac{\partial F_3}{\partial v^2}(0,0) \cdot v^2 \\ &\quad + \frac{1}{2} \sum_{i,j=1}^{2} \frac{\partial^2 F_3}{\partial v^i \partial v^j}(0,0) \cdot v^i \cdot v^j + O(\|v\|^3) \\ &= v^1 \cdot X_1 + v^2 \cdot X_2 + \frac{1}{2} \sum_{i,j=1}^{2} v^i v^j \left\langle \frac{\partial^2 F_3}{\partial v^i \partial v^j}(0,0), N(p) \right\rangle N(p) \\ &\quad + \frac{1}{2} \sum_{i,j,k=1}^{2} v^i v^j \left\langle \frac{\partial^2 F_3}{\partial v^i \partial v^j}(0,0), X_k \right\rangle X_k + O(\|v\|^3).\end{aligned}$$

We now want to reparametrise (one last time) in order to make the term $\sum_{i,j,k=1}^{2} v^i v^j \langle (\partial^2 F/\partial v^i \partial v^j)(0,0), X_k \rangle X_k$ vanish. For this purpose we consider the map

$$\psi : U_3 \to \mathbb{R}^2,$$

$$\psi(v^1, v^2) = \begin{pmatrix} v^1 + \frac{1}{2} \cdot \sum_{i,j=1}^{2} v^i v^j \left\langle \frac{\partial^2 F_3}{\partial v^i \partial v^j}(0,0), X_1 \right\rangle \\ v^2 + \frac{1}{2} \cdot \sum_{i,j=1}^{2} v^i v^j \left\langle \frac{\partial^2 F_3}{\partial v^i \partial v^j}(0,0), X_2 \right\rangle \end{pmatrix}.$$

For this map we have that

$$\psi(0,0) = (0,0)^\top,$$

$$D_{(0,0)}\psi = \begin{pmatrix} 1 & 0 \\ 0 & 1 \end{pmatrix}.$$

By the inverse function theorem there exist neighbourhoods U'_3 and U of $(0,0)^\top$, $U'_3 \subset U_3$, such that $\psi : U'_3 \to U$ is a diffeomorphism. By reducing the size of $V'_3 \subset V_3$ and restricting $F'_3 = F_3|_{U'_3}$, we obtain another local parametrisation $\left(U'_3, F'_3, V'_3\right)$ of S around p. We set $F := F'_3 \circ \psi^{-1}$, $V := V'_3$. We can now show that (U, F, V) has properties (i)–(iv).

For (i), $F(0,0) = F'_3(\psi^{-1}(0,0)) = F'_3(0,0) = F_3(0,0) = p$.

For (iv), with the abbreviation

$$u^k := v^k + \frac{1}{2}\sum_{i,j=1}^{2} v^i v^j \left\langle \frac{\partial^2 F_3}{\partial v^i \partial v^j}(0,0), X_k \right\rangle,$$

we obtain

$$\begin{aligned}
F(u^1,u^2) - p &= F(\psi(v^1,v^2)) - p \\
&= F_3(v^1,v^2) - p \\
&= \sum_{k=1}^{2} \left\{ v^k + \frac{1}{2}\sum_{i,j=1}^{2} v^i v^j \left\langle \frac{\partial^2 F_3}{\partial v^i \partial v^j}(0,0), X_k \right\rangle \right\} X_k \\
&\quad + \frac{1}{2}\sum_{i,j=1}^{2} v^i v^j \left\langle \frac{\partial^2 F_3}{\partial v^i \partial v^j}(0,0), N(p) \right\rangle N(p) + \mathrm{O}(\|v\|^3) \\
&= \sum_{k=1}^{2} u^k X_k + \frac{1}{2}\sum_{i,j=1}^{2} v^i v^j h_{ij}^{(F_3)}(0,0) \cdot N(p) + \mathrm{O}(\|v\|^3).
\end{aligned}$$

Here $h_{ij}^{(F_3)} = \left\langle (\partial^2 F_3/\partial v^i \partial v^j) N \circ F_3 \right\rangle$ denotes the components of the second fundamental form with respect to the local parametrisation (U_3, F_3, V_3). Because of $\psi(0,0) = (0,0)^\top$ and $D_{(0,0)^\top}\psi = \begin{pmatrix} 1 & 0 \\ 0 & 1 \end{pmatrix}$ we also have that $\psi^{-1}(0,0) = (0,0)^\top$ and that $D_{(0,0)^\top}(\psi^{-1}) = \begin{pmatrix} 1 & 0 \\ 0 & 1 \end{pmatrix}$ and hence

$$v^j = u^j + \mathrm{O}(\|u\|^2)$$

as well as

$$\mathrm{O}(\|v\|^3) = \mathrm{O}(\|u\|^3).$$

It therefore follows that

$$F(u^1,u^2)-p=\sum_{k=1}^{2}u^kX_k+\frac{1}{2}\sum_{i,j=1}^{2}(u^iu^j+O(\|u\|^3))\cdot h_{ij}^{(F_3)}(0,0)\cdot N(p)+O(\|u\|^3)$$

$$=\sum_{k=1}^{2}u^kX_k+\frac{1}{2}\sum_{i,j=1}^{2}u^iu^jh_{ij}^{(F_3)}(0,0)\cdot N(p)+O(\|u\|^3). \tag{3.7}$$

Differentiating (3.7) gives

$$h_{ij}^{(F)}(0,0)=\left\langle\frac{\partial^2F}{\partial u^i\partial u^j}(0,0),N(p)\right\rangle=\left\langle h_{ij}^{(F_3)}(0,0)\cdot N(p),N(p)\right\rangle=h_{ij}^{(F_3)}(0,0). \tag{3.8}$$

Substituting (3.8) into (3.7) gives (iv).

For (ii) and (iii), because of (iv) we have that

$$\frac{\partial F}{\partial u^i}(u)=X_i+\sum_{k=1}^{2}h_{ik}(0,0)u^k\cdot N(p)+O(\|u\|^2).$$

Multiplying out gives

$$\begin{aligned}g_{ij}(u)&=\Bigg\langle X_i+\sum_{k=1}^{2}h_{ik}(0,0)u^k\cdot N(p)+O(\|u\|^2),\\&\qquad X_j+\sum_{l=1}^{2}h_{jl}(0,0)u^l\cdot N(p)+O(\|u\|^2)\Bigg\rangle\\&=\langle X_i,X_j\rangle+\sum_{k=1}^{2}h_{ik}(0,0)u^k\langle N(p),X_j\rangle\\&\qquad+\sum_{l=1}^{2}h_{jl}(0,0)u^l\langle X_i,N(p)\rangle+O(\|u\|^2)\\&=\delta_{ij}+O(\|u\|^2),\end{aligned}$$

as $\langle N(p),X_j\rangle=\langle X_i,N(p)\rangle=0$. This proves (ii) and (iii). □

Corollary 3.6.16 *Every regular surface S can locally be given as a graph on its affine tangent plane T_pS+p.*

Proof Let $S \subset \mathbb{R}^3$ be a regular surface, $p \in S$. For simplicity we rotate and translate the surface in $\mathbb{R}^3$ in such a way that $p = (0,0,0)^\top$ and that the tangent plane T_pS is spanned precisely by the first two unit vectors e_1 and e_2. By the preceding theorem there exists a local parametrisation (U, F, V) of S and p of the form

$$F(u^1, u^2) = \left(u^1, u^2, \frac{1}{2} \sum_{i,j=1}^{2} h_{ij}(0,0) u^i u^j \right)^\top + \mathrm{O}(\|u\|^3).$$

Let $\pi : \mathbb{R}^3 \to T_pS = \mathbb{R}^2 \times \{0\} \cong \mathbb{R}^2$ be the orthogonal projection onto the tangent plane.

As $\pi \circ F(u^1, u^2) = (u^1, u^2)^\top + \mathrm{O}(\|u\|^3)$ we have that

$$D_{(0,0)}(\pi \circ F) = \begin{pmatrix} 1 & 0 \\ 0 & 1 \end{pmatrix}.$$

It follows that by the inverse function theorem the map $\pi \circ F$ can be inverted on a possibly smaller neighbourhood of p, i.e. there exists a smooth map

$$\varphi : \tilde{U} \subset T_pS \to \mathbb{R}^2$$

with

$$\pi \circ F \circ \varphi = \mathrm{Id}.$$

Then $F(\varphi(v^1, v^2)) = \left(v^1, v^2, (F \circ \varphi)^3(v^1, v^2)\right)^\top$, i.e. near p S is precisely the graph of the third component of the function of $F \circ \varphi$. □

We can now begin to interpret the Gauss curvature geometrically. Disregarding terms of order 3, we can approximately present the regular surface S near a point $p \in S$ above the tangent plane T_pS as the graph of the function $(u^1, u^2)^\top \mapsto \frac{1}{2} \sum_{i,j=1}^{2} h_{ij}(0,0) u^i u^j$.

First case Let $K(p) > 0$. Then $(h_{ij}(0,0))_{ij}$ is positive or negative definite and hence S is approximated by a paraboloid.

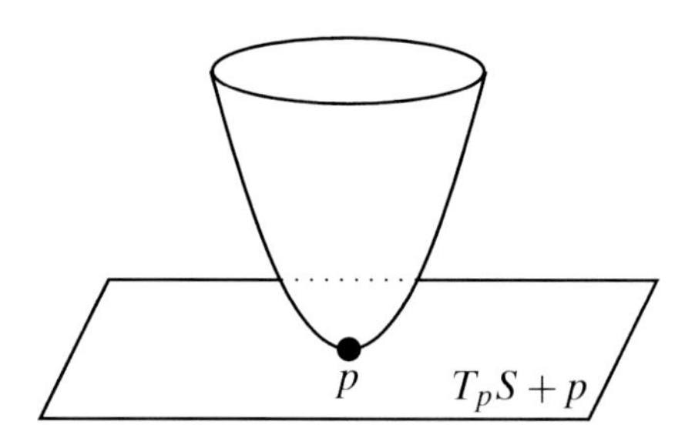

Second case Let $K(p) < 0$. Then $(h_{ij}(0,0))_{ij}$ indefinite, but not degenerate. Hence S is approximated by a hyperbolic paraboloid (or saddle surface) near p.

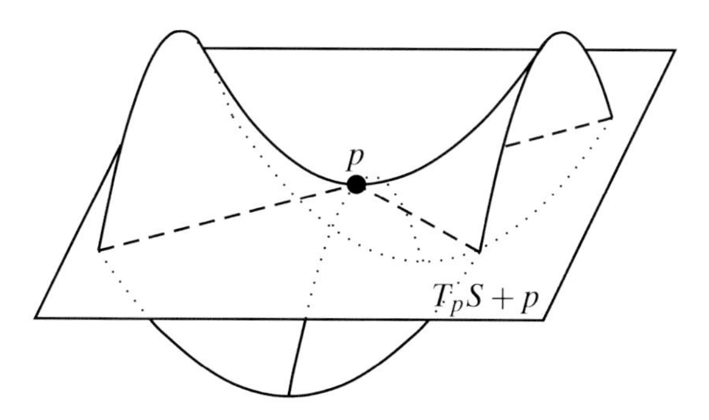

Third case p is parabolic. Then $(h_{ij}(0,0))_{ij}$ is degenerate, but not 0. Near p the surface S looks like the cylindrical surface over a parabola.

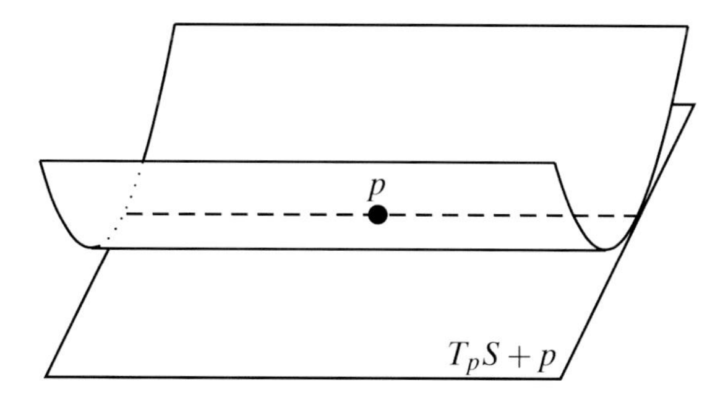

Fourth case p is a planar point. Then $(h_{ij}(0,0)) = 0$ for all $i,j = 1,2$ and hence the surface S agrees with its tangent plane up to terms of order 3.

Exercise 3.11 Let $S \subset \mathbb{R}^3$ be a regular surface, $p \in S$. Show the following:

(a) If $K(p) > 0$, then a neighbourhood of p in S lies entirely on one side of the affine tangent plane $T_pS + p$.
(b) If $K(p) < 0$, then every neighbourhood of p in S meets both sides of the affine tangent plane.

What can be said in the case $K(p) = 0$?

We already know compact surfaces with positive Gauss curvature, e.g. $S = S^2$. The following theorem says that a regular surface $S \subset \mathbb{R}^3$ with $K \leq 0$ cannot be compact.

Theorem 3.6.17 *Let $S \subset \mathbb{R}^3$ be a compact non-empty regular surface. Then there is a point $p \in S$ with $K(p) > 0$.*

Proof As S is compact, we have that S is bounded and hence there exists $R > 0$ such that

$$S \subset \overline{B}(0,R) = \left\{ x \in \mathbb{R}^3 \,\middle|\, \|x\| \leq R \right\}.$$

We choose the minimal radius R_0 which satisfies this, i.e.

$$R_0 := \inf \left\{ R \,\middle|\, S \subset \overline{B}(0,R) \right\}.$$

We have that $S \subset \overline{B}(0,R_0)$. We first see that $S \cap S^2(R_0) \neq \emptyset$, where

$$S^2(R_0) = \partial \overline{B}(0,R_0) = \left\{ x \in \mathbb{R}^3 \middle| \|x\| = R_0 \right\}.$$

If S and $S^2(R_0)$ had no point in common, then by compactness of S and $S^2(R_0)$

$$\varepsilon := \operatorname{dist}(S, S^2(R_0)) = \min \left\{ \|x - y\| \,\middle|\, x \in S, y \in S^2(R_0) \right\}$$

would be strictly positive. But then $S \subset \overline{B}(0, R_0 - \varepsilon)$ would contradict minimality of R_0.

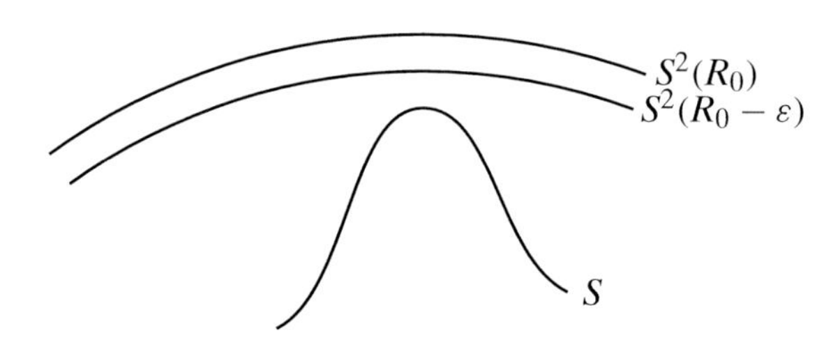

Hence there exists $p \in S \cap S^2(R_0)$; $S^2(R_0)$ also is a regular surface, being the surface of the sphere of radius R_0. We now argue that S and $S^2(R_0)$ have the same tangent plane at p:

$$T_p S = T_p S^2(R_0).$$

The tangent plane of $S^2(R_0)$ at p is exactly the orthogonal complement of the vector p. If the tangent planes were not the same, then $T_p S$ would *not* be the orthogonal complement of p and there would be an $X \in T_p S$ with $\langle X, p \rangle \neq 0$. After possibly having exchanged X for $-X$, we can now assume that $\langle X, p \rangle > 0$.

Let $c : (-\varepsilon, \varepsilon) \to S$ be a curve with $c(0) = p$ and $\dot{c}(0) = X$. The Taylor expansion of c gives

$$c(t) = p + X \cdot t + \mathrm{O}(t^2).$$

Hence

$$\begin{aligned}\|c(t)\|^2 &= \|p\|^2 + 2\langle p, X\rangle t + \mathrm{O}(t^2)\\ &= R_0^2 + 2\langle p, X\rangle t + \mathrm{O}(t^2).\end{aligned}$$

Because of $\langle p, X\rangle > 0$ we have that

$$\frac{\|c(t)\|^2 - R_0^2}{t} = 2\langle p, X\rangle + \mathrm{O}(t)$$

is positive for small $t > 0$. But this contradicts $\|c(t)\| \leq R_0$, which must be true because of $S \subset \overline{B}(0, R_0)$. We have therefore shown that $T_pS = T_pS^2(R_0)$.

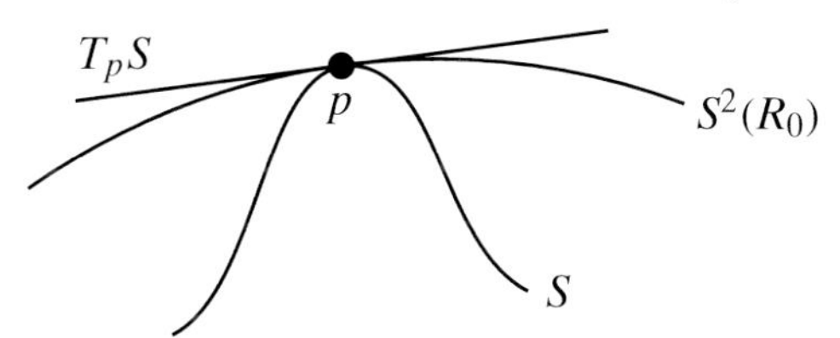

We now look at normal sections of S and $S^2(R_0)$ at the point p. Let E be a plane spanned by $N(p)$ and by a tangent vector from $T_pS = T_pS^2(R_0)$. We see that $S \cap E$ always lies within the circular line $S^2(R_0) \cap E$ and that it touches the circular line at p. Hence the normal curvature in this direction satisfies $|\kappa_{\mathrm{nor}}| \geq 1/R_0$, see exercise 2.11.

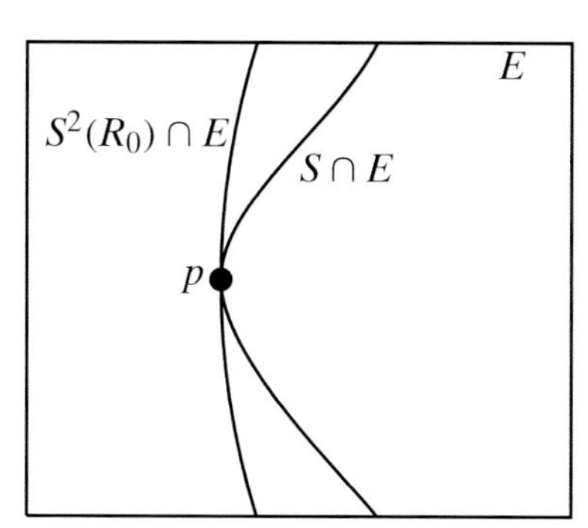

In particular, no normal curvature vanishes at the point p, hence the second fundamental form at p is definite and $K(p) > 0$. □

For the point p with positive Gauss curvature we took a point on S with maximal distance to the origin. In particular, we have shown that the difference vector from the origin, i.e. the vector p itself, is perpendicular to T_pS. This part of the proof would also have worked for points with minimal distance. Let us put this down as a corollary of the proof for later use.

Corollary 3.6.18 *Let S be a compact regular surface, let $q \in \mathbb{R}^3$, let $p \in S$ be a point on S with minimal distance from q. Then $q - p$ is perpendicular to T_pS.*

We will later generalise the concept of the first fundamental form and move on to surfaces with Riemannian metrics. Those surfaces also have a Gauss curvature. It will then be possible to find compact surfaces with $K \leq 0$, even $K \equiv -1$ is possible. Theorem 3.6.17 should be understood as follows: compact surfaces with $K \leq 0$ do exist in an abstract sense that remains to be made precise, but they cannot be embedded in $\mathbb{R}^3$ without bending it somewhere to such an extent that Gauss curvature becomes positive.

Exercise 3.12 Let $S \subset \mathbb{R}^3$ be a compact non-empty regular surface.

(a) Show that the Gauss map $S \to S^2$ is onto.
(b) Show that if the Gauss map is injective, then $K \geq 0$.
(c) Improve (a) and show that the Gauss map restricted to $S_+ := \{x \in S \mid K(x) \geq 0\}$ is onto.

3.7 Surface area and integration on surfaces

Before having a closer look at some special classes of surfaces, we want to prepare ourselves by studying the integration of functions on surfaces and the area of a surface in particular.

For this purpose let $S \subset \mathbb{R}^3$ be a regular surface and (U, F, V) a local parametrisation of S. We first only consider functions $f : S \to \mathbb{R}$ that vanish everywhere outside V, i.e. $f|_{S-V} \equiv 0$. Denoting the coordinates in U as u^1, u^2 and the components of the first fundamental form as usual as g_{ij}, we make the following definition.

Definition 3.7.1 A function $f : S \to \mathbb{R}$ with $f|_{S-V} \equiv 0$ is called (Lebesgue-) ***integrable*** if the function

$$
\begin{aligned}
U &\to \mathbb{R}, \\
(u^1, u^2)^\top &\mapsto f(F(u^1, u^2)) \cdot \sqrt{\det(g_{ij}(u^1, u^2))},
\end{aligned}
$$

is (Lebesgue-)integrable. The value of the integral is

$$\int_S f\, dA := \int_U f(F(u^1, u^2))\sqrt{\det(g_{ij}(u^1, u^2))}\, du^1 du^2.$$

We call the formal expression

$$dA = \sqrt{\det(g_{ij})}\, du^1 du^2$$

the ***surface element***.

We do not simply integrate the function f or $f \circ F$, but multiply by the factor $\sqrt{\det(g_{ij})}$, which accounts for the distortion of the surface that is caused by the parametrisation F. This makes the definition independent of the choice of parametrisation.

Lemma 3.7.2 *Let $S \subset \mathbb{R}^3$ be a regular surface, let (U, F, V) and $(\tilde{U}, \tilde{F}, \tilde{V})$ be local parametrisations of S. Let $f : S \to \mathbb{R}$ be a function satisfying $f|_{S-(V\cap\tilde{V})} \equiv 0$.*

$$(f \circ F) \cdot \sqrt{\det(g_{ij})} : U \to \mathbb{R}$$

is integrable if and only if

$$(f \circ \tilde{F}) \cdot \sqrt{\det(\tilde{g}_{ij})} : \tilde{U} \to \mathbb{R}$$

is integrable, and in this case

$$\int_U (f \circ F) \cdot \sqrt{\det(g_{ij})}\, du^1 du^2 = \int_{\tilde{U}} (f \circ \tilde{F}) \cdot \sqrt{\det(\tilde{g}_{ij})}\, d\tilde{u}^1 d\tilde{u}^2.$$

Proof Let $\varphi := \tilde{F}^{-1} \circ F$ be the parameter transformation. By (3.2) we have

$$(g_{ij}) = (D\varphi)^\top \cdot (\tilde{g}_{ij} \circ \varphi) \cdot D\varphi,$$

hence

$$\det(g_{ij}) = \det(\tilde{g}_{ij} \circ \varphi) \cdot (\det(D\varphi))^2,$$

and thus

$$\sqrt{\det(g_{ij})} = \sqrt{\det(\tilde{g}_{ij} \circ \varphi)} \cdot |\det(D\varphi)|.$$

Let $(f \circ \tilde{F})\sqrt{\det(\tilde{g}_{ij})}$ be integrable. By the change of variable formula [18, p. 593, theorem 4.7]

$$(f \circ \tilde{F} \circ \varphi) \cdot \sqrt{\det(\tilde{g}_{ij} \circ \varphi)} \cdot |\det(D\varphi)| = (f \circ F) \cdot \sqrt{\det(g_{ij})}$$

is integrable as well, and

$$\int_{\tilde{U}} f(\tilde{F}(\tilde{u}^1, \tilde{u}^2))\sqrt{\det\left(\tilde{g}_{ij}(\tilde{u}^1, \tilde{u}^2)\right)}\, d\tilde{u}^1 d\tilde{u}^2$$
$$= \int_U f(F(u^1, u^2))\sqrt{\det\left(g_{ij}(u^1, u^2)\right)}\, du^1 du^2. \qquad \square$$

Example 3.7.3 If $S = \mathbb{R}^2 \times \{0\} \subset \mathbb{R}^3$ is the x–y plane, then we choose the Cartesian coordinates $U = \mathbb{R}^2$, $V = \mathbb{R}^3$, $F(x,y) = (x,y,0)^\top$. Then $(g_{ij}(x,y)) = \begin{pmatrix} 1 & 0 \\ 0 & 1 \end{pmatrix}$ and hence

$$dA = dx\,dy.$$

We obtain the original integral over $\mathbb{R}^2$:

$$\int_S f\,dA = \int_{-\infty}^{\infty}\int_{-\infty}^{\infty} f(x,y)\,dx\,dy.$$

If one wants to integrate the function in polar coordinates r and φ, given by the local parametrisation

$$\tilde{F} : (0,\infty) \times (0,2\pi) \to \mathbb{R}^3,$$
$$\tilde{F}(r,\varphi) = (r \cdot \cos\varphi, r \cdot \sin\varphi, 0)^\top,$$

then by (3.1)

$$dA = rdrd\varphi$$

and therefore

$$\int_S f\,dA = \int_0^{\infty}\int_0^{2\pi} f(\tilde{F}(r,\varphi))d\varphi rdr.$$

It remains to think about what to do if the function to be integrated does not have its support on one coordinate chart.

Definition 3.7.4 Let $S \subset \mathbb{R}^3$ be a regular surface. A function $f : S \to \mathbb{R}$ is called ***integrable*** if f can be written as a finite sum

$$f = f_1 + \cdots + f_k, \tag{3.9}$$

where the $f_i : S \to \mathbb{R}$ are integrable functions that vanish outside a coordinate chart (which depends on i). In this case we set

$$\int_S f\,dA := \sum_{i=1}^{k} \int_S f_i\,dA.$$

How do we know whether a function f can be written in the form (3.9)? Simply writing the function as a finite sum $f = f_1 + \cdots + f_k$, where each f_i vanishes outside a coordinate chart, and then checking whether the f_i are integrable in the sense of definition 3.7.1, is *not* sufficient, since the sum of non-integrable functions may be integrable.

If, for example, S is the x–y plane, then we can take the parametrisation via Cartesian coordinates, which has the pleasing property that it parametrises the whole plane. So in this case $V = \mathbb{R}^3$, and the condition that f_i must vanish on $S - V$ is satisfied trivially. We could write the zero function $f \equiv 0$, which is certainly integrable, in a clumsy way as a sum $f = f_1 + f_2$, where $f_1 \equiv 1$ and $f_2 \equiv -1$. Then f_1 and f_2 are not integrable, but f is.

To find a decomposition of the form (3.9) in which every f_i is integrable and vanishes outside of a coordinate chart for an integrable function f, one can proceed as follows. Cover the entire surface with finitely many coordinate charts $(U_1, F_1, V_1), \ldots, (U_k, F_k, V_k)$, i.e. $S \subset \bigcup_{i=1}^k V_k$. This is always possible; indeed, we will sketch an argument in chapter 6 which shows that every regular surface can even be covered by only *three* coordinate charts, see corollary 6.2.18.

For a subset $A \subset \mathbb{R}^3$ let $\chi_A : \mathbb{R}^3 \to \mathbb{R}$ be the *characteristic function*, i.e.

$$\chi_A(x) = \begin{cases} 1, & \text{if } x \in A, \\ 0, & \text{otherwise.} \end{cases}$$

We now define

$$\begin{aligned} f_1 &:= \chi_{V_1} \cdot f, \\ f_2 &:= \chi_{V_2 - V_1} \cdot f, \\ &\vdots \\ f_k &:= \chi_{V_k - (V_1 \cup \cdots \cup V_{k-1})} \cdot f. \end{aligned}$$

Then $f = f_1 + \cdots + f_k$, and every f_i vanishes outside of V_i.

Exercise 3.13 Show that all f_i are integrable in the sense of definition 3.7.1, if f is integrable in the sense of definition 3.7.4.

Exercise 3.14 Show that the value of the integral $\int_S f\, dA$ in definition 3.7.4 is independent of the choice of the decomposition $f = f_1 + \cdots + f_k$.

The usual properties of the integral translate directly to the integral of functions over surfaces, for example

(1) if $f, g : S \to \mathbb{R}$ are integrable and $\alpha, \beta \in \mathbb{R}$, then $\alpha f + \beta g : S \to \mathbb{R}$ is integrable as well, and

$$\int_S (\alpha f + \beta g)\, dA = \alpha \int_S f\, dA + \beta \int_S g\, dA;$$

(2) if $f, g : S \to \mathbb{R}$ are integrable and $f \leq g$, then

$$\int_S f\, dA \leq \int_S g\, dA.$$

Definition 3.7.5 A subset $N \subset S$ of a regular surface is called a ***null set*** if for every local parametrisation (U, F, V) of S the set

$$F^{-1}(V \cap N)$$

is a null set in $U \subset \mathbb{R}^2$.

Exercise 3.15 Let $S \subset \mathbb{R}^3$ be a regular surface, let $(U_i, F_i, V_i)_{i=1,\dots,\infty}$ be local parametrisations that cover S, i.e. $S \subset \bigcup_{i=1}^{\infty} V_i$. Show that a subset $N \subset S$ is already a null set if the $F_i^{-1}(N) \subset U_i$ are null sets.

Hint To show that for every other local parametrisation (U, F, V) the set $F^{-1}(N)$ is a null set in $U \subset \mathbb{R}^2$, use that countable unions of null sets are again null sets and that smooth maps map null sets to null sets.

Remark Suppose that $S \subset \mathbb{R}^3$ is a regular surface, $N \subset S$ a null set and $f, g : S \to \mathbb{R}^3$ are functions that agree on $S - N$. If f is integrable, then the same is true for g and

$$\int_S f\, dA = \int_S g\, dA.$$

Definition 3.7.6 Let $S \subset \mathbb{R}^3$ be a regular surface. If the constant function $f \equiv 1$ is integrable, then we call

$$A[S] := \int_S dA$$

the ***area*** of S.

Example 3.7.7 On the x–y plane $S = \mathbb{R}^2 \times \{0\}$ the integral

$$\int_S dA = \int_{-\infty}^{\infty} \int_{-\infty}^{\infty} dx\, dy$$

diverges, i.e. the function $f \equiv 1$ is not integrable. Hence S does not have a (finite) area.

Example 3.7.8 Let $I \subset \mathbb{R}$ be an open interval, and $c : I \to \mathbb{R}^2$ a plane parametrised regular curve. Let $h > 0$. We consider the ***generalised cylinder*** over c:

$$S = \{c(t) + se_3 \mid t \in I, 0 < s < h\}.$$

We can cover S with a single parametrisation (U, F, V), where

$$U = I \times (0, h) \subset \mathbb{R}^2, \qquad V = \mathbb{R}^3, \qquad F(t, s) = c(t) + se_3.$$

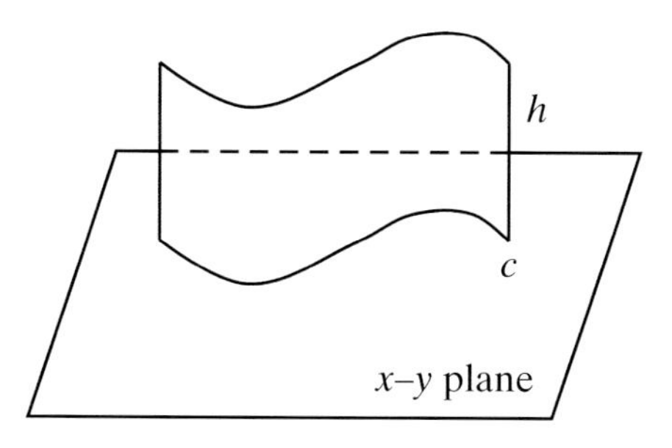

For the differential of F we have

$$D_{(t,s)}F = (\dot{c}(t), e_3).$$

It follows that

$$\begin{aligned} g_{11}(t,s) &= \langle \dot{c}(t), \dot{c}(t) \rangle, \\ g_{12}(t,s) &= g_{21}(t,s) = \langle \dot{c}(t), e_3 \rangle = 0, \\ g_{22}(t,s) &= \langle e_3, e_3 \rangle = 1 \end{aligned}$$

and thus

$$dA = \sqrt{g_{11} \cdot g_{22} - g_{12} \cdot g_{21}}\, dt\, ds = \|\dot{c}(t)\|\, dt\, ds.$$

It might be necessary to reduce the size of the interval I in order to make S a regular surface, since c could, for example, intersect itself. For the area we obtain

$$A[S] = \int_I \int_0^h \|\dot{c}(t)\| ds\, dt = h \cdot L[c].$$

Example 3.7.9 We calculate the area of the sphere $S = S^2$. We use polar coordinates

$$U = (0, 2\pi) \times \left(-\frac{\pi}{2}, \frac{\pi}{2}\right), \quad V = \mathbb{R}^3 - \left\{(x, y, z)^\top \mid x \geq 0, y = 0\right\},$$

$$F(\varphi, \vartheta) = (\cos(\varphi)\cos(\vartheta), \sin(\varphi)\cos(\vartheta), \sin(\vartheta))^\top.$$

The part of S^2 which is not covered by this parametrisation, which is $N - S^2 \cap \{(x, 0, z)^\top | x \geq 0\}$, is a null set in S^2 and can therefore be ignored for the calculation of the area.

We have

$$D_{(\varphi,\vartheta)}F = \begin{pmatrix} -\sin(\varphi)\cos(\vartheta) & -\cos(\varphi)\sin(\vartheta) \\ \cos(\varphi)\cos(\vartheta) & -\sin(\varphi)\sin(\vartheta) \\ 0 & \cos(\vartheta) \end{pmatrix}$$

and thus

$$g_{11} = \cos^2(\vartheta), \quad g_{12} = g_{21} = 0, \quad g_{22} = 1.$$

The surface element is then

$$dA = \sqrt{g_{11}g_{22} - g_{12}g_{21}}\, d\varphi d\vartheta = \cos(\vartheta)\, d\varphi d\vartheta.$$

For the area we obtain

$$\begin{aligned} A[S] &= A[S - N] \\ &= \int_0^{2\pi} \int_{-\pi/2}^{\pi/2} \cos(\vartheta)\, d\vartheta\, d\varphi \\ &= 2\pi \int_{-\pi/2}^{\pi/2} \cos(\vartheta)\, d\vartheta \\ &= 4\pi. \end{aligned}$$

Exercise 3.16 Let $S \subset \mathbb{R}^3$ be a compact non-empty regular surface. Put $S_+ := \{x \in S \mid K(x) \geq 0\}$. Prove that

$$\int_{S_+} K\, dA \geq 4\pi.$$

Hint Use exercise 3.12(c).

3.8 Some classes of surfaces

We now want to study some special classes of surfaces. These are the ruled surfaces, which consist of straight lines, surfaces of revolution, which result from a rotation of a plane curve around an axis, tubular surfaces, which are defined by space curves, as well as minimal surfaces, which are important in the theory of soap films.

3.8.1 Ruled surfaces Let $I \subset \mathbb{R}$ be an open interval and let $c : I \to \mathbb{R}^3$ be a parametrised space curve. We now want to attach a straight line to each point of this curve, which will give us a surface. For this purpose let $v : I \to \mathbb{R}^3$ be a smooth map with $v(t) \neq (0,0,0)^\top$ for all $t \in I$. Let $J \subset \mathbb{R}$ be another open interval. We set

$$F : I \times J \to \mathbb{R}^3, \quad F(t,s) = c(t) + sv(t). \tag{3.10}$$

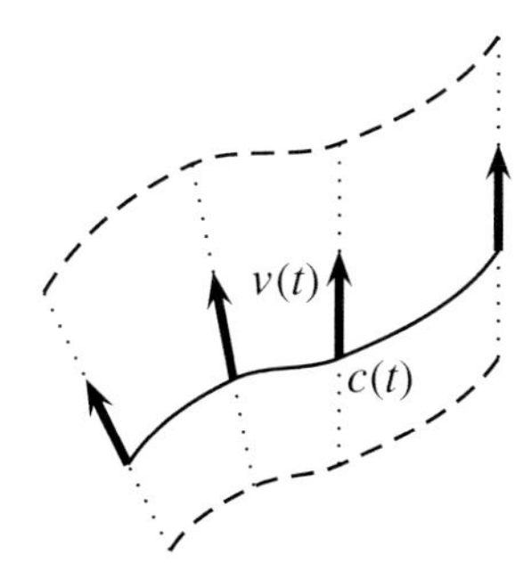

To check whether this gives us a regular surface we find the differential of F:

$$D_{(t,s)}F = (\dot{c}(t) + s\dot{v}(t), v(t)).$$

If we require that $v(t)$ and $\dot{c}(t)$ are linearly independent, then the matrix $D_{(t,0)}F$ has full rank for a fixed t. Hence there exists an open neighbourhood of $(t,0)$ in $I \times J$ such that (U, F, S) is a parametrisation of the regular surface $S = F(U)$.

Definition 3.8.1 A regular surface $S \subset \mathbb{R}^3$ which can be covered by parametrisations of the form (3.10) is called a ***ruled surface***.

Example 3.8.2 If $c : I \to \mathbb{R}^3$ is a plane parametrised curve that does not intersect itself, $c(t) = (c_1(t), c_2(t), 0)$ and $v(t) = (0, 0, 1)$, then the corresponding ruled surface

$$F(t,s) = \begin{pmatrix} c_1(t) \\ c_2(t) \\ s \end{pmatrix}$$

is the ***generalised cylinder over*** c. We can take $U = I \times \mathbb{R}$ as the domain. Compare example 3.7.8.

Example 3.8.3 Let us consider a plane parametrised curve $c : I \to \mathbb{R}^3$ that does not intersect itself, $c(t) = (c^1(t), c^2(t), 0)$. For a fixed point $p \in \mathbb{R}^3 - (\mathbb{R}^2 \times \{0\})$ we set $v(t) = p - c(t)$. Then

$$F : I \times (-\infty, 1) \to \mathbb{R}^3, \quad F(t,s) = (1-s)c(t) + sp$$

is the ***generalised cone*** over c with apex p.

Example 3.8.4 The ***Möbius strip*** is a ruled surface as well. We consider

$$F : \mathbb{R} \times (-1, 1) \to \mathbb{R}^3,$$

$$F(t,s) = \begin{pmatrix} \cos(t) + s \cdot \cos(t)\cos(t/2) \\ \sin(t) + s \cdot \sin(t)\cos(t/2) \\ s \cdot \sin(t/2) \end{pmatrix}.$$

The vector field $v(t) = (\cos(t)\cos(t/2), \sin(t)\cos(t/2), \sin(t/2))^\top$ rotates with half the speed from $(1,0,0)^\top$ to $(-1,0,0)^\top$ while t runs through one period of $c(t) = (\cos(t), \sin(t), 0)^\top$, e.g. $[0, 2\pi]$. Of course, $(1,0,0)^\top$ and $(-1,0,0)^\top$ generate the same straight line, thus the surface closes up.

Example 3.8.5 It may be surprising that the ***hyperboloid of revolution***, also called the ***one-sheeted hyperboloid*** or ***hyperboloid of one sheet***,

$$S = \left\{ (x,y,z)^\top \in \mathbb{R}^3 \,\middle|\, 1 + z^2 = x^2 + y^2 \right\},$$

is a ruled surface.

However, it is easy to check that the ruled surface given by

$$c(t) = (\cos(t), \sin(t), 0)^\top,$$
$$v(t) = \dot{c}(t) + e_3 = (-\sin(t), \cos(t), 1)^\top,$$

agrees with S. Alternatively, one could take $\tilde{v}(t) = \dot{c}(t) - e_3$ instead of $v(t)$, see plate 1.

Example 3.8.6 The ***hyperbolic paraboloid*** or ***saddle surface***

$$S = \left\{ (x,y,z)^\top \in \mathbb{R}^3 \,\middle|\, z = xy \right\}$$

is a ruled surface as well. This is the case since

$$c(t) = (t, 0, 0)^\top,$$
$$v(t) = \frac{1}{\sqrt{1+t^2}}(0, 1, t)^\top.$$

See plate 2.

We have seen that the property of being a ruled surface is not always obvious for a given surface. How can we now show that a given surface is *not* a ruled surface?

Here is a condition.

Theorem 3.8.7 *Let $S \subset \mathbb{R}^3$ be a ruled surface. Then the Gauss curvature satisfies*

$$K \leq 0.$$

Proof Let us remember that the Gauss curvature can be given in terms of the determinants of the first and second fundamental forms,

$$K = \frac{\det(h_{ij})}{\det(g_{ij})}.$$

As the first fundamental form is (g_{ij}) positive definite, we have in particular that $\det(g_{ij}) > 0$. We now need to show that $\det(h_{ij}) \le 0$.

For this purpose let us consider a parametrisation of the form (3.10)

$$F(t,s) = c(t) + sv(t).$$

In particular, it follows that

$$\frac{\partial^2 F}{(\partial s)^2} = 0.$$

For the component h_{22} of the second fundamental form we therefore obtain

$$h_{22} = \left\langle \frac{\partial^2 F}{(\partial s)^2}, N \right\rangle = 0,$$

where N is the normal to the surface. It follows that

$$\det(h_{ij}) = h_{11}h_{22} - h_{12}h_{21} \;=\; h_{11} \cdot 0 - h_{12}^2 \;\le\; 0,$$

where we used the symmetry of the second fundamental form, $h_{12} = h_{21}$. □

Exercise 3.17 Let $F(t,s) = c(t) + sv(t)$ be a parametrisation of a ruled surface. Show that

$$K(F(t,s)) < 0$$

if and only if $\dot{c}(t), v(t)$ and $\dot{v}(t)$ are linearly independent.

Exercise 3.18 Show that for the generalised cylinder and the generalised cone (examples 3.8.2 and 3.8.3) we have that

$$K \equiv 0.$$

Exercise 3.19 Show that for the Möbius strip, the hyperboloid of revolution and the hyperbolic paraboloid (examples 3.8.4, 3.8.5 and 3.8.6) we have that

$$K < 0.$$

3.8.2 Minimal surfaces

Minimal surfaces are, as the name suggests, related to minimal areas. To understand this better, we first need to investigate how area changes under the deformation of a surface.

Theorem 3.8.8 (variation of area) *Let S be a regular surface with finite area. Let $\mathcal{H}$ be the mean curvature field. Let $\Phi : S \to \mathbb{R}^3$ be a smooth normal field on S with compact support.*

Then for all sufficiently small $|t|$ the set $S_t := \{p + t\Phi(p) | \; p \in S\}$ is a regular surface with finite area and

$$\frac{d}{dt} A[S_t]|_{t=0} = -2 \int_S \langle \Phi, \mathcal{H} \rangle \, dA.$$

Proof Let us first consider the case that the support of Φ is fully contained in a coordinate neighbourhood, i.e. for a local parametrisation (U, F, V) we have that $\text{supp}(\Phi) \subset S \cap V$. The corresponding parametrisation for S_t is then given by (U, F_t, V), where

$$F_t(u^1, u^2) = F(u^1, u^2) + t \cdot \Phi(F(u^1, u^2)).$$

As before the differential of F_t has maximal rank and is for reasons of continuity injective for $|t|$ sufficiently small. Hence $F_t : U \to \mathbb{R}^3$ is a local parametrisation of the regular surface S_t.

Let N be the unit normal field on $S \cap V$ given by the parametrisation. Then Φ may be written as

$$\Phi = f \cdot N$$

with a smooth function $f : S \to \mathbb{R}$ with support in $S \cap V$. We calculate

$$\frac{\partial F_t}{\partial u^i} = \frac{\partial F}{\partial u^i} + t \cdot \frac{\partial (f \circ F)}{\partial u^i} \cdot (N \circ F) + t \cdot (f \circ F) \cdot \frac{\partial (N \circ F)}{\partial u^i}.$$

It follows for the first fundamental form of S_t that

$$\begin{aligned} g_{t,ij} &= \left\langle \frac{\partial F_t}{\partial u^i}, \frac{\partial F_t}{\partial u^j} \right\rangle \\ &= g_{ij} + t \frac{\partial (f \circ F)}{\partial u^j} \overbrace{\left\langle \frac{\partial F}{\partial u^i}, N \circ F \right\rangle}^{=0} + t (f \circ F) \overbrace{\left\langle \frac{\partial F}{\partial u^i}, \frac{\partial (N \circ F)}{\partial u^j} \right\rangle}^{=-h_{ij}} \\ &\quad + t \frac{\partial (f \circ F)}{\partial u^i} \overbrace{\left\langle N \circ F, \frac{\partial F}{\partial u^j} \right\rangle}^{=0} + t (f \circ F) \overbrace{\left\langle \frac{\partial (N \circ F)}{\partial u^i}, \frac{\partial F}{\partial u^j} \right\rangle}^{=-h_{ji}} + O(t^2) \\ &= g_{ij} - 2t (f \circ F) h_{ij} + O(t^2). \end{aligned}$$

Multiplying this matrix equation by the inverse matrix (g^{ij}) of (g_{ij}), we obtain after using the formula $w_i^k = \sum_j h_{ij} g^{jk}$ for the coefficients of the Weingarten map that

$$g_{t,ij} = \sum_k (\delta_i^k - 2t(f \circ F)w_i^k + O(t^2))g_{kj}.$$

Equating the determinants of both sides of the equation and using the Taylor expansion of the determinant $\det(\mathrm{Id} + X) = 1 + \mathrm{Trace}(X) + O(\|X\|^2)$, see lemma 5.1.6, we obtain

$$\begin{aligned}
\det(g_{t,ij}) &= \det(g_{ij}) \cdot \det(\mathrm{Id} - 2t(f \circ F) \cdot (w_i^k) + O(t^2)) \\
&= \det(g_{ij}) \cdot (1 + \mathrm{Trace}(-2t(f \circ F)(w_i^k) + O(t^2)) + O(t^2)) \\
&= \det(g_{ij}) \cdot (1 - 4t(f \circ F)(H \circ F) + O(t^2)).
\end{aligned}$$

Taking the square root and using the Taylor expansion of the root-function $\sqrt{1+x} = 1 + \frac{1}{2}x + O(x^2)$, we get

$$\begin{aligned}
\sqrt{\det(g_{t,ij})} &= \sqrt{\det(g_{ij})} \cdot \sqrt{1 - 4t(f \circ F)(H \circ F) + O(t^2)} \\
&= \sqrt{\det(g_{ij})} \cdot (1 - 2t(f \circ F)(H \circ F) + O(t^2)).
\end{aligned}$$

Integrating over S gives that

$$\begin{aligned}
A[S_t] &= \int_S (1 - 2tfH + O(t^2))dA \\
&= A[S] - 2t \int_S \langle \Phi, \mathscr{H} \rangle \, dA + O(t^2).
\end{aligned}$$

This proves the theorem for the case that the support of Φ is fully contained in a coordinate neighbourhood. Let us now consider the general case. As the support is compact by assumption, it can be covered by finitely many systems of local coordinates $(U_1, F_1, V_1), \ldots, (U_k, F_k, V_k)$, $\mathrm{supp}(\Phi) \subset \bigcup_{j=1}^k V_j$. We choose smooth functions $\rho_j : \mathbb{R}^3 \to \mathbb{R}$ with $0 \le \rho_j \le 1$, $\mathrm{supp}\, \rho_j \subset V_j$ and $\sum_{j=1}^k \rho_j \equiv 1$ in a neighbourhood of $\mathrm{supp}(\Phi)$. We set $\Phi_j := \rho_j \cdot \Phi$. Then $\sum_{j=1}^k \Phi_j = \Phi$ and $\mathrm{supp}(\Phi_j) \subset V_j$.

We now obtain a k-parameter family of surfaces

$$S_{(t_1,\ldots,t_k)} = \Big\{ p + \sum_{j=1}^k t_j \Phi_j(p) \,\Big|\, p \in S \Big\}.$$

By the statement proved above we know that

$$\frac{\partial}{\partial t_j} A[S_{(t_1,\ldots,t_k)}]\Big|_{(t_1,\ldots,t_k)=(0,\ldots,0)} = -2\int_S \langle \Phi_j, \mathscr{H}\rangle \, dA.$$

By the chain rule it follows that

$$\begin{aligned}\frac{d}{dt} A[S_t]|_{t=0} &= \sum_{j=1}^{k} \frac{\partial}{\partial t_j} A[S_{(t_1,\ldots,t_k)}]\Big|_{(t_1,\ldots,t_k)=(0,\ldots,0)} \\ &= -2\sum_{j=1}^{k}\int_S \langle \Phi_j, \mathscr{H}\rangle \, dA \\ &= -2\int_S \langle \Phi, \mathscr{H}\rangle \, dA.\end{aligned}$$

□

Corollary 3.8.9 *Let $S \subset \mathbb{R}^3$ be a regular surface with compact closure $\bar{S}$. We assume that S has minimal area among all regular surfaces $\tilde{S}$ with the same boundary $\partial\tilde{S} = \partial S$. Then the mean curvature field of S satisfies*

$$\mathscr{H} \equiv (0,0,0)^\top.$$

Proof Suppose that $\mathscr{H}(p) \neq (0,0,0)^\top$ for a point $p \in S$. In a neighbourhood of p we consider the smooth unit normal field N, for which $\langle \mathscr{H}(p), N(p)\rangle > 0$. For reasons of continuity $\langle \mathscr{H}, N\rangle > 0$ also in a neighbourhood V of p in S. We choose a smooth function $f : S \to \mathbb{R}$ with compact support $\operatorname{supp} f \subset V$, $f \geq 0$ and $f(p) > 0$. Then

$$\Phi(q) := \begin{cases} f(q)N(q), & q \in S \cap V, \\ (0,0,0)^\top, & q \in S - V \end{cases}$$

defines a smooth normal field on S with compact support. Hence

$$\int_S \langle \mathscr{H}, \Phi\rangle \, dA > 0.$$

On the other hand, the deformed surfaces S_t have the same boundary as S. As S has minimal area among all such surfaces, we have

$$\frac{d}{dt} A[S_t]\Big|_{t=0} = 0.$$

This contradicts the variation formula for area. □

If we want to find a surface S that is enclosed by a closed space curve c, $\partial S = c$, and has minimal surface area, then this surface necessarily needs to satisfy $\mathscr{H} \equiv (0,0,0)^\top$. This leads to the following definition.

Definition 3.8.10 A regular surface $S \subset \mathbb{R}^3$ is called a ***minimal surface*** if

$$\mathscr{H} \equiv (0,0,0)^\top.$$

Note that minimal surfaces do not necessarily minimise area. $\mathscr{H} \equiv (0,0,0)^\top$ is only a necessary condition.

The question of whether we can, for a given closed space curve c, find a surface of minimum area with boundary c is known as the *Plateau's problem*. The answer is "yes" for a very general class of boundary curves, see, for example, [1] for a discussion of this problem. We can physically create minimal surfaces by dipping a closed wire (our closed space curve) into a soap solution. The soap film created is then a minimal surface.

Remark If the surface S is orientable, then there exists a smooth unit normal field N on S and we can write the mean normal curvature field in the form $\mathscr{H} = H \cdot N$. The minimum area condition is then

$$H \equiv 0.$$

It is now time for some examples.

Example 3.8.11 The simplest and most uninteresting example is certainly the affine plane $S \subset \mathbb{R}^3$. For this surface we obviously have

$$K \equiv H \equiv 0.$$

Example 3.8.12 ***Enneper's surface*** can be given by one single parametrisation:

$$F : \mathbb{R}^2 \to \mathbb{R}^3,$$

$$F(u^1, u^2) = \begin{pmatrix} u^1 - \dfrac{(u^1)^3}{3} + u^1(u^2)^2 \\ u^2 - \dfrac{(u^2)^3}{3} + u^2(u^1)^2 \\ (u^1)^2 - (u^2)^2 \end{pmatrix}.$$

Plate 3 shows that Enneper's surface intersects itself. To obtain a regular surface, we need to restrict the domain of F in a suitable way.

Exercise 3.20 Show that Enneper's surface is a minimal surface.

Example 3.8.13 The ***catenoid*** is given by the parametrisation

$$F(u^1, u^2) = \begin{pmatrix} \cosh(u^1)\cos(u^2) \\ \cosh(u^1)\sin(u^2) \\ u^1 \end{pmatrix}.$$

The catenoid is an example of a surface of revolution, a class of surfaces that we will consider after the minimal surfaces. Indeed, the catenoid is essentially a single surface of revolution, which is at the same time a minimal surface, see exercise 3.30.

Exercise 3.21 Show that the catenoid is a minimal surface.

Example 3.8.14 The ***helicoid*** is given by the parametrisation

$$F(u^1, u^2) = \begin{pmatrix} u^1 \sin(u^2) \\ -u^1 \cos(u^2) \\ u^2 \end{pmatrix}.$$

Exercise 3.22 Show that the helicoid is a minimal surface and at the same time a ruled surface.

Theorem 3.8.15 *For every regular surface we have*

$$K \leq H^2.$$

In particular, the Gauss curvature of minimal surfaces satisfies

$$K \leq 0.$$

Proof Expressing the mean curvature and the Gauss curvature in terms of the principal curvatures, $H = (\kappa_1 + \kappa_2)/2$, $K = \kappa_1\kappa_2$, we observe that

$$4 \cdot (H^2 - K) = (\kappa_1 + \kappa_2)^2 - 4 \cdot \kappa_1\kappa_2 = (\kappa_1 - \kappa_2)^2 \geq 0.$$

The claim follows. □

Corollary 3.8.16 *Compact minimal surfaces do not exist.*

Proof By theorem 3.6.17 every compact surface $S \subset \mathbb{R}^3$ has a point with positive Gauss curvature. By theorem 3.8.15 it follows that S cannot be a minimal surface. □

Exercise 3.23 Show that the mean curvature H, the Gauss curvature K and the two principal curvatures $\kappa = \kappa_1$ and $\kappa = \kappa_2$ of a regular surface satisfy

$$\kappa^2 - 2H\kappa + K = 0$$

and thus

$$\kappa = H \pm \sqrt{H^2 - K}.$$

Exercise 3.24 Let S be the graph of the function $\varphi : U \to \mathbb{R}$, $U \subset \mathbb{R}^2$ open. Show that S is minimal if and only if φ satisfies the following differential equation:

$$\left(1 + \left(\frac{\partial \varphi}{\partial y}\right)^2\right)\frac{\partial^2 \varphi}{\partial x^2} - 2\frac{\partial \varphi}{\partial x}\frac{\partial \varphi}{\partial y}\frac{\partial^2 \varphi}{\partial x \partial y} + \left(1 + \left(\frac{\partial \varphi}{\partial x}\right)^2\right)\frac{\partial^2 \varphi}{\partial y^2} = 0.$$

Exercise 3.25 Let S be the graph of the function $\varphi : U \to \mathbb{R}$, $U \subset \mathbb{R}^2$ open. Derive a formula for the Gauss curvature of S and show that the Gauss curvature is positive if and only if the Hessian of φ is definite while K is negative if and only if the Hessian is indefinite and non-degenerate.

Exercise 3.26 Show that the graph of the function $\varphi : (-\pi/2, \pi/2) \times (-\pi/2, \pi/2) \to \mathbb{R}$, $\varphi(x, y) = \ln(\cos(y)) - \ln(\cos(x))$, called ***Scherk's minimal surface***, is a minimal surface. See plate 4.

Exercise 3.27 The ***rescaled catenoid*** is given by the parametrisation

$$F(u^1, u^2) = \begin{pmatrix} R \cdot \cosh(u^1/R)\cos(u^2) \\ R \cdot \cosh(u^1/R)\sin(u^2) \\ u^1 \end{pmatrix}.$$

(a) Show that the rescaled catenoid is a minimal surface for all $R > 0$.

Hint Use the results of exercise 3.21 instead of computing the mean curvature from scratch.

(b) Calculate the height h for $0 < R < 1$, at which the $(x$–$y)$ plane translated up by h along the z-axis intersects the rescaled catenoid such that the resulting circle has radius 1.

(c) Show that this height is bounded above.

Statement (c) can be illustrated experimentally. If we hold two wire circles of radius 1 parallel to the x–y plane, one at a height h above, the other one at a height $-h$ below, such that the centre of one circle lies above that of the other, then we can, if h is not too large, fit a rescaled catenoid as a soap film to those wires. If we now carefully pull the two circles away from each other, i.e. if we increase h, then the soap film will burst at the latest when the above postulated maximal height is reached.

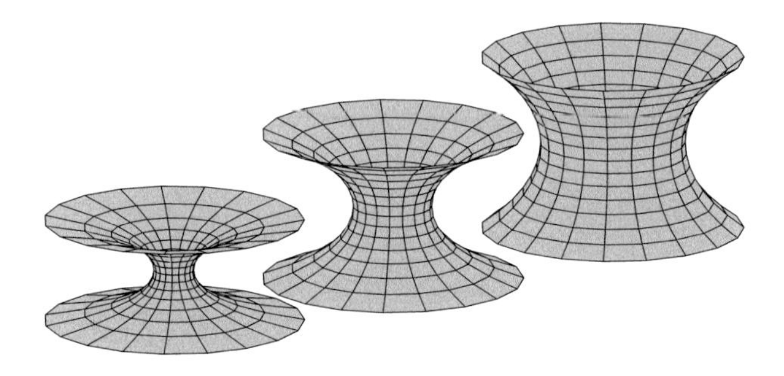

Exercise 3.28 Let S_α be given by

$$F_\alpha(u^1, u^2) = \begin{pmatrix} \sin(\alpha)\cosh(u^1)\cos(u^2) + \cos(\alpha)\sinh(u^1)\sin(u^2) \\ \sin(\alpha)\cosh(u^1)\sin(u^2) - \cos(\alpha)\sinh(u^1)\cos(u^2) \\ \sin(\alpha)u^1 + \cos(\alpha)u^2 \end{pmatrix}.$$

Show that S_α is a minimal surface for all $\alpha \in \mathbb{R}$, and that $S_{\pi/2}$ is the catenoid while S_0 is the helicoid.

The surfaces S_α are therefore deformations of the helicoid into the catenoid. This is illustrated in plate 5.

We will conclude this subsection about minimal surfaces with a slightly more involved exercise.

Exercise 3.29 Let $c : I \to \mathbb{R}^3$ be a space curve parametrised by arc-length, let $v : I \to \mathbb{R}^3$ be smooth with $\|v\| \equiv 1$. We assume that $v(t)$ is perpendicular to $\dot{c}(t)$ for all $t \in I$. Show that the ruled surface generated by c and v is a minimal surface if and only if

(a) it is contained in a plane, or
(b) there exist $\omega > 0$ and $A \in [-1, 1]$, such that after an application of a Euclidean motion

$$c(t) = \frac{A}{\omega}\begin{pmatrix} \sin(\omega t) \\ -\cos(\omega t) \\ 0 \end{pmatrix} + \sqrt{1 - A^2}\begin{pmatrix} 0 \\ 0 \\ t \end{pmatrix},$$

$$v(t) = \begin{pmatrix} \sin(\omega t) \\ -\cos(\omega t) \\ 0 \end{pmatrix}.$$

If $A=0$, then c is a straight line and we are dealing with the helicoid. If $|A|=1$, then c is a circular line, and the surface is the x–y plane. For other A, curve c is a helix. As A increases from 0 to 1, the helix is deformed into the plane, and it remains a minimal ruled surface in the whole process of the deformation.

3.8.3 Surfaces of revolution

Surfaces of revolution are the result of rotating a plane curve that lies, in the x–z plane, say, around the z-axis. We can imagine the x–y plane as a "potter's wheel". If the afore-mentioned plane curve is given by the parametrisation $t \mapsto (r(t), t)^\top$, $t \in I$, then we obtain a local parametrisation of the corresponding surface of revolution by

$$F(t,\varphi) = \begin{pmatrix} r(t)\cos(\varphi) \\ r(t)\sin(\varphi) \\ t \end{pmatrix}, \quad t \in I, \varphi \in (\varphi_0, \varphi_0 + 2\pi).$$

Choosing two values of φ_0, e.g. $\varphi_0=0$ and $\varphi_0=\pi$, we obtain two local parametrisations that cover the entire surface of revolution.

We calculate the first fundamental form:

$$\frac{\partial F}{\partial t}(t,\varphi) = \begin{pmatrix} \dot r(t)\cos(\varphi) \\ \dot r(t)\sin(\varphi) \\ 1 \end{pmatrix}, \quad \frac{\partial F}{\partial \varphi}(t,\varphi) = \begin{pmatrix} -r(t)\sin(\varphi) \\ r(t)\cos(\varphi) \\ 0 \end{pmatrix},$$

hence

$$\begin{aligned} g_{11} &= 1 + \dot r(t)^2, \\ g_{12} &= g_{21} = 0, \\ g_{22} &= r(t)^2. \end{aligned}$$

For the surface element we obtain

$$dA = r(t)\sqrt{1+\dot r(t)^2}\, dt\, d\varphi.$$

For the calculation of the second fundamental form we need a unit normal field. We choose the inner unit normal field

$$N(F(t,\varphi)) = \frac{1}{\sqrt{1+\dot r(t)^2}} \begin{pmatrix} \cos(\varphi) \\ \sin(\varphi) \\ -\dot r(t) \end{pmatrix}.$$

Further,

$$\frac{\partial^2 F}{\partial t^2}(t,\varphi) = \begin{pmatrix} \ddot r(t)\cos(\varphi) \\ \ddot r(t)\sin(\varphi) \\ 0 \end{pmatrix},$$

$$\frac{\partial^2 F}{\partial t \partial \varphi}(t,\varphi) = \begin{pmatrix} -\dot{r}(t)\sin(\varphi) \\ \dot{r}(t)\cos(\varphi) \\ 0 \end{pmatrix},$$

$$\frac{\partial^2 F}{\partial \varphi^2}(t,\varphi) = \begin{pmatrix} -r(t)\cos(\varphi) \\ -r(t)\sin(\varphi) \\ 0 \end{pmatrix}.$$

The second fundamental form is therefore

$$h_{11} = \frac{\ddot{r}(t)}{\sqrt{1+\dot{r}(t)^2}},$$
$$h_{12} = h_{21} = 0,$$
$$h_{22} = \frac{-r(t)}{\sqrt{1+\dot{r}(t)^2}}.$$

From the first and second fundamental form we obtain the Weingarten map

$$W = \frac{1}{\sqrt{1+\dot{r}(t)^2}} \begin{pmatrix} \frac{\ddot{r}(t)}{1+\dot{r}(t)^2} & 0 \\ 0 & -\frac{1}{r(t)} \end{pmatrix}.$$

Conveniently the Weingarten map w.r.t. the coordinates t and φ is in diagonal form already, so we can therefore read off the principal curvatures directly:

$$\kappa_1 = \frac{\ddot{r}(t)}{(1+\dot{r}(t)^2)^{3/2}}, \quad \kappa_2 = -\frac{1}{r(t)\sqrt{1+\dot{r}(t)^2}}.$$

We also obtain Gauss curvature and mean curvature immediately:

$$K = -\frac{\ddot{r}(t)}{r(t)(1+\dot{r}(t)^2)^2},$$

$$H = \frac{1}{2}\frac{r(t)\ddot{r}(t) - 1 - \dot{r}(t)^2}{r(t)(1+\dot{r}(t)^2)^{3/2}}.$$

Example 3.8.17 The paraboloid of revolution $S = \{(x,y,z)^\top \in \mathbb{R}^3 \mid z = x^2 + y^2\}$ is a surface of revolution with function $r(t) = \sqrt{t}$, $t > 0$. Application of the formulae derived above gives:

$$K = \frac{4}{(1+4t)^2}, \quad H = \frac{2+4t}{(1+4t)^{3/2}}.$$

We will come back to surfaces of revolution in the next chapter in the context of geodesics, see theorem 4.5.13.

Exercise 3.30 A surface of revolution with $r(t) = c_1 \cosh((t + c_2)/c_1)$, $c_1 > 0, c_2 \in \mathbb{R}$, is a catenoid, see example 3.8.13 and exercise 3.27. Show that a surface of revolution is a minimal surface if and only if it is a catenoid.

Exercise 3.31 Let S be the surface of revolution of the tractrix, see example 2.1.5. A parametrisation is given by

$$F : (0, \pi/2) \times (\varphi_0, \varphi_0 + 2\pi) \to \mathbb{R}^3,$$

$$F(t, \varphi) = \begin{pmatrix} \sin(t)\sin(\varphi) \\ \sin(t)\cos(\varphi) \\ \cos(t) + \ln(\tan(t/2)) \end{pmatrix}.$$

Show that the surface has constant Gauss curvature $K \equiv -1$. Because of this property it is somewhat analogous to the sphere, which has constant Gauss curvature $K \equiv 1$, and is therefore called a ***pseudo-sphere***. See plate 6.

3.8.4 Tubular surfaces

We have so far always considered curves as one-dimensional objects. If we model a curve in three-dimensional space, e.g. using wire, then this wire has positive thickness $2r > 0$. The surface of this wire forms a ***tubular surface***. We will now analyse the geometry of this tubular surface.

Let $c : I \to \mathbb{R}^3$ be a curve parametrised by arc-length that has non-vanishing curvature, $\kappa(t) \neq 0$ for all $t \in I$. Then the torsion τ and the Frenet dreibein $(\dot{c}, n, b)$ are defined. Let $r > 0$. We consider

$$F : I \times \mathbb{R} \to \mathbb{R}^3,$$
$$F(t, \varphi) = c(t) + r \cdot (\cos(\varphi) \cdot n(t) + \sin(\varphi) \cdot b(t)).$$

To check whether this gives us a parametrisation of a regular surface, we calculate the partial derivatives of F:

$$\begin{aligned}
\frac{\partial F}{\partial t}(t, \varphi) &= \dot{c}(t) + r \cdot (\cos(\varphi) \cdot \dot{n}(t) + \sin(\varphi) \cdot \dot{b}(t)) \\
&= \dot{c}(t) + r(\cos(\varphi)(-\kappa(t)\dot{c}(t) + \tau(t)b(t)) + \sin(\varphi)(-\tau(t)n(t))) \\
&= (1 - r\cos(\varphi)\kappa(t))\dot{c}(t) - r\sin(\varphi)\tau(t)n(t) + r\cos(\varphi)\tau(t)b(t), \\
\frac{\partial F}{\partial \varphi}(t, \varphi) &= r(-\sin(\varphi)n(t) + \cos(\varphi)b(t)).
\end{aligned}$$

We used the Frenet formulae (proposition 2.3.7). For the first fundamental form we obtain

$$\begin{aligned}
g_{11}(t,\varphi) &= (1 - r\cos(\varphi)\kappa(t))^2 + r^2\tau(t)^2,\\
g_{12}(t,\varphi) &= g_{21}(t,\varphi) = r^2\tau(t),\\
g_{22}(t,\varphi) &= r^2.
\end{aligned}$$

The determinant of the matrix g_{ij} is therefore $r^2(1 - r\cos(\varphi)\kappa(t))^2$. If r is so small that we have $r < 1/\kappa(t)$ for all $t \in I$, then this determinant is never equal to 0, and F is a parametrisation of a regular surface after a restriction of the domain. For the surface element we obtain

$$dA = r(1 - r\cos(\varphi)\kappa(t))dt\,d\varphi$$

and for the inverse matrix of the first fundamental form

$$\left(g^{ij}(t,\varphi)\right)_{ij} = \begin{pmatrix} \dfrac{1}{(1 - r\cos(\varphi)\kappa(t))^2} & -\dfrac{\tau(t)}{(1 - r\cos(\varphi)\kappa(t))^2} \\ -\dfrac{\tau(t)}{(1 - r\cos(\varphi)\kappa(t))^2} & \dfrac{1}{r^2} + \dfrac{\tau(t)^2}{(1 - r\cos(\varphi)\kappa(t))^2} \end{pmatrix}.$$

The unit normal field of F is then

$$N = -\cos(\varphi)\cdot n(t) - \sin(\varphi)\cdot b(t).$$

To find the second fundamental form, we find the second partial derivatives of F, again using the Frenet formulae:

$$\begin{aligned}
\frac{\partial^2 F}{\partial t^2}(t,\varphi) &= -r\cos(\varphi)\dot{\kappa}(t)\dot{c}(t) + (1 - r\cos(\varphi)\kappa(t))\ddot{c}(t) - r\sin(\varphi)\dot{\tau}(t)n(t)\\
&\quad - r\sin(\varphi)\tau(t)\dot{n}(t) + r\cos(\varphi)\dot{\tau}(t)b(t) + r\cos(\varphi)\tau(t)\dot{b}(t)\\
&= r(-\cos(\varphi)\dot{\kappa}(t) + \sin(\varphi)\tau(t)\kappa(t))\dot{c}(t)\\
&\quad + [\kappa(t)(1 - r\cos(\varphi)\kappa(t)) - r\sin(\varphi)\dot{\tau}(t) - r\cos(\varphi)\tau(t)^2]n(t)\\
&\quad + r(-\sin(\varphi)\tau(t)^2 + \cos(\varphi)\dot{\tau}(t))b(t),\\
\frac{\partial^2 F}{\partial t\partial\varphi}(t,\varphi) &= r\sin(\varphi)\kappa(t)\dot{c}(t) - r\cos(\varphi)\tau(t)n(t) - r\sin(\varphi)\tau(t)b(t),\\
\frac{\partial^2 F}{\partial \varphi^2}(t,\varphi) &= -r(\cos(\varphi)n(t) + \sin(\varphi)b(t)).
\end{aligned}$$

By (3.4) the second fundamental form is

$$\begin{aligned}
h_{11}(t,\varphi) &= \left\langle \frac{\partial^2 F}{\partial t^2}(t,\varphi), N(t,\varphi) \right\rangle \\
&= \left(\kappa(t)(1 - r\cos(\varphi)\kappa(t)) - r\sin(\varphi)\dot{\tau}(t) - r\cos(\varphi)\tau(t)^2\right)(-\cos(\varphi)) \\
&\quad + (-\sin(\varphi)\tau(t)^2 + \cos(\varphi)\dot{\tau}(t))(-\sin(\varphi)) \\
&= r\tau(t)^2 - \cos(\varphi)\kappa(t) + r\cos(\varphi)^2\kappa(t)^2, \\
h_{12}(t,\varphi) &= r\cos(\varphi)^2\tau(t) + r\sin(\varphi)^2\tau(t) \\
&= r\tau(t), \\
h_{22}(t,\varphi) &= r.
\end{aligned}$$

From (3.5) it follows that the matrix of the Weingarten map is

$$\begin{aligned}
\left(w_i^j(t,\varphi)\right)_{ij} &= (h_{ik}(t,\varphi))(g^{kj}(t,\varphi))) \\
&= \begin{pmatrix} r\tau(t)^2 - \cos(\varphi)\kappa(t) + r\cos(\varphi)^2\kappa(t)^2 & r\tau(t) \\ r\tau(t) & r \end{pmatrix} \cdot \\
&\quad \cdot \begin{pmatrix} \frac{1}{(1 - r\cos(\varphi)\kappa(t))^2} & -\frac{\tau(t)}{(1 - r\cos(\varphi)\kappa(t))^2} \\ -\frac{\tau(t)}{(1 - r\cos(\varphi)\kappa(t))^2} & \frac{1}{r^2} + \frac{\tau(t)^2}{(1 - r\cos(\varphi)\kappa(t))^2} \end{pmatrix} \\
&= \begin{pmatrix} -\frac{\kappa(t)\cos(\varphi)}{1 - r\cos(\varphi)\kappa(t)} & * \\ 0 & \frac{1}{r} \end{pmatrix}.
\end{aligned}$$

As this matrix is triangular, we can read off the eigenvalues and obtain the principal curvatures:

$$\begin{aligned}
\kappa_1 &= \frac{-\kappa(t)\cos(\varphi)}{1 - r\cos(\varphi)\kappa(t)}, \\
\kappa_2 &= \frac{1}{r}.
\end{aligned}$$

It follows that the Gauss curvature is

$$K = -\frac{1}{r}\frac{\kappa(t)\cos(\varphi)}{1 - r\cos(\varphi)\kappa(t)},$$

and the mean curvature is

$$H = \frac{1 - 2r\cos(\varphi)\kappa(t)}{2r(1 - r\cos(\varphi)\kappa(t))}.$$

Example 3.8.18 We consider the ***torus***, which is defined as the tubular surface around a circular line

$$c(t) = (\cos(t), \sin(t), 0)^\top$$

with thickness $2r < 2$ (plate 7). We then have

$$\kappa(t) \equiv 1, \quad \tau(t) \equiv 0$$

and thus

$$K(t, \varphi) = -\frac{1}{r}\frac{\cos(\varphi)}{1 - r\cos(\varphi)}, \qquad H(t, \varphi) = \frac{\cos(\varphi) - r/2}{1 - r\cos(\varphi)}.$$

Exercise 3.32 Let c be a closed space curve parametrised by arc-length. Show that for every tubular surface S around c we have

$$\int_S K\,dA = 0.$$

We will understand this fact better against the background of the Gauss–Bonnet theorem.

Exercise 3.33 Let $c : I \to \mathbb{R}^3$ be a closed space curve. Using the results of this section and exercise 3.16 give a new simple proof of the total curvature estimate in Fenchel's theorem 2.3.19:

$$\kappa(c) \geq 2\pi.$$

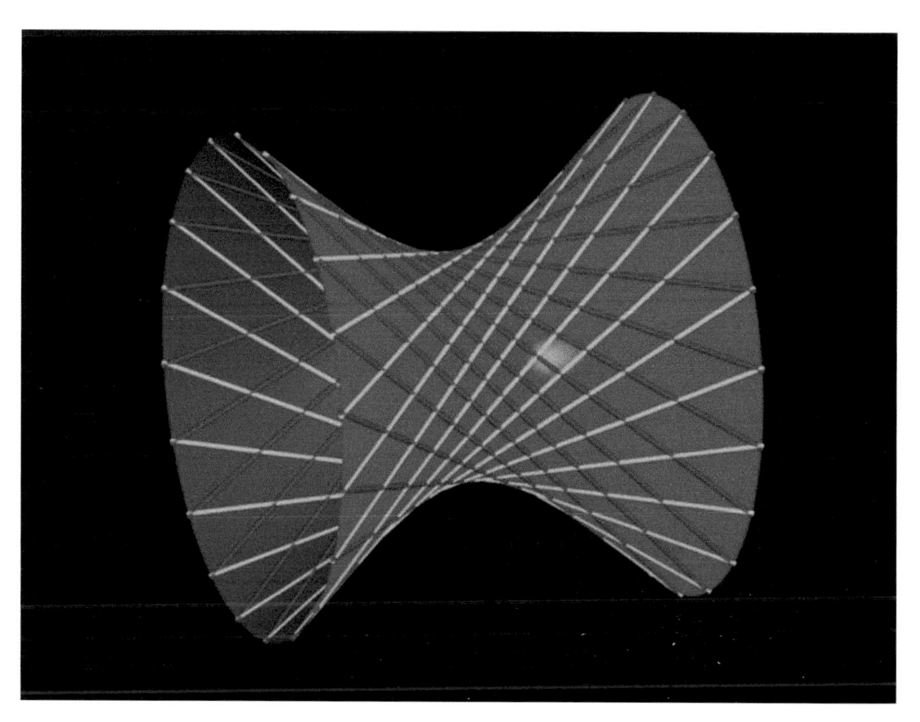

Plate 1. Hyperboloid of revolution

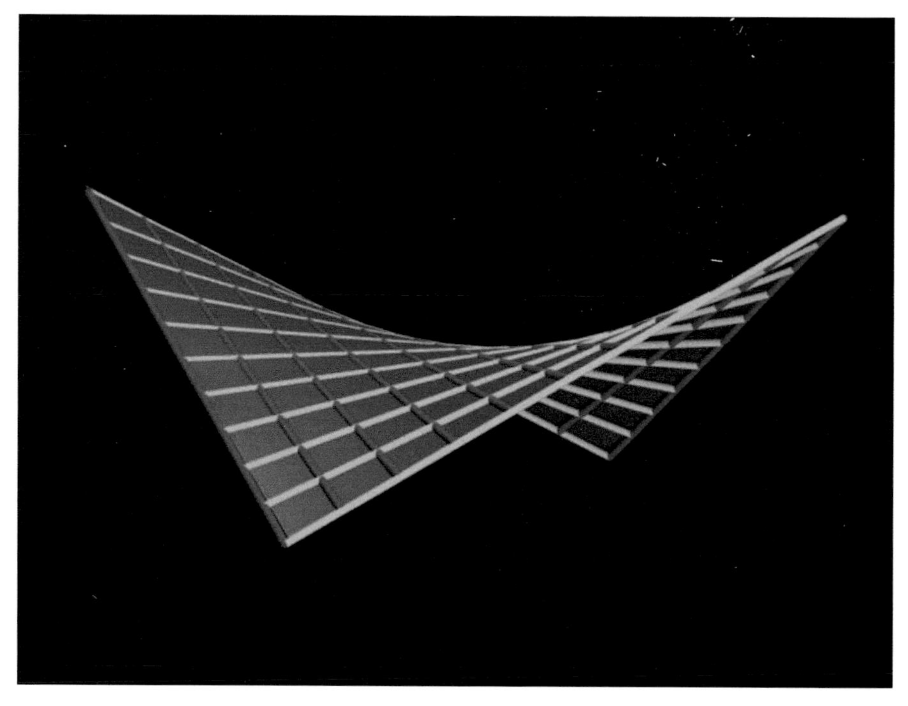

Plate 2. Hyperbolic paraboloid

Plate 3. Enneper's surface

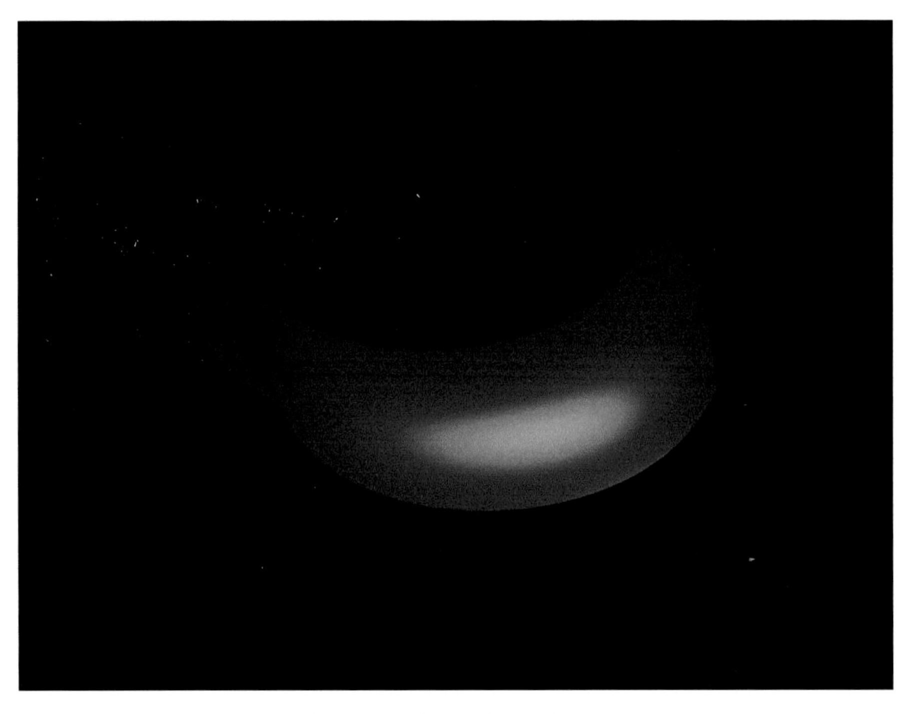

Plate 4. Scherk's surface

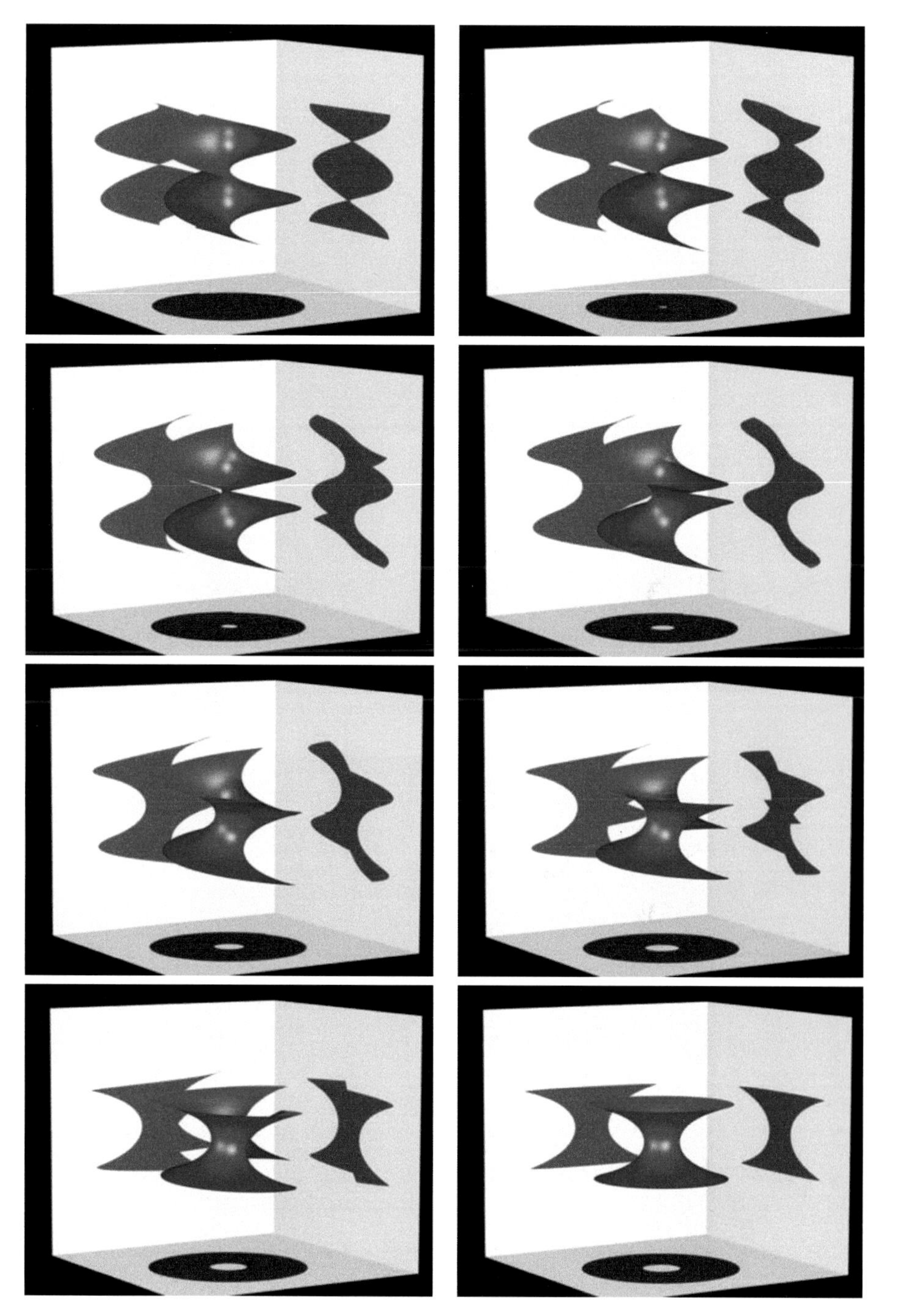

Plate 5. Deformation of helicoid into catenoid

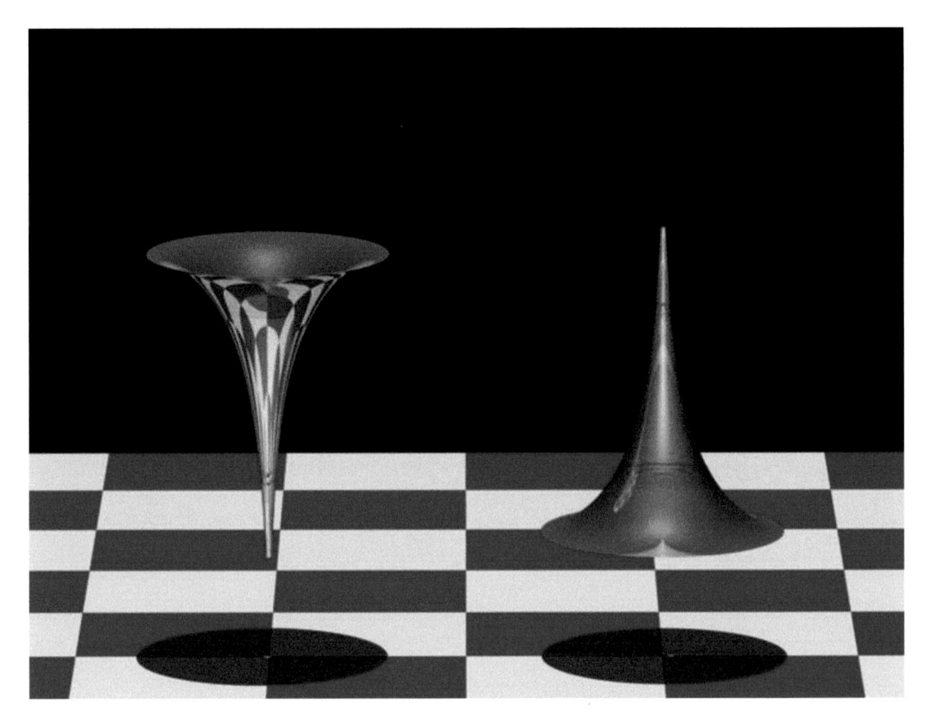

Plate 6. Pseudo-sphere

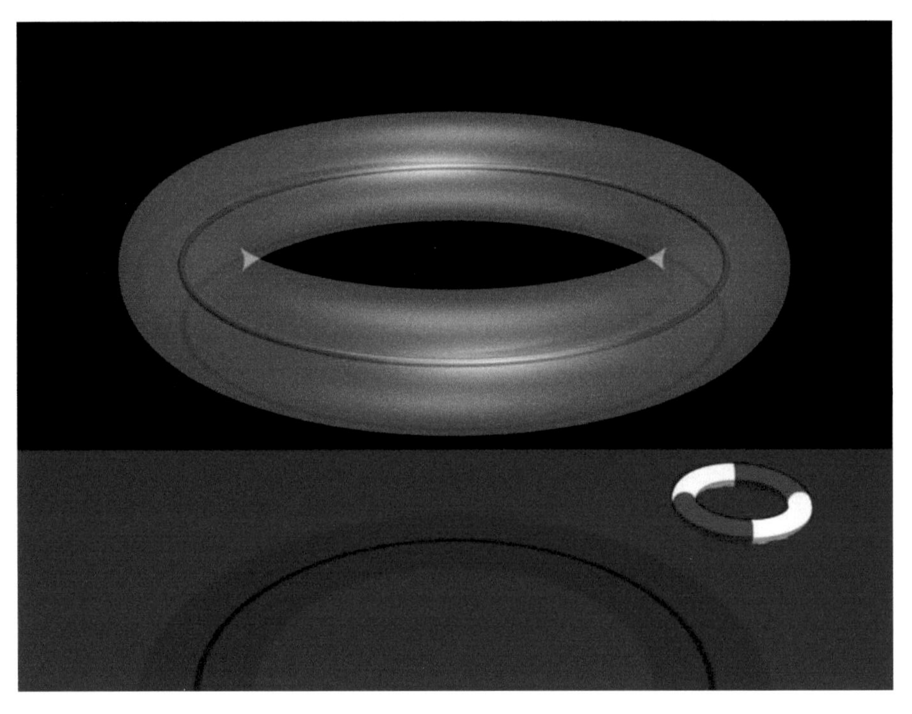

Plate 7. Torus of revolution

4 The inner geometry of surfaces

We use the notion of an isometry to make the concept of inner geometry of surfaces more precise. Vector fields and their first and second covariant derivatives are introduced. The Theorema Egregrium ('remarkable theorem') expresses the Gauss curvature in terms of the curvature tensor and shows the Gauss curvature belongs to the inner geometry of the surface. General Riemann metrics generalise the first fundamental form. The problem of the shortest way from one point to another leads to the concept of the geodesic and the Riemann exponential mapping. In this way it is particularly straightforward to obtain coordinates that are convenient in geometry, like Riemann normal coordinates, geodesic polar coordinates and Fermi coordinates. Jacobi fields illustrate the inner geometric importance of the Gauss curvature. Spherical and hyperbolic geometry are investigated in more detail. Their trigonometry is derived and applications to cartography are discussed. The hyperbolic plane satisfies all axioms of Euclidean geometry except for the parallel axiom.

4.1 Isometries

When we consider surfaces in $\mathbb{R}^3$ we tend to pay special attention to their relative geometries, i.e. to how the surface is embedded into the surrounding space. We quasi look at them from outside. One could also try to imagine oneself in the position of a (two-dimensional) inhabitant of the surface, and examine those properties of the surface that can be observed by a being who cannot peek out of the surface. For instance, viewed from outside, the cylindrical surface and the plane seem very different. But we will see that it would not be easy for our two-dimensional being to decide whether it lives on a cylindrical surface or on a plane. If we took small pieces of plane and cylinder, it would be impossible to make a decision by means of measurements within the surface. By contrast, it would be relatively easy to distinguish between a piece of a sphere and a piece of a plane.

We will now start to say more precisely which quantities can be observed by an inhabitant of the plane.

Definition 4.1.1 Let S_1 and S_2 be regular surfaces in $\mathbb{R}^3$. A smooth map $f : S_1 \to S_2$ is a ***local isometry***, if for every point $p \in S_1$ the differential

$$d_p f : T_p S_1 \to T_{f(p)} S_2$$

is a linear isometry regarding the first fundamental form, i.e.

$$\big\langle d_p f(X), d_p f(Y)\big\rangle = \langle X, Y\rangle$$

for all $X, Y \in T_p S_1$.

Exercise 4.1 Let $S_1 = \mathbb{R}^2 \times \{0\}$ be the x–y plane, and $S_2 = S^1 \times \mathbb{R}$ the cylindrical surface. Show that the map

$$f : S_1 \to S_2, \qquad f(x, y, 0) = (\cos(x), \sin(x), y)^\top$$

is a local isometry.

Exercise 4.2 Let $S_1 = \mathbb{R}^2 \times \{0\}$ be the x–y plane, and $S_2 = \{(\xi, \eta, \zeta)^\top \mid \xi^2 + \eta^2 = \frac{1}{3}\zeta^2, \zeta > 0\}$ the conical surface. Show that the map

$$f : S_1 \to S_2, \qquad f(x, y, 0) = \frac{1}{2\sqrt{x^2 + y^2}} \left(x^2 - y^2, 2xy, \sqrt{3}(x^2 + y^2)\right)^\top$$

is a local isometry.

Everything which can be "measured" within the surface, e.g. the lengths of curves which lie on the surface or the angle between two tangent vectors, depends on the first fundamental form. Thus, if such a local isometry f exists, then the angle between the two vectors in the image is the same as the one between the two original tangent vectors, for example. "Small" open subsets $U \subset S_1$ cannot be distinguished from their isometric images $f(U) \subset S_2$ by means of such measurements. Because of this we call geometric quantities which do not change under local isometries ***quantities of the inner geometry***.

What it exactly means when a geometric quantity does not change under local isometries depends on the type of the mathematical object. If the geometric quantity is a function $F_S : S \to \mathbb{R}$ in the plane, e.g. the Gauss curvature or the mean curvature, then this means that every local isometry $f : S_1 \to S_2$ satisfies

$$F_{S_1} = F_{S_2} \circ f.$$

Example 4.1.2 The mean curvature H is *not* a quantity of the inner geometry, since in the plane $H_{\text{plane}} \equiv 0$, whereas in the cylindrical surface $H_{\text{cylinder}} \equiv \frac{1}{2}$. Since plane and cylinder are locally isometric,

$H_{\text{cylinder}} = H_{\text{plane}} \circ f$ would have to apply if the mean curvature was a quantity of the inner geometry.

Since the mean curvature is not a quantity of the inner geometry, the principal curvatures, from which the mean curvature is calculated, cannot be quantities of the inner geometry either. We will see later that the Gauss curvature by contrast is a quantity of the inner geometry, although it is also calculated from the principal curvatures.

Remark Since the differential of a local isometry in particular always has full rank, a local isometry is always a local diffeomorphism by the inverse mapping theorem. However, in general it is not a global diffeomorphism, i.e. not bijective, as the example of the plane and cylinder has already shown.

Exercise 4.3 Let $f : S_1 \to S_2$ be a local isometry. Let (U, F, V) be a local parametrisation of S_1. Without loss of generality assume that $V \cap S_1 \subset S_1$ is so small that $f|_{V \cap S_1} : V \cap S_1 \to f(V \cap S_1)$ is a diffeomorphism. Then $f \circ F$ is a local parametrisation of S_2.

Show that the coefficient functions of the matrix representations $g_{ij} : U \to \mathbb{R}$ of S_1 with respect to F and of S_2 with respect to $f \circ F$ agree.

Definition 4.1.3 A local isometry $f : S_1 \to S_2$ which is in addition bijective, is called an ***isometry***. If there exists such an isometry $f : S_1 \to S_2$, then the surfaces S_1 and S_2 are said to be ***isometric***. The surfaces S_1 and S_2 are called ***locally isometric*** if for every point $p \in S_1$ there exists an open neighbourhood $U_1 \subset S_1$ of p, an open subset $U_2 \subset S_2$ and an isometry $f : U_1 \to U_2$ and conversely for every point $q \in S_2$ an open neighbourhood $U_2 \subset S_2$ of q, an open subset $U_1 \subset S_1$ and an isometry $f : U_2 \to U_1$.

Exercise 4.4 Show that if $f : S_1 \to S_2$ is an isometry, then $f^{-1} : S_2 \to S_1$ is also an isometry.

Exercise 4.5 Show that if there exists a surjective local isometry $f : S_1 \to S_2$, then S_1 and S_2 are locally isometric. Is this also true if f is not surjective?

For example the cylinder and the plane are locally isometric. But they are not isometric, since they are not even diffeomorphic. The relations "isometric" and "locally isometric" are obviously equivalence relations on the set of regular surfaces.

Exercise 4.6 Let $F : \mathbb{R}^3 \to \mathbb{R}^3$ be a Euclidean motion, i.e. $F(x) = Ax + b$, where $A \in \mathrm{O}(3)$ is an orthogonal map and $b \in \mathbb{R}^3$ the translational component. Let $S \subset \mathbb{R}^3$ be a regular surface. Show that $f := F|_S : S \to F(S)$ is an isometry.

Exercise 4.7 Let E_1 and $E_2 \subset \mathbb{R}^3$ be affine planes. Show that E_1 and E_2 are isometric.

4.2 Vector fields and the covariant derivative

Definition 4.2.1 Let $S \subset \mathbb{R}^3$ be a regular surface. A ***vector field*** on S is a map $v : S \to \mathbb{R}^3$, such that $v(p) \in T_pS$ for all $p \in S$.

A vector field assigns each point on the surface a vector, which at that point is tangential to the surface. We also talk about continuous, differentiable, smooth, ... vector fields, if map v has the corresponding property.

Example 4.2.2 Let $f : S \to \mathbb{R}$ be a smooth function. Since the first fundamental form is non-degenerate, there exists, for a fixed point p, exactly one vector $v(p) \in T_pS$ with the property

$$d_pf(X) = I(v(p), X)$$

for all $X \in T_pS$. In this way the ***gradient vector field*** $v =: \operatorname{grad} f$ is defined. We will see soon that the gradient vector field is a smooth vector field.

The differentiability of a vector field is best verified using a local parametrisation. Let (U, F, V) be a local parametrisation of the regular surface S. Then for every point $p \in S \cap V$ the vectors $(\partial F/\partial u^1)(F^{-1}(p))$ and $(\partial F/\partial u^2)(F^{-1}(p))$ form a basis of T_pS. Hence, a vector field v over S is for all $p \in S \cap V$ uniquely representable in the following form:

$$v(p) = \sum_{j=1}^{2} \xi^j(p) \frac{\partial F}{\partial u^j}(F^{-1}(p))$$

Since the basis fields $(\partial F/\partial u^j)(F^{-1}(p))$ are smooth, we have that v on $V \cap S$ is continuous, differentiable, smooth, etc., if and only if the coefficient functions

$$\xi^j : V \to \mathbb{R}$$

have the corresponding properties.

Example 4.2.3 Let us check that the gradient vector field of a smooth function $f : S \to \mathbb{R}$ is smooth. For this purpose let (U, F, V) be a local parametrisation. Then $\tilde{f} := f \circ F : U \to \mathbb{R}$ is also a smooth function. We need to find the coefficient functions of the gradient vector field with respect to the basis given by the parametrisation, i.e. the functions ξ^j in the representation $\operatorname{grad} f = \sum_{j=1} \xi^j \cdot (\partial F/\partial u^1)$. We calculate

$$
\begin{aligned}
\frac{\partial \tilde{f}}{\partial u^k}\left(F^{-1}(p)\right) &= d_p f\left(\frac{\partial F}{\partial u^k}\left(F^{-1}(p)\right)\right) \\
&= I\left(\operatorname{grad} f(p), \frac{\partial F}{\partial u^k}\left(F^{-1}(p)\right)\right) \\
&= I\left(\sum_{j=1}^{2} \xi^j(p)\frac{\partial F}{\partial u^j}\left(F^{-1}(p)\right), \frac{\partial F}{\partial u^k}\left(F^{-1}(p)\right)\right) \\
&= \sum_{j=1}^{2} \xi^j(p) g_{jk}\left(F^{-1}(p)\right).
\end{aligned}
$$

It follows that

$$
\xi^j \circ F = \sum_{k=1}^{2} g^{jk} \frac{\partial \tilde{f}}{\partial u^k}.
$$

Thus the functions ξ^j are smooth, i.e. $\operatorname{grad} f$ is a smooth vector field.

The gradient of a smooth map is closely connected to the directional derivatives of the map.

Definition 4.2.4 Let S be a regular surface, $p \in S$ a point, $X_p \in T_pS$ a tangent vector and $f : S \to \mathbb{R}$ a smooth map. Then

$$
\partial_{X_p} f := d_p f(X_p) = I(\operatorname{grad} f(p), X_p) \in \mathbb{R}
$$

is called the ***directional derivative of f in the direction X_p***. If X is a vector field on S, then the function

$$
\partial_X f : S \to \mathbb{R}, \quad \partial_X f(p) := \partial_{X(p)} f
$$

is the ***directional derivative of f in direction of the vector field X***.

Exercise 4.8 Let $S = S^1 \times \mathbb{R} \subset \mathbb{R}^3$ be the cylindrical surface with the vector fields $X(x, y, z) = (-y, x, 0)^\top$ and $Y(x, y, z) = (0, 0, 1)^\top$. Find the directional derivatives in the directions X and Y for the functions $f_1(x, y, z) = x$, $f_2(x, y, z) = y$ and $f_3(x, y, z) = z$ on S.

Exercise 4.9 Let S be a regular surface and X and Y two smooth vector fields on S. Show that there exists exactly one vector field Z on S that satisfies

$$
\partial_X(\partial_Y f) - \partial_Y(\partial_X f) = \partial_Z f
$$

for all smooth maps $f : S \to \mathbb{R}$.

Further show that if X and Y are, with respect to a local parametrisation (U, F, V), given by

$$X = \sum_{i=1}^{2} \xi_i \frac{\partial F}{\partial u^i},$$

$$Y = \sum_{i=1}^{2} \eta_i \frac{\partial F}{\partial u^i},$$

then the following relation holds:

$$Z = \sum_{i,j=1}^{2} \left(\xi_i \frac{\partial \eta_j}{\partial u^i} - \eta_i \frac{\partial \xi_j}{\partial u^i} \right) \frac{\partial F}{\partial u^j}.$$

Definition 4.2.5 The vector field

$$Z =: [X, Y]$$

is called ***Lie bracket*** of X and Y.

The vector field $[X, Y]$ is therefore characterised by

$$\partial_X(\partial_Y f) - \partial_Y(\partial_X f) = \partial_{[X,Y]} f$$

for all f.

The formula for $[X, Y]$ w.r.t. a local parametrisation from the above exercise also shows that if X and Y are coordinate fields, i.e. $X = \partial F/\partial u^i$ and $Y = \partial F/\partial u^j$, then the Lie bracket vanishes:

$$\left[\frac{\partial F}{\partial u^i}, \frac{\partial F}{\partial u^j} \right] = 0.$$

This is nothing other than Schwarz's theorem from analysis [18, p. 372, theorem 1.1].

For instance, if one wants to study the velocity field of a curve which lies on a plane, then the concept of vector fields used up to now is not suitable for the following two reasons. Firstly, the velocity field is not defined on the whole surface, but only along the curve. Secondly, the curve may intersect itself, leading to the problem that the velocity vector has two different values at this point of intersection. Considered as a vector field on the surface it would not be uniquely determined at such points. Hence we introduce the following definition.

Definition 4.2.6 Let $S \subset \mathbb{R}^3$ be a regular surface, and let $c: I \to S$ be a parametrised curve. A ***vector field on S along c*** is a map $v: I \to \mathbb{R}^3$, such that $v(t) \in T_{c(t)}S$ for all $t \in I$.

Example 4.2.7 The velocity field $v(t) = \dot{c}(t)$ is such a vector field along c.

Example 4.2.8 Let S be a ruled surface given by the parametrisation $F(t,s) = c(t) + sv(t)$ as in (3.10). Then v is a vector field on S along the curve c.

Now we need a useful concept of the derivative of such vector fields. If one naively differentiates a differentiable vector field v on S along c, then one obtains a map $\dot{v}: I \to \mathbb{R}^3$. The problem now is that $\dot{v}$ is generally not tangential to the surface. As often in mathematics, we solve the problem by imposing the desired property. We replace $\dot{v}$ by the projection of $\dot{v}$ on the tangential plane at the corresponding point of the curve.

Definition 4.2.9 Let $S \subset \mathbb{R}^3$ be a regular surface, let $c: I \to S$ be a parametrised curve, and let $v: I \to \mathbb{R}^3$ be a differentiable vector field on S along c. For every point $p \in S$, let $\Pi_p: \mathbb{R}^3 \to T_pS$ be the orthogonal projection, i.e. if $N(p)$ is one of the two unit normal vectors on S at the point p, then

$$\Pi_p(X) = X - \langle X, N(p)\rangle\, N(p).$$

Then

$$\frac{\nabla}{dt} v(t) := \Pi_{c(t)}\left(\dot{v}(t)\right),$$

$t \in I$, is called the ***covariant derivative*** of v.

Hence, $(\nabla/dt)v$ is also a vector field on S along c.

Example 4.2.10 Let $S = \mathbb{R}^2 \times \{0\}$ be the x–y plane and c a parametrised plane curve, $c(t) = (c_1(t), c_2(t), 0)^\top$. A vector field v on S along c is then of the form $v(t) = (v^1(t), v^2(t), 0)^\top$. It follows that

$$\begin{aligned}
\frac{\nabla}{dt} v(t) &= \Pi_{c(t)}\left(\dot{v}(t)\right) \\
&= \Pi_{c(t)}\left((\dot{v}_1(t), \dot{v}_2(t), 0)^\top\right) \\
&= (\dot{v}_1(t), \dot{v}_2(t), 0)^\top \\
&= \dot{v}(t).
\end{aligned}$$

Thus the usual derivative and the covariant derivative agree in the plane.

Example 4.2.11 Let $S = S^2$ be the sphere. We calculate the covariant derivative of the velocity field of the curve

$$c : \mathbb{R} \to S^2,$$

$$c(t) = (\cos(t), \sin(t), 0)^\top.$$

The curve c passes through the equator of S^2, i.e. the intersection of S^2 with the x–y plane. The usual derivative of c is $\dot{c}(t) = (-\sin(t), \cos(t), 0)^\top$ and

$$\ddot{c}(t) = (-\cos(t), -\sin(t), 0)^\top = -c(t).$$

We observe that for every $t \in \mathbb{R}$ the vector $\ddot{c}(t)$ is perpendicular to $T_{c(t)}S^2$. It follows that

$$\frac{\nabla}{dt}\dot{c}(t) \equiv 0.$$

Exercise 4.10 Let $S = S^2$ be the sphere and

$$c : \mathbb{R} \to S, \quad c(t) = (\cos(t)\cos(\theta), \sin(t)\cos(\theta), \sin(\theta))^\top,$$

with $\theta \in (-\pi/2, \pi/2)$ fixed. The curve c describes a circle of latitude. Show that the covariant derivative of $\dot{c}$ vanishes if and only if $\theta = 0$.

The following calculation rules for covariant derivatives result directly from the definition.

Lemma 4.2.12 *Let S be a regular surface, let $c : I \to S$ be a parametrised curve, let $f : I \to \mathbb{R}$ be a differentiable function and let $\varphi : J \to I$ be a change of parametrisation of c. Further let v and w be differentiable vector fields on S along c. Then $v + w$ and fv are also differentiable vector fields on S along c and we have:*

(a) *additivity:*

$$\frac{\nabla}{dt}(v + w)(t) = \frac{\nabla}{dt}v(t) + \frac{\nabla}{dt}w(t);$$

(b) *product rule I:*

$$\frac{\nabla}{dt}(fv)(t) = \dot{f}(t)v(t) + f(t)\frac{\nabla}{dt}v(t);$$

(c) *product rule II:*

$$\frac{d}{dt}I(v(t), w(t)) = I\left(\frac{\nabla}{dt}v(t), w(t)\right) + I\left(v(t), \frac{\nabla}{dt}w(t)\right);$$

(d) *change of parametrisation:*

$$\frac{\nabla}{dt}(v \circ \varphi) = \dot{\varphi} \cdot \left(\left(\frac{\nabla}{dt} v \right) \circ \varphi \right). \qquad \square$$

One can also find the covariant derivative using local parametrisations. For this purpose we express for a local parametrisation (U, F, V) of S the vectors $(\partial^2 F/\partial u^i \partial u^j)(u) \in \mathbb{R}^3$ in terms of the basis $(\partial F/\partial u^1)(u)$, $(\partial F/\partial u^2)(u)$ and $N(F(u))$:

$$\frac{\partial^2 F}{\partial u^i \partial u^j}(u) = \Gamma^1_{ij}(u) \frac{\partial F}{\partial u^1}(u) + \Gamma^2_{ij}(u) \frac{\partial F}{\partial u^2}(u) + h_{ij}(u) N(F(u)). \tag{4.1}$$

Definition 4.2.13 The coefficient functions

$$\Gamma^k_{ij} : U \to \mathbb{R},$$

$1 \leq i, j, k \leq 2$, are called ***Christoffel symbols***.

From $(\partial^2 F/\partial u^i \partial u^j) = (\partial^2 F/\partial u^j \partial u^i)$ it follows directly that the Christoffel symbols are symmetric in the lower indices:

$$\Gamma^k_{ij} = \Gamma^k_{ji}.$$

Let us now consider the local formula for the covariant derivative. Let (U, F, V) be a local parametrisation of the regular surface S. Let $c : I \to S$ be a parametrised curve. Of course, with the coordinates given by the parametrisation we can only deal with the part of the curve c that lies in V. After possibly reducing the size of I we assume that $c(I) \subset V$. We can now set $\tilde{c} := F^{-1} \circ c : I \to U$. Let $v : I \to \mathbb{R}^3$ be a smooth vector field on S along c. We express v in terms of the basis given by the parametrisation,

$$v(t) = \xi^1(t) \frac{\partial F}{\partial u^1}(\tilde{c}(t)) + \xi^2(t) \frac{\partial F}{\partial u^2}(\tilde{c}(t)).$$

We calculate

$$\begin{aligned} \frac{\nabla}{dt} v(t) &= \Pi_{c(t)} \left(\dot{v}(t) \right) \\ &= \Pi_{c(t)} \Bigg(\sum_{i=1}^{2} \Bigg(\dot{\xi}^i(t) \frac{\partial F}{\partial u^i}(\tilde{c}(t)) + \xi^i(t) \sum_{j=1}^{2} \frac{\partial^2 F}{\partial u^i \partial u^j}(\tilde{c}(t)) \dot{\tilde{c}}^j(t) \Bigg) \Bigg) \end{aligned}$$

$$
\begin{aligned}
&= \sum_{i=1}^{2} \dot{\xi}^i(t) \frac{\partial F}{\partial u^i}(\tilde{c}(t)) + \sum_{i,j,k=1}^{2} \Gamma_{ij}^k(\tilde{c}(t)) \xi^i(t) \dot{\tilde{c}}^j(t) \frac{\partial F}{\partial u^k}(\tilde{c}(t)) \\
&= \sum_{k=1}^{2} \left(\dot{\xi}^k(t) + \sum_{i,j=1}^{2} \Gamma_{ij}^k(\tilde{c}(t)) \xi^i(t) \dot{\tilde{c}}^j(t) \right) \frac{\partial F}{\partial u^k}(\tilde{c}(t)). \qquad (4.2)
\end{aligned}
$$

Expressed in the coefficient functions ξ^1 and ξ^2, the covariant derivative corresponds to the map

$$
\begin{pmatrix} \xi^1 \\ \xi^2 \end{pmatrix} \mapsto \begin{pmatrix} \dot{\xi}^1 + \sum_{i,j=1}^2 \Gamma_{ij}^1(\tilde{c}) \xi^i \dot{\tilde{c}}^j \\ \dot{\xi}^2 + \sum_{i,j=1}^2 \Gamma_{ij}^2(\tilde{c}) \xi^i \dot{\tilde{c}}^j \end{pmatrix}.
$$

The Christoffel symbols determine exactly those correction terms that make the difference between the usual derivative $(\xi^1, \xi^2)^\top \mapsto (\dot{\xi}^1, \dot{\xi}^2)^\top$ of the coefficient functions and the covariant derivative. Next we want to see that the Christoffel symbols can be determined from the first fundamental form and thus that the covariant derivative is a quantity of the inner geometry.

Lemma 4.2.14 *The Christoffel symbols satisfy the following:*

$$
\Gamma_{ij}^k = \frac{1}{2} \sum_{m=1}^{2} \left(\frac{\partial g_{jm}}{\partial u^i} + \frac{\partial g_{im}}{\partial u^j} - \frac{\partial g_{ij}}{\partial u^m} \right) g^{mk}
$$

Proof We calculate

$$
\begin{aligned}
\frac{\partial g_{jm}}{\partial u^i} &= \frac{\partial}{\partial u^i} \left\langle \frac{\partial F}{\partial u^j}, \frac{\partial F}{\partial u^m} \right\rangle \\
&= \left\langle \frac{\partial^2 F}{\partial u^i \partial u^j}, \frac{\partial F}{\partial u^m} \right\rangle + \left\langle \frac{\partial F}{\partial u^j}, \frac{\partial^2 F}{\partial u^i \partial u^m} \right\rangle \\
&= \left\langle \sum_{k=1}^{2} \Gamma_{ij}^k \frac{\partial F}{\partial u^k}, \frac{\partial F}{\partial u^m} \right\rangle + \left\langle \frac{\partial F}{\partial u^j}, \sum_{k=1}^{2} \Gamma_{im}^k \frac{\partial F}{\partial u^k} \right\rangle \\
&= \sum_{k=1}^{2} \left(\Gamma_{ij}^k g_{km} + \Gamma_{im}^k g_{kj} \right). \qquad (4.3)
\end{aligned}
$$

Analogously we obtain

$$
\frac{\partial g_{im}}{\partial u^j} = \sum_{k=1}^{2} \left(\Gamma_{ji}^k g_{km} + \Gamma_{jm}^k g_{ki} \right)
$$

and

$$\frac{\partial g_{ij}}{\partial u^m} = \sum_{k=1}^{2} \left(\Gamma^k_{mi} g_{kj} + \Gamma^k_{mj} g_{ki} \right).$$

It follows that

$$\frac{\partial g_{jm}}{\partial u^i} + \frac{\partial g_{im}}{\partial u^j} - \frac{\partial g_{ij}}{\partial u^m} = 2 \sum_{k=1}^{2} \Gamma^k_{ij} g_{km}.$$

By multiplying with the inverse matrix (g^{km}) one solves for Γ^k_{ij} and obtains the equation of the lemma. □

Now we know how to work with vector fields along curves in a covariant way. How can we now differentiate traditional vector fields on a surface? For this we also need to fix the direction in which we want to differentiate.

Definition 4.2.15 Let S be a regular surface, v a differentiable vector field on S and $w_p \in T_pS$ a tangent vector. Then the ***covariant derivative*** $\nabla_{w_p} v \in T_pS$ of v in direction w_p is defined as follows:

choose a parametrised curve $c : (-\varepsilon, \varepsilon) \to S$ with $\dot{c}(0) = w_p$ and set

$$\nabla_{w_p} v := \frac{\nabla}{dt}(v \circ c)(0).$$

Exercise 4.11 Show that the definition does not depend on the choice of the parametrised curve c with $\dot{c}(0) = w_p$.

Hint Show that the covariant derivative in direction $w_p = \sum_k \eta^k (\partial F / \partial u^k)(u)$, expressed in a local parametrisation (U, F, V), corresponds to the map

$$\begin{pmatrix} \xi^1 \\ \xi^2 \end{pmatrix} \mapsto \begin{pmatrix} d\xi^1 \cdot (\eta^1, \eta^2)^\top + \sum_{i,j=1}^{2} \Gamma^1_{ij} \xi^i \eta^j \\ d\xi^2 \cdot (\eta^1, \eta^2)^\top + \sum_{i,j=1}^{2} \Gamma^2_{ij} \xi^i \eta^j \end{pmatrix},$$

where the parametrisation maps u to p. More precisely,

$$\nabla_{w_p} \left(\sum_k \xi^k \frac{\partial F}{\partial u^k} \right) = \sum_k \Big(\sum_\ell \frac{\partial \xi^k}{\partial u^\ell}(u) \eta^\ell + \sum_{i,j=1}^{2} \Gamma^k_{ij}(u) \xi^i(u) \eta^j \Big) \frac{\partial F}{\partial u^k}(u).$$

Definition 4.2.16 If v and w are two vector fields on S, then we define a new vector field $\nabla_w v$ by

$$(\nabla_w v)(p) := \nabla_{w(p)} v.$$

Lemma 4.2.17 *Let S be a regular surface, let $c_1, c_2 \in \mathbb{R}$, v, v_1, v_2, w, w_1 and w_2 be differentiable vector fields on S, and let $f : S \to \mathbb{R}$ be a differentiable function.*

Then the following hold:

(a) *linearity in the vector field that is differentiated:*

$$\nabla_w(c_1 v_1 + c_2 v_2) = c_1 \nabla_w v_1 + c_2 \nabla_w v_2;$$

(b) *product rule I:*

$$\nabla_w(fv) = df(w)v + f\nabla_w v;$$

(c) *product rule II:*

$$\partial_w(I(v_1, v_2)) = I(\nabla_w v_1, v_2) + I(v_1, \nabla_w v_2);$$

(d) *linearity in the vector field w.r.t. which we differentiate:*

$$\nabla_{(c_1 w_1 + c_2 w_2)} v = c_1 \nabla_{w_1} v + c_2 \nabla_{w_2} v;$$

(e) *linearity w.r.t. functions on the vector field w.r.t. which we differentiate:*

$$\nabla_{fw} v = f\nabla_w v.$$

Proof Properties (a), (b) and (c) follow directly from the corresponding properties of the covariant derivatives of vector fields along curves in lemma 4.2.12. Points (d) and (e) are most easily deduced from the formula in terms of a local parametrisation, given in the above exercise. □

4.3 Riemann curvature tensor and *Theorema Egregium*

In chapter 3 we met several notions of curvature, such as the principal curvatures, the mean curvature and the Gauss curvature. In this section we will introduce a new quantity, the Riemann curvature tensor. We will begin by looking at the second covariant derivative. The value of the second covariant derivative depends on the order in which we differentiate, i.e. the theorem of Schwarz about the interchangeability of directional derivatives does not hold for the covariant derivative! The Riemann curvature tensor measures exactly this error that arises when interchanging the derivatives.

To begin with, we will examine how the second covariant derivative should be defined correctly. If v, w and z are vector fields on a regular surface S, then one can of course differentiate the vector field $\nabla_w z$ once more with respect to v in a covariant way. But this will introduce derivatives of w with respect to v. Hence, if we are only interested in the second derivative of z in direction

v and w, then we have to compensate for this effect and make the following definition.

Definition 4.3.1 The ***second covariant derivative*** of z with respect to v and w is defined via

$$\nabla^2_{v,w} z := \nabla_v(\nabla_w z) - \nabla_{\nabla_v w} z.$$

Let us now look at the second covariant derivative in a local parametrisation.

Lemma 4.3.2 *Let S be a regular surface and v, w and z be vector fields on S. Let (U, F, V) be a local parametrisation of S. As usual we express v in the basis given by the parametrisation, $v = \sum_{i=1}^2 v^i(\partial F/\partial u^i)$, and analogously for all other vector fields.*

Then, w.r.t. the basis $\partial F/\partial u^m$, $\nabla^2_{v,w} z$ is given by the coefficients

$$\left(\sum_{i,j} \frac{\partial^2 z^m}{\partial u^i \partial u^j} v^i w^j + \sum_{i,j,k} \Gamma^m_{ij} \frac{\partial z^i}{\partial u^k} (v^j w^k + v^k w^j) - \sum_{i,j,k} \Gamma^k_{ij} \frac{\partial z^m}{\partial u^k} v^i w^j \right.$$

$$\left. + \sum_{i,j,k} \left(\frac{\partial \Gamma^m_{kj}}{\partial u^i} + \sum_{\ell} \left(\Gamma^m_{\ell i} \Gamma^\ell_{kj} - \Gamma^m_{k\ell} \Gamma^\ell_{ij} \right) v^i w^j z^k \right) \right)_{m=1,2}.$$

The first sum contains exactly the usual second covariant derivative in directions $(v^1, v^2)^\top$ and $(w^1, w^2)^\top$. The other sums are correction terms involving lower-order derivatives of $(z^1, z^2)^\top$. We also observe that there are indeed no derivatives of $(v^1, v^2)^\top$ and $(w^1, w^2)^\top$. Thus

Corollary 4.3.3 *The value of the second covariant derivative $\nabla^2_{v,w} z$ at a point $p \in S$ depends only on $v(p)$, $w(p)$ and the derivatives of z at p up to order 2.*

For a vector field z on S we can therefore define the second covariant differential of z as

$$\nabla^2 z : T_pS \times T_pS \to T_pS,$$
$$(v_p, w_p) \mapsto \left(\nabla^2_{v,w} z\right)(p),$$

where v and w are arbitrary vector fields on S with $v(p) = v_p$ and $w(p) = w_p$.

Proof of lemma 4.3.2 The proof consists of thorough calculations, in which it is most important to keep a level head. The vector field $\nabla_w z$ has the components

$$\Big(\sum_{\ell}\frac{\partial z^k}{\partial u^\ell}w^\ell+\sum_{ij}\Gamma_{ij}^k z^i w^j\Big)_{k=1,2}=:\big(\nu^k\big)_{k=1,2}.$$

For $\nabla_v(\nabla_w z)$ we obtain

$$\begin{aligned}
&\Big(\sum_m \frac{\partial \nu^\alpha}{\partial u^m}v^m+\sum_{\beta\gamma}\Gamma_{\beta\gamma}^\alpha \nu^\beta v^\gamma\Big)_{\alpha=1,2}\\
&=\Big(\sum_{m\ell}\Big(\frac{\partial^2 z^\alpha}{\partial u^\ell\partial u^m}w^\ell v^m+\frac{\partial z^\alpha}{\partial u^\ell}\frac{\partial w^\ell}{\partial u^m}v^m\Big)\\
&\quad+\sum_{ijm}\Big(\frac{\partial\Gamma_{ij}^\alpha}{\partial u^m}z^i w^j v^m+\Gamma_{ij}^\alpha\frac{\partial z^i}{\partial u^m}w^j v^m+\Gamma_{ij}^\alpha z^i\frac{\partial w^j}{\partial u^m}v^m\Big)\\
&\quad+\sum_{\ell\beta\gamma}\frac{\partial z^\beta}{\partial u^\ell}w^\ell\Gamma_{\beta\gamma}^\alpha v^\gamma+\sum_{ij\beta\gamma}\Gamma_{ij}^\beta\Gamma_{\beta\gamma}^\alpha z^i w^j v^\gamma\Big)_{\alpha=1,2}.
\end{aligned}\tag{4.4}$$

For $\nabla_v w$ we write

$$\Big(\sum_{\ell}\frac{\partial w^k}{\partial u^\ell}v^\ell+\sum_{ij}\Gamma_{ij}^k v^i w^j\Big)_{k=1,2}=:\big(\mu^k\big)_{k=1,2}$$

and for $\nabla_{\nabla_v w}z$ we obtain

$$\begin{aligned}
&\Big(\sum_m \frac{\partial z^\alpha}{\partial u^m}\mu^m+\sum_{\beta\gamma}\Gamma_{\beta\gamma}^\alpha z^\beta \mu^\gamma\Big)_{\alpha=1,2}\\
&=\Big(\sum_{m\ell}\frac{\partial z^\alpha}{\partial u^m}\frac{\partial w^m}{\partial u^\ell}v^\ell+\sum_{mij}\frac{\partial z^\alpha}{\partial u^m}\Gamma_{ij}^m v^i w^j\\
&\quad+\sum_{\ell\beta\gamma}\frac{\partial w^\gamma}{\partial u^\ell}v^\ell\Gamma_{\beta\gamma}^\alpha z^\beta+\sum_{ij\beta\gamma}\Gamma_{ij}^\gamma\Gamma_{\beta\gamma}^\alpha v^i w^j z^\beta\Big)_{\alpha=1,2}.
\end{aligned}\tag{4.5}$$

Subtraction of (4.4) and (4.5) cancels terms involving derivatives of v and w, and we obtain the claim. □

As the first covariant derivative is a concept that depends only on the inner geometry, the same is true for the second covariant derivative.

As announced before we will now define the curvature tensor as the error term that measures the non-interchangeability of the two arguments in the second covariant derivative.

Definition 4.3.4 Let S be a regular surface, $p \in S$ a point, $v_p, w_p \in T_pS$ tangent vectors, and z a vector field on S. Then the ***Riemann (curvature) tensor*** R, also called just the ***curvature tensor***, is defined by

$$R(v_p, w_p)z := \nabla^2_{v_p,w_p} z - \nabla^2_{w_p,v_p} z.$$

We can express the Riemann curvature tensor in terms of a local parametrisation as well. For this purpose let (U, F, V) be a local parametrisation of our regular surface, $p = F(u_0)$. We express v_p, w_p and z in terms of a local parametrisation:

$$v_p = \sum_i v^i \frac{\partial F}{\partial u^i}(u_0),$$

$$w_p = \sum_j w^j \frac{\partial F}{\partial u^j}(u_0),$$

$$z = \sum_k z^k \frac{\partial F}{\partial u^k}.$$

Lemma 4.3.5 *The Riemann curvature tensor w.r.t. a local parametrisation has the form*

$$R(v_p, w_p)z = \sum_{ijk\ell=1}^{2} R^\ell_{ijk}(u_0) v^i w^j z^k \frac{\partial F}{\partial u^\ell}(u_0),$$

where

$$R^\ell_{ijk} = \frac{\partial \Gamma^\ell_{kj}}{\partial u^i} - \frac{\partial \Gamma^\ell_{ki}}{\partial u^j} + \sum_m \left(\Gamma^\ell_{mi}\Gamma^m_{kj} - \Gamma^\ell_{mj}\Gamma^m_{ki} \right).$$

Proof All terms from lemma 4.3.2 that involve derivatives of z are symmetric in v_p and w_p and therefore cancel. Considering further that the Christoffel symbols Γ^k_{ij} are symmetric in the lower indices i and j proves the claim. □

As a direct implication we have the following corollary.

Corollary 4.3.6 (a) *The tangent vector $R(v_p, w_p)z$ at $p \in S$ depends only on $z(p)$, not on the values of the vector field z on $S - \{p\}$. Then the map*

$$R_p : T_pS \times T_pS \times T_pS \to T_pS,$$

$$R_p(v_p, w_p)z_p := R(v_p, w_p)z,$$

is well defined, where z is an arbitrary vector field on S with $z(p) = z_p$.

(b) R_p *is linear in each argument.*

(c) R_p *is skew-symmetric in the first two arguments,*

$$R_p(v_p, w_p)z_p = -R_p(w_p, v_p)z_p.$$

It can at times be quite involved to calculate the curvature tensor using the Christoffel symbols. It is therefore useful to know that this can also be done using the second fundamental form and the Weingarten map.

Theorem 4.3.7 (Gauss's equation) *Let $S \subset \mathbb{R}^3$ be an oriented regular surface, $p \in S$. Then any $v, w, z \in T_pS$ satisfy the following:*

$$R(v, w)z = II(w, z) \cdot W(v) - II(v, z) \cdot W(w).$$

With respect to a local parametrisation this is given by

$$R_{ijk}^{\ell} = h_{jk} w_i^{\ell} - h_{ik} w_j^{\ell}.$$

Proof We prove the local version. Let (U, F, V) be a local parametrisation. We recall (4.1):

$$\frac{\partial^2 F}{\partial u^i \partial u^j} = \sum_k \Gamma_{ij}^k \frac{\partial F}{\partial u^k} + h_{ij} \cdot (N \circ F).$$

Differentiating this equation with respect to u^ℓ we obtain

$$\begin{aligned}
\frac{\partial^3 F}{\partial u^\ell \partial u^i \partial u^j} &= \sum_k \left(\frac{\partial \Gamma_{ij}^k}{\partial u^\ell} \frac{\partial F}{\partial u^k} + \Gamma_{ij}^k \frac{\partial^2 F}{\partial u^\ell \partial u^k} \right) \\
&\quad + \frac{\partial h_{ij}}{\partial u^\ell} \cdot (N \circ F) + h_{ij} \cdot \frac{\partial (N \circ F)}{\partial u^\ell} \\
&= \sum_k \left(\frac{\partial \Gamma_{ij}^k}{\partial u^\ell} \frac{\partial F}{\partial u^k} + \Gamma_{ij}^k \sum_m \Gamma_{\ell k}^m \frac{\partial F}{\partial u^m} + \text{normal component} \right) \\
&\quad + \text{normal component} + h_{ij} \cdot \left(-W \left(\frac{\partial F}{\partial u^\ell} \right) \right) \\
&= \sum_m \left(\frac{\partial \Gamma_{ij}^m}{\partial u^\ell} + \sum_k \Gamma_{ij}^k \Gamma_{\ell k}^m - h_{ij} w_\ell^m \right) \frac{\partial F}{\partial u^m} \\
&\quad + \text{normal component}. \qquad (4.6)
\end{aligned}$$

By the theorem of Schwarz we can interchange the derivations with respect to u^i and u^ℓ and obtain by (4.6) that

$$\begin{aligned}0 &= \frac{\partial^3 F}{\partial u^\ell \partial u^i \partial u^j} - \frac{\partial^3 F}{\partial u^i \partial u^\ell \partial u^j} \\ &= \sum_m \left(\frac{\partial \Gamma_{ij}^m}{\partial u^\ell} - \frac{\partial \Gamma_{\ell j}^m}{\partial u^i} + \sum_k \left(\Gamma_{ij}^k \Gamma_{\ell k}^m - \Gamma_{\ell j}^k \Gamma_{ik}^m \right) - h_{ij} w_\ell^m + h_{\ell j} w_i^m \right) \frac{\partial F}{\partial u^m} \\ &\quad + \text{normal component} \\ &= \sum_m \left(R_{\ell ij}^m - h_{ij} w_\ell^m + h_{\ell j} w_i^m \right) \frac{\partial F}{\partial u^m} + \text{normal component.}\end{aligned}$$

It follows that

$$R_{\ell ij}^m - h_{ij} w_\ell^m + h_{\ell j} w_i^m = 0.$$

This is the claim up to renaming of the indices. □

A remarkable feature of Gauss's equation is that the left hand side, i.e. the Riemann curvature tensor, is a quantity of the inner geometry, while the quantities on the right hand side of the equation, i.e. the second fundamental form and the Weingarten map, are quantities that do not only depend on the inner geometry of the surface. This now allows us to prove that Gauss curvature, unlike mean curvature, is a quantity of the inner geometry of the surface.

Theorem 4.3.8 (Theorema Egregium) *Gauss curvature can be calculated form the Riemann curvature tensor as follows: let $p \in S$ be a point. Choose an orthonormal basis v, w of T_pS. Then*

$$K(p) = I(R_p(v, w)w, v).$$

In particular, Gauss curvature depends on the inner geometry of the surface only.

Proof According to the Gauss equation we have

$$\begin{aligned}I(R(v, w)w, v) &= I(II(w, w) \cdot W(v) - II(v, w) \cdot W(w), v) \\ &= II(w, w)II(v, v) - II(v, w)II(w, v) \\ &= \det(W) \\ &= K.\end{aligned}$$

□

Example 4.3.9 The two regular surfaces $S_1 = \{(x, y, 0)^\top \mid x^2 + y^2 < 1\}$ (circular disc) and $S_2 = \{(x, y, z)^\top \mid x^2 + y^2 < 1, z = \sqrt{1 - x^2 - y^2}\}$ (hemisphere) are diffeomorphic. The projection $(x, y, z)^\top \mapsto (x, y, 0)^\top$ gives a diffeomorphism from S_2 to S_1. However, there cannot be an isometry between those two surfaces, since S_1 has Gauss curvature $K_{S_1} \equiv 0$, while for S_2 we have $K_{S_2} \equiv 1$. Since, as we now know, Gauss curvature is preserved under isometries, such an isometry from S_1 to S_2 cannot exist. Our "inhabitant of the surface" from the beginning of this chapter can therefore use the Gauss curvature of the surface in order to determine whether he lives on a circular disc or on a sphere.

This example is of practical significance, since it shows that in principle it is impossible to draw absolutely correct maps. Such a map would have to transfer ratios of lengths of a part of the sphere to scale to a part of the plane. But this would be (up to a scaling) an isometry between two surfaces of different curvatures. For this reason every map will distort some lengths. The larger the area of the earth represented by the map, the larger the distortion. See section 4.10 for more on this.

Lemma 4.3.10 *Let S be a regular surface, let $p \in S$, let $v, w, x, y \in T_pS$. The curvature tensor has the following symmetries:*

(a) $R(v,w)x = -R(w,v)x$;
(b) $I(R(v,w)x,y) = -I(R(v,w)y,x)$;
(c) $I(R(v,w)x,y) = I(R(x,y)v,w)$;
(d) *Bianchi identity:*

$$R(v,w)x + R(x,v)w + R(w,x)v = 0.$$

Proof Part (a) is trivial. Statement (c) follows from Gauss's equation:

$$\begin{aligned} I(R(v,w)x,y) &= I(II(w,x)\cdot W(v) - II(v,x)\cdot W(w), y) \\ &= II(w,x)\cdot II(v,y) - II(v,x)\cdot II(w,y), \end{aligned}$$

since this expression does not change when interchanging the pairs (v,w) and (x,y). Part (b) follows directly from (a) and (c). Statement (d) can also be derived from Gauss's equation:

$$\begin{aligned} &R(v,w)x + R(x,v)w + R(w,x)v \\ &\quad = II(w,x)W(v) - II(v,x)W(w) + II(v,w)W(x) \\ &\qquad - II(x,w)W(v) + II(x,v)W(w) - II(w,v)W(x) \\ &\quad = 0. \end{aligned}$$

□

Exercise 4.12 Prove that Gauss curvature with respect to a local parametrisation is given by

$$K = \frac{1}{2}\sum_{ijk} g^{jk} R^i_{ijk}.$$

The *Theorema Egregium* and the last statement tell us how to calculate Gauss curvature from the curvature tensor. Conversely, the curvature tensor can be obtained from the Gauss curvature.

Lemma 4.3.11 *Let S be a regular surface, let $p \in S$. For all $v, w, x \in T_pS$ we have*

$$R(v,w)x = K(p) \cdot (I(w,x)v - I(v,x)w).$$

In local coordinates,

$$R_{ijk}^{\ell} = K \cdot \left(g_{jk}\delta_i^{\ell} - g_{ik}\delta_j^{\ell} \right).$$

Proof (a) Let V be a two-dimensional real vector space. We first show that the vector space of all multi-linear maps

$$\mathscr{R} : V \times V \times V \times V \to \mathbb{R}$$

that satisfy the symmetries

$$\mathscr{R}(v,w,x,y) = -\mathscr{R}(w,v,x,y) \quad \text{and} \quad \mathscr{R}(v,w,x,y) = -\mathscr{R}(v,w,y,x)$$

is one-dimensional. We choose a basis e_1, e_2 of V and write $v = v^1e_1 + v^2e_2$, $w = w^1e_1 + w^2e_2$, $x = x^1e_1 + x^2e_2$ and $y = y^1e_1 + y^2e_2$. It follows that

$$\begin{aligned}\mathscr{R}(v,w,x,y) &= (v^1w^2 - v^2w^1)\mathscr{R}(e_1,e_2,x,y) \\ &= (v^1w^2 - v^2w^1)(x^1y^2 - x^2y^1)\mathscr{R}(e_1,e_2,e_1,e_2).\end{aligned}$$

Hence $\mathscr{R}$ is determined by the coefficient $\mathscr{R}(e_1,e_2,e_1,e_2)$.

(b) By lemma 4.3.10

$$\mathscr{R}_1(v,w,x,y) := I(R(v,w)x,y)$$

satisfies those symmetries. One sees directly that this is also true for

$$\mathscr{R}_2(v,w,x,y) := K(p) \cdot (I(w,x)I(v,y) - I(v,x)I(w,y)).$$

$\mathscr{R}_1$ and $\mathscr{R}_2$ are linearly dependent by (a). By the *Theorema Egregium*, an orthonormal basis e_1, e_2 of T_pS satisfies

$$\mathscr{R}_1(e_1,e_2,e_1,e_2) = -K(p) = \mathscr{R}_2(e_1,e_2,e_1,e_2).$$

Hence $\mathscr{R}_1$ and $\mathscr{R}_2$ agree. The claim follows. □

We see that Gauss curvature and curvature tensor can be calculated from each other and hence contain the same information about the surface.

Table 4.1 *Quantities that depend on the inner geometry only*

geometric quantity	symbol	expression w.r.t. local param.
first fundamental form	I	g_{ij}
surface element	dA	$dA = \sqrt{g_{11}g_{22} - (g_{12})^2}du^1 du^2$
covariant derivative	∇	$\Gamma_{ij}^k = \frac{1}{2}\sum_m \left(\frac{\partial g_{jm}}{\partial u^i} + \frac{\partial g_{im}}{\partial u^j} - \frac{\partial g_{ij}}{\partial u^m}\right) g^{mk}$
Riemann curvature tensor	R	$R_{ijk}^\ell = \frac{\partial \Gamma_{kj}^\ell}{\partial u^i} - \frac{\partial \Gamma_{ki}^\ell}{\partial u^j} + \sum_m \left(\Gamma_{mi}^\ell \Gamma_{kj}^m - \Gamma_{mj}^\ell \Gamma_{ki}^m\right)$
Gauss curvature	K	$K = \frac{1}{2}\sum_{ijk} g^{jk} R_{ijk}^i$

Table 4.2 *Quantities that are not invariant under local isometries*

geometric quantity	symbol	expression w.r.t. local param.
second fundamental form	II	h_{ij}
Weingarten map	W	$w_i^j = \sum_k h_{ik} g^{kj}$
principal curvatures	κ_i	
mean curvature	H	$H = \frac{\kappa_1 + \kappa_2}{2} = \frac{w_1^1 + w_2^2}{2}$

The most important geometric quantities of a regular surface are summarised in tables 4.1 and 4.2.

4.4 Riemannian metrics

Since we will as of now just look at quantities that depend on the inner geometry only, we will at this point generalise the concept of the first fundamental form. What actually is the first fundamental form? It assigns a Euclidean scalar product to each tangent plane. That this scalar product is the restriction of the standard scalar product on $\mathbb{R}^3$ is of no importance for quantities that depend on the inner geometry only. We therefore make the following definition.

Definition 4.4.1 Let $S \subset \mathbb{R}^3$ be a regular surface. A ***Riemannian metric*** g on S assigns a Euclidean scalar product g_p on the tangent plane T_pS to each point $p \in S$, such that for every local parametrisation (U, F, V) of S the functions

$$g_{ij} : U \to \mathbb{R},$$

$$g_{ij}(u) := g_{F(u)}\left(\frac{\partial F}{\partial u^i}(u), \frac{\partial F}{\partial u^j}(u)\right)$$

are smooth.

The first fundamental form is, of course, an example of a Riemannian metric. But there are important examples of Riemannian metrics that are not the first fundamental form, as we will see later. One of them is the hyperbolic plane. The condition that the functions g_{ij} must be smooth ensures the scalar product g_p does not "wildly" depend on p, but is differentiable w.r.t. p. In the case of the first fundamental form this holds automatically.

All quantities that depend on the inner geometry only, such as the surface element, the covariant derivative, the Gauss curvature and the Riemann curvature tensor, are also defined for regular surfaces with a Riemannian metric that is not the first fundamental form.

We therefore *define*, for example, the Christoffel symbols via the formula from lemma 4.2.14 as

$$\Gamma_{ij}^k := \frac{1}{2} \sum_{m=1}^{2} \left(\frac{\partial g_{jm}}{\partial u^i} + \frac{\partial g_{im}}{\partial u^j} - \frac{\partial g_{ij}}{\partial u^m} \right) g^{mk} \tag{4.7}$$

and then the covariant derivative for vector fields

$$v = \sum_i v^i \frac{\partial F}{\partial u^i}, w = \sum_j w^j \frac{\partial F}{\partial u^j}$$

via

$$\nabla_w v(F(u)) := \sum_k \left(d_u v^k \begin{pmatrix} w^1(u) \\ w^2(u) \end{pmatrix} + \sum_{ij} \Gamma_{ij}^k(u) v^i(u) w^j(u) \right) \frac{\partial F}{\partial u^k}(u).$$

Of course, it needs to be checked that this definition does not depend on the local parametrisation chosen; this is left as an exercise for the reader. In the case of the first fundamental form this formula holds automatically. We can analogously translate the other quantities that depend on the inner geometry, such as the Riemann curvature tensor and the Gauss curvature. We see that it was good to have required the smoothness of the g_{ij}. For the definition of the covariant derivative we need the first derivatives of the g_{ij}, for the curvature we also need the second derivatives.

Exercise 4.13 Show that lemma 4.2.12 and lemma 4.2.17 also hold for general Riemannian metrics.

Exercise 4.14 Show that for any two smooth vector fields v and w on a regular surface equipped with a Riemannian metric one has

$$\nabla_v w - \nabla_w v = [v, w].$$

Exercise 4.15 Show that lemma 4.3.10 and lemma 4.3.11 also hold for general Riemannian metrics.

Example 4.4.2 We consider the torus S from example 3.8.18 with the parametrisation

$$F(t,\varphi) = \begin{pmatrix} (1 - r\cos(\varphi))\cos(t) \\ (1 - r\cos(\varphi))\sin(t) \\ r\sin(\varphi) \end{pmatrix}.$$

The vector fields $\partial F/\partial t$ and $\partial F/\partial \varphi$ are defined on the whole of S. The two vectors $(\partial F/\partial t)(t,\varphi)$ and $(\partial F/\partial \varphi)(t,\varphi)$ are a basis of the tangent plane $T_{F(t,\varphi)}S$, but they are generally not orthonormal with respect to the first fundamental form. We now *define* a new Riemannian metric, determining the Euclidean scalar product $g_{F(t,\varphi)}$ on $T_{F(t,\varphi)}S$ by requiring that $(\partial F/\partial t)(t,\varphi)$ and $(\partial F/\partial \varphi)(t,\varphi)$ must be orthonormal. For $p \in S$ this uniquely determines a Euclidean scalar product on T_pS, since although $p = F(t,\varphi) = F(t',\varphi')$ may be possible for different parameter values, the corresponding tangent vectors will agree, $(\partial F/\partial t)(t,\varphi) = (\partial F/\partial t)(t',\varphi')$ and $(\partial F/\partial \varphi)(t,\varphi) = (\partial F/\partial \varphi)(t',\varphi')$.

In this way we obtain a well-defined Riemannian metric on the torus. What is the Gauss curvature for this Riemannian metric? Well, the definition is formulated such that with the parametrisation from above the Riemannian metric has components

$$(g_{ij})_{ij} = \begin{pmatrix} 1 & 0 \\ 0 & 1 \end{pmatrix}.$$

The fact that all the g_{ij} are constant causes the Christoffel symbols to vanish, and hence

$$K \equiv 0.$$

Note that this is not possible for the first fundamental form on a compact regular surface, since because of theorem 3.6.17 the Gauss curvature would then have to be positive somewhere.

This construction of a Riemannian metric with $K \equiv 0$ can be done on all regular surfaces on which there are two smooth vector fields that, at each point, form a basis of the corresponding tangent plane.

A very different approach for the construction of Riemannian metrics consists of *pulling back* a metric that is defined on a different, diffeomorphic surface.

Definition 4.4.3 Let S_1 and S_2 be regular surfaces, let $\Phi : S_1 \to S_2$ be a diffeomorphism. Let g be a Riemannian metric on S_2. The ***pullback Riemannian metric*** Φ^*g on S_1 is defined by

$$(\Phi^*g)_p(X,Y) := g_{\Phi(p)}(d_p\Phi(X), d_p\Phi(Y))$$

for all $p \in S_1$, $X, Y \in T_pS_1$.

One easily sees that Φ^*g is a Riemannian metric on S_1. Indeed, Φ^*g is the unique Riemannian metric on S_1 for which $\Phi : S_1 \to S_2$ is an *isometry*. If F is a local parametrisation of S_1, then $\Phi \circ F$ is a local parametrisation of S_2. From the definition of Φ^*g it follows immediately that Φ^*g in the coordinates given by F looks exactly like g with respect to $\Phi \circ F$:

$$(\Phi^* g)_{ij} = g_{ij}.$$

Accordingly, all quantities that only depend on the inner geometry agree, e.g. for the Gauss curvature we have

$$K_{\Phi^* g} = K_g \circ \Phi.$$

Example 4.4.4 As we have seen in example 3.1.16, there is a diffeomorphism $\Phi : S^2 \to S$, where S is an ellipsoid. We can then pull back the first fundamental form of the ellipsoid to the sphere. This gives a Riemannian metric on S^2 that does not agree with the first fundamental form of S^2, since the Gauss curvature of the ellipsoid is not constant, and hence the one of the pulled-back metric is not constant either.

Exercise 4.16 Let $\kappa \in \mathbb{R}$ be a constant. In the case $\kappa \geq 0$ let $S = \mathbb{R}^2 \times \{0\}$ be the x–y plane; in the case $\kappa < 0$; on the other hand, let $S = \{(x, y, 0)^\top \mid x^2 + y^2 < -4/\kappa\}$. On S let the Riemannian metric g in Cartesian coordinates be given by

$$(g_{ij}(x, y))_{ij} = \frac{1}{\left(1 + \kappa(x^2 + y^2)/4\right)^2} \cdot \begin{pmatrix} 1 & 0 \\ 0 & 1 \end{pmatrix}.$$

Show that S with this Riemannian metric has constant Gauss curvature $K \equiv \kappa$.

4.5 Geodesics

We now want to investigate the following problem. In the plane the shortest connecting curve between two points is the corresponding segment of a straight line. What do shortest connecting curves look like on a regular surface?

Definition 4.5.1 Let S be a regular surface with Riemannian metric g. Let $c : I \to S$ be a parametrised curve. Then the ***length*** of c (w.r.t. (S, g)) is defined by

$$L[c] := \int_I \sqrt{g_{c(t)}(\dot{c}(t), \dot{c}(t))}dt.$$

If g is the first fundamental form, then this notion of length agrees with the one that we defined for $c : I \to S \subset \mathbb{R}^3$ as a space curve in chapter 2. Otherwise there is no connection. It is beneficial to look not only at the length, but also at the energy of curves.

Definition 4.5.2 Let S be a regular surface with Riemannian metric g. Let $c : I \to S$ be a parametrised curve. Then the ***energy*** of c (w.r.t. (S, g)) is defined by

$$E[c] := \frac{1}{2} \int_I g_{c(t)}(\dot{c}(t), \dot{c}(t))dt.$$

Exercise 4.17 Show that the length of a parametrised curve does not change under parameter transformations, while the energy of a curve does.

We can compare those two concepts as follows.

Lemma 4.5.3 *Let S be a regular surface with Riemannian metric g. Let $c : [a, b] \to S$ be a parametrised curve. Then*

$$L[c]^2 \leq 2(b-a)E[c]$$

where the equality applies if and only if c is parametrised proportional to arc-length, i.e. if

$$g_{c(t)}(\dot{c}(t), \dot{c}(t)) \equiv \text{const.}$$

Proof We set $f : [a, b] \to \mathbb{R}$, $f(t) := \sqrt{g_{c(t)}(\dot{c}(t), \dot{c}(t))}$. By the Cauchy–Schwarz inequality we have

$$\begin{aligned} L[c]^2 &= \Big(\int_a^b f(t) \cdot 1 dt\Big)^2 \\ &\leq \int_a^b f(t)^2 dt \cdot \int_a^b 1^2 dt \\ &= 2 \cdot E[c] \cdot (b-a). \end{aligned}$$

We have equality precisely if f and 1 are linearly dependent, i.e. if f is constant. This means exactly that c is parametrised proportional to arc-length. □

In conclusion we can note that a connecting curve has minimal energy if and only if it has minimal length and is parametrised proportional to arc-length. It is for this reason that cyclists like to jump red traffic lights. Although the distance covered would not be increased by stopping, the energy needed would be higher.

We will in the following work with energy instead of with length. When studying minimal surfaces, we derived a variation formula for surface area (theorem 3.8.8). We will now derive a similar variation formula for the energy of curves. Let us first prove a lemma.

Lemma 4.5.4 *Let S be a regular surface with Riemannian metric g. Let $c : I \times J \to S$, $(s,t) \mapsto c(s,t)$, be a smooth map. Then*

$$\frac{\nabla}{\partial s}\frac{\partial c}{\partial t} = \frac{\nabla}{\partial t}\frac{\partial c}{\partial s}.$$

Proof From the formula

$$\Gamma_{ij}^k = \frac{1}{2}\sum_{m=1}^{2}\left(\frac{\partial g_{jm}}{\partial u_i} + \frac{\partial g_{im}}{\partial u_j} - \frac{\partial g_{ij}}{\partial u_m}\right) g^{mk}$$

we see that in the case of general Riemannian metrics the Christoffel symbols are also symmetric in the lower indices i and j. Let (U,F,V) be a local parametrisation. We set

$$u : c^{-1}(V) \subset I \times J \to U, \quad u := F^{-1} \circ c.$$

We have

$$c = F \circ u$$

and hence

$$\frac{\partial c}{\partial t} = \sum_k \frac{\partial u^k}{\partial t}\frac{\partial F}{\partial u^k}.$$

It follows that

$$\frac{\nabla}{\partial s}\frac{\partial c}{\partial t} = \sum_k \frac{\partial^2 u^k}{\partial t \partial s}\frac{\partial F}{\partial u^k} + \sum_{ijk} \Gamma_{ij}^k \frac{\partial u^i}{\partial t}\frac{\partial u^j}{\partial s}\frac{\partial F}{\partial u^k}.$$

This expression is symmetric in s and t because of the symmetry of the Christoffel symbols in the lower indices i and j. □

We can now move on to the variation formula.

Theorem 4.5.5 (Variation of energy) *Let S be a regular surface with Riemannian metric g. Let $p, q \in S$. Let $c : (-\varepsilon, \varepsilon) \times [a,b] \to S$ be a smooth map such that for $c_s : [a,b] \to S$, $c_s(t) := c(s,t)$ we have*

$$c_s(a) = p, \quad c_s(b) = q.$$

Let $V(t) := (\partial c/\partial s)(0,t)$ be the so-called variation vector field. Then:

$$\frac{d}{ds}E[c_s]\bigg|_{s=0} = -\int_a^b g_{c_0(t)}\left(V(t), \frac{\nabla}{dt}\dot{c}_0(t)\right) dt.$$

Proof We differentiate inside the integral and use lemma 4.5.4:

$$\begin{aligned}
&\frac{d}{ds}E[c_s]\bigg|_{s=0} \\
&= \frac{1}{2}\int_a^b \frac{d}{ds} g_{c_s(t)}(\dot{c}_s(t), \dot{c}_s(t))\big|_{s=0}\, dt \\
&= \frac{1}{2}\int_a^b \left(g_{c_0(t)}\left(\frac{\nabla}{\partial s}\dot{c}_s(t)\bigg|_{s=0}, \dot{c}_0(t)\right) + g_{c_0(t)}\left(\dot{c}_0(t), \frac{\nabla}{\partial s}\dot{c}_s(t)\bigg|_{s=0}\right)\right) dt \\
&= \int_a^b g_{c_0(t)}\left(\frac{\nabla}{\partial s}\frac{\partial c}{\partial t}(0,t), \dot{c}_0(t)\right) dt \\
&= \int_a^b g_{c_0(t)}\left(\frac{\nabla}{\partial t}\frac{\partial c}{\partial s}(0,t), \dot{c}_0(t)\right) dt \\
&= \int_a^b g_{c_0(t)}\left(\frac{\nabla}{dt}V(t), \dot{c}_0(t)\right) dt. \qquad (4.8)
\end{aligned}$$

Because $V(a) = V(b) = (0,0,0)^\top$, the fundamental theorem of calculus gives that

$$\begin{aligned}
0 &= g_q(V(b), \dot{c}_0(b)) - g_p(V(a), \dot{c}_0(a)) \\
&= \int_a^b \frac{d}{dt} g_{c_0(t)}(V(t), \dot{c}_0(t))\, dt \\
&= \int_a^b \left(g_{c_0(t)}\left(\frac{\nabla}{dt}V(t), \dot{c}_0(t)\right) + g_{c_0(t)}\left(V(t), \frac{\nabla}{dt}\dot{c}_0(t)\right)\right) dt, \qquad (4.9)
\end{aligned}$$

thus

$$\int_a^b g_{c_0(t)}\left(\frac{\nabla}{dt}V(t), \dot{c}_0(t)\right) dt = -\int_a^b g_{c_0(t)}\left(V(t), \frac{\nabla}{dt}\dot{c}_0(t)\right) dt. \qquad (4.10)$$

Substitution of (4.10) in (4.8) gives the claim. □

Corollary 4.5.6 *Let S be a regular surface with Riemannian metric g. Let $p, q \in S$. If $c : [a,b] \to S$ is a connecting curve from p to q with minimal energy, then*

$$\frac{\nabla}{dt}\dot{c}_0(t) = 0$$

for all $t \in [a,b]$.

Proof For reasons of continuity it suffices to prove the claim for all $t \in (a,b)$. Suppose that we had $(\nabla/dt)\dot{c}_0(t_0) \neq 0$ for some $t_0 \in (a,b)$.

We choose a local parametrisation (U, F, V), such that $c(t_0) \in V$. We further choose a $\delta > 0$ with

- $[t_0 - \delta, t_0 + \delta] \subset (a,b)$,
- $c(t) \in V$ for all $t \in [t_0 - \delta, t_0 + \delta]$.

We define

$$u : [t_0 - \delta, t_0 + \delta] \to U \qquad u(t) := F^{-1}(c(t))$$

and

$$X : [t_0 - \delta, t_0 + \delta] \to \mathbb{R}^2,$$

$$X(t) := (D_{u(t)}F)^{-1}\left(\frac{\nabla}{dt}\dot{c}_0(t)\right).$$

Then

$$\frac{\nabla}{dt}\dot{c}(t) = D_{u(t)}F(X(t)).$$

We choose a smooth function $\varphi : [t_0 - \delta, t_0 + \delta] \to \mathbb{R}$ with

- $\varphi \geq 0$,
- $\varphi(t_0) > 0$,
- $\operatorname{supp}(\varphi) \subset [t_0 - \delta, t_0 + \delta]$.

For a sufficiently small $\varepsilon > 0$ we have that $u(t) + s \cdot \varphi(t) \cdot X(t) \subset U$ for all $t \in [t_0 - \delta, t_0 + \delta]$ and all $s \in (-\varepsilon, \varepsilon)$. We can therefore define

$$c_s(t) := F(u(t) + s \cdot \varphi(t) \cdot X(t)) \subset V \subset S$$

for all $t \in [t_0 - \delta, t_0 + \delta]$ and all $s \in (-\varepsilon, \varepsilon)$. For $s \in (-\varepsilon, \varepsilon)$ and $t \in [a,b] - [t_0 - \delta, t_0 + \delta]$ we set

$$c_s(t) := c(t).$$

Then $c_s(t)$ is smooth at $(s,t) \in (-\varepsilon, \varepsilon) \times [a,b]$. We now find the variation vector field. For $t \in [a,b] - [t_0 - \delta, t_0 + \delta]$ we obviously have

$$V(t) = \left.\frac{\partial}{\partial s}\right|_{s=0} c_s(t) = 0,$$

while for $t \in [t_0 - \delta, t_0 + \delta]$ we have

$$\begin{aligned} V(t) &= \left.\frac{\partial}{\partial s}\right|_{s=0} c_s(t) \\ &= \left.\frac{\partial}{\partial s}\right|_{s=0} F(u(t) + s \cdot \varphi(t) \cdot X(t)) \\ &= D_{u(t)}F(\varphi(t) \cdot X(t)) \\ &= \varphi(t) \cdot \frac{\nabla}{dt}\dot{c}(t). \end{aligned}$$

We substitute this into the variation formula for energy and obtain

$$\begin{aligned} \left.\frac{d}{ds}E[c_s]\right|_{s=0} &= -\int_a^b g_{c(t)}\left(V(t), \frac{\nabla}{dt}\dot{c}(t)\right) dt \\ &= -\int_{t_0-\delta}^{t_0+\delta} g_{c(t)}\left(\varphi(t) \cdot \frac{\nabla}{dt}\dot{c}(t), \frac{\nabla}{dt}\dot{c}(t)\right) dt \\ &= -\int_{t_0-\delta}^{t_0+\delta} \varphi(t) \cdot g_{c(t)}\left(\frac{\nabla}{dt}\dot{c}(t), \frac{\nabla}{dt}\dot{c}(t)\right) dt \\ &< 0. \end{aligned}$$

Since the curve $c = c_0$ minimises energy we have

$$\left.\frac{d}{ds}E[c_s]\right|_{s=0} = 0,$$

a contradiction. □

Definition 4.5.7 Let S be a regular surface and I an interval. A parametrised curve $c : I \to S$ is a ***geodesic*** if

$$\frac{\nabla}{dt}\dot{c}(t) = 0$$

for all $t \in I$.

Example 4.5.8 Let $S \subset \mathbb{R}^3$ be the x–y plane with the first fundamental form as Riemannian metric. As we already know, the covariant derivative agrees with the usual derivative in this case:

$$\frac{\nabla}{dt}\dot{c}(t) = \ddot{c}(t).$$

The geodesics are therefore precisely those straight lines that have a constant speed:

$$c(t) = p + t \cdot v.$$

Example 4.5.9 Let $S = S^2 \subset \mathbb{R}^3$ be the sphere. We have already seen that among the lines of latitude

$$c(t) = \begin{pmatrix} \cos(t)\cos(\theta) \\ \sin(t)\cos(\theta) \\ \sin(\theta) \end{pmatrix},$$

θ fixed, only the equator $\theta = 0$ satisfies the geodesic equation.

Exercise 4.18 Let S_1 and S_2 be regular surfaces and let $f : S_1 \to S_2$ be a local isometry. Show that if $c : I \to S_1$ is a geodesic, then the same is true for $f \circ c : I \to S_2$.

Exercise 4.19 Let S be a regular surface and let $c : I \to S$ be a geodesic. Further let $\alpha, \beta \in \mathbb{R}$. Show that $\tilde{c}(t) := c(\alpha t + \beta)$ is a geodesic as well. A linear reparametrisation of a geodesic is thus again a geodesic.

Exercise 4.20 Use the last two exercises to show that on the sphere $S = S^2$ all great circles with constant speed are geodesics.

Lemma 4.5.10 *Geodesics are parametrised proportional to arc-length.*

Proof Let c be a geodesic. We differentiate and use the product rule II from lemma 4.2.17:

$$\frac{d}{dt} g_{c(t)}(\dot{c}(t), \dot{c}(t)) = g_{c(t)}\left(\frac{\nabla}{dt}\dot{c}(t), \dot{c}(t)\right) + g_{c(t)}\left(\dot{c}(t), \frac{\nabla}{dt}\dot{c}(t)\right) = 0.$$

Thus $g_{c(t)}(\dot{c}(t), \dot{c}(t))$ is constant. $\square$

So for we know the following about geodesics

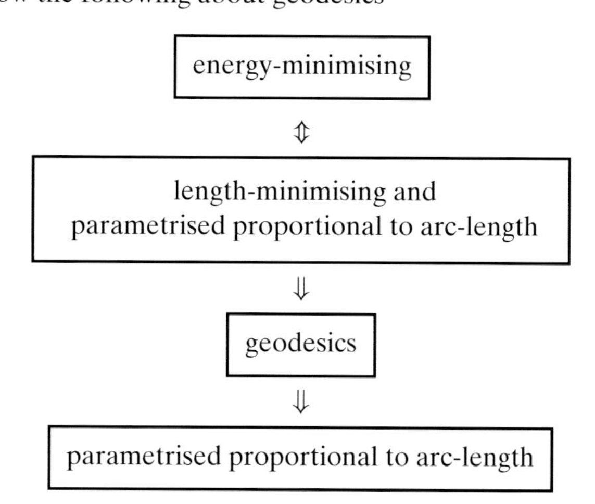

In general, the converses of the lower two arrows are not true. In the above example with the sphere $S = S^2$ the lines of latitude are parametrised proportional to arc-length for all θ, yet only the equator is a geodesic. Thus the converse of the lower arrow is not true. Further, the equator, traversed once, $c(t) = (\cos(t), \sin(t), 0)^\top$, $t \in [0, 2\pi]$, is a geodesic. However, the length- and energy-minimising connecting curve of $p = (1,0,0)^\top$ with itself is, of course, the constant curve, $\tilde{c}(t) = (1,0,0)^\top$. Hence, in general, the converse of the middle arrow is also not true.

To find out how many geodesics there are on a regular surface, we look at the geodesic equation w.r.t. a local parametrisation. For a local parametrisation (U, F, V) and a curve c we write, where defined, $u := F^{-1} \circ c$, i.e. $c = F \circ u$. The geodesic equation is then

$$\begin{aligned} 0 &= \frac{\nabla}{dt}\dot{c}(t) \\ &= \sum_k \left(\ddot{u}^k(t) + \sum_{ij} \Gamma_{ij}^k(u(t))\dot{u}^i(t)\dot{u}^j(t) \right) \frac{\partial F}{\partial u^k}(u(t)). \end{aligned}$$

The geodesic equation is hence equivalent to a system of (non-linear) ordinary differential equations in $u(t) = (u^1(t), u^2(t))$:

$$\ddot{u}^k(t) + \sum_{ij} \Gamma_{ij}^k(u(t))\dot{u}^i(t)\dot{u}^j(t) = 0, \tag{4.11}$$

$k = 1, 2$. Existence and uniqueness theorems for ordinary differential equations [18, Ch. XIX] now give corresponding statements for geodesics.

Theorem 4.5.11 (Existence of geodesics) *Let $S \subset \mathbb{R}^3$ be a regular surface with Riemannian metric g. Let $p \in S$, $v \in T_pS$ and $t_0 \in \mathbb{R}$.*

Then there is an interval $I \subset \mathbb{R}$ with $t_0 \in I$ and a geodesic $c : I \to S$ with the "initial conditions"

$$c(t_0) = p \quad \textit{and} \quad \dot{c}(t_0) = v.$$

Proof We choose a local parametrisation (U, F, V) such that $p \in V$. We set $u_0 := F^{-1}(p) \in U$ and $X := (D_{u_0}F)^{-1}(v) \in \mathbb{R}^2$. By the existence theorem for ordinary differential equations we can now solve (4.11) with the initial conditions $u(t_0) = u_0$ and $\dot{u}(t_0) = X$. With $c := F \circ u$ we have found a geodesic with the desired properties. □

Remark It is in some cases possible to solve the geodesic equation completely in $\mathbb{R}$, i.e. we can choose $I = \mathbb{R}$. In the case that S is the x–y plane or the sphere, the geodesics (straight line and great circle, respectively) indeed have a maximal domain $I = \mathbb{R}$. If we consider, for example, the unit disc

$S = \{(x, y, 0)^\top \mid x^2 + y^2 < 1\}$ in the x–y plane, then the geodesic equation is the same as for the x–y plane itself, i.e. the geodesics are the straight lines that are traversed at constant speed, but those leave the unit disc at a certain point. The maximal domain I of a geodesic is therefore bounded on both sides, unless the geodesic is a constant curve.

Theorem 4.5.12 (Uniqueness of geodesics) *Let $S \subset \mathbb{R}^3$ be a regular surface with Riemannian metric g. Let I be an interval, $t_0 \in I$. Let $c : I \to S$ be a geodesic. Then c is uniquely determined by $c(t_0) \in S$ and $\dot{c}(t_0) \in T_{c(t_0)}S$.*

Proof If we knew that the trace of c lay entirely in one open neighbourhood of the local parametrisation, then we could argue in a way similar to that used for theorem 4.5.11. We would simply cite the uniqueness statement instead of the existence statement from the theory of ordinary differential equations.

But since we cannot assume this, we need to argue a bit more carefully. Let c_1 and c_2 be geodesics with the same initial conditions $c_1(t_0) = c_2(t_0)$ and $\dot{c}_1(t_0) = \dot{c}_2(t_0)$. Suppose that there exists a $t \in I, t > t_0$, with $c_1(t) \neq c_2(t)$. We set

$$t_1 := \sup\{t \in I \mid t > t_0, \text{ such that } c_1(\tau) = c_2(\tau) \text{ for all } \tau \in [t_0, t]\}.$$

In words, t_1 is exactly the point at which c_1 and c_2 cease to agree. We now choose a local parametrisation (U, F, V) with $c_1(t_1) \in V$. Since $c_1(t) = c_2(t)$ for all $t < t_1$ (and $t \geq t_0$) we have that

$$c_1(t_1) = c_2(t_1) \quad \text{as well as} \quad \dot{c}_1(t_1) = \dot{c}_2(t_1).$$

The uniqueness theorem for ordinary differential equations now tells us that $c_1(t) = c_2(t)$ as long as $c_1(t) \in V$ and $c_2(t) \in V$. For a sufficiently small $\varepsilon > 0$ this is the case for all $t \in (t_1 - \varepsilon, t_1 + \varepsilon)$. This contradicts the maximality of t_1.

It follows that $c_1(t) = c_2(t)$ for all $t \geq t_0$. The proof for $t \leq t_0$ is analogous. □

Exercise 4.21 Let $S \subset \mathbb{R}^3$ be a regular surface with Riemannian metric g, let $p \in S$ and $v \in T_pS$. Let c be the geodesic with initial conditions $c(0) = p$ and $\dot{c}(0) = v$. Let $\delta \in \mathbb{R}$ be a constant. Show that the curve $\tilde{c}(t) := c(\delta t)$ is the geodesic with the initial conditions $\tilde{c}(0) = p$ and $\dot{\tilde{c}}(0) = \delta v$.

The following theorem allows us to get a qualitative idea about the course of geodesics on surfaces of revolution.

Theorem 4.5.13 (Clairaut's theorem) *Let S be a surface of revolution, given by the parametrisation $F(t, \varphi) = (r(t)\cos(\varphi), r(t)\sin(\varphi), t)^\top$. We take the first fundamental form as the Riemannian metric. Let $c : I \to S$ be a geodesic,*

$c(t) = F(r(t), \varphi(t))$. Let $\theta(t)$ be the angle between $\dot{c}(t)$ and the line of latitude through $c(t)$.

Then

$$r(t)\cos(\theta(t)) = \text{const.}$$

If $r(t)$ becomes bigger, for example, then $\cos(\theta(t))$ must become smaller, and hence $\theta(t)$ bigger.

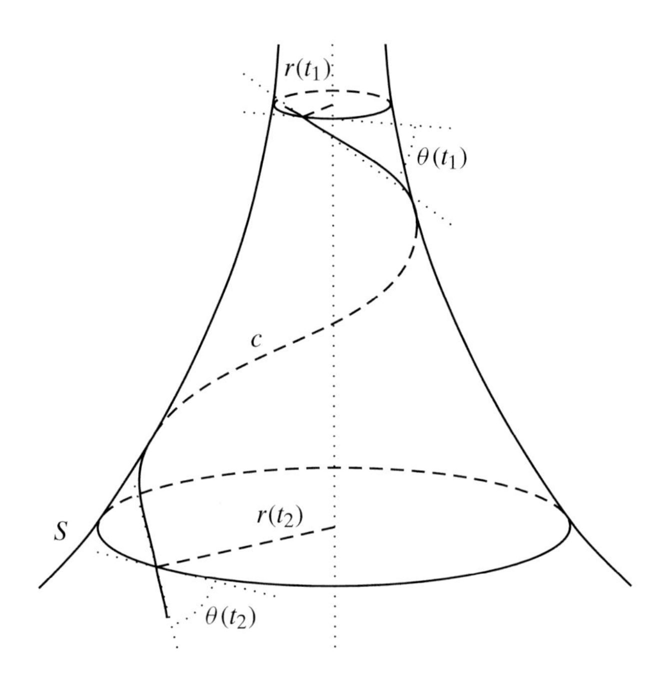

Proof The first fundamental form was calculated in section 3.8.3. The result is

$$\left(g_{ij}(t,\varphi)\right)_{ij} = \begin{pmatrix} 1+\dot{r}(t)^2 & 0 \\ 0 & r(t)^2 \end{pmatrix}.$$

We set

$$v := \frac{\partial F}{\partial t} \quad \text{and} \quad w := \frac{\partial F}{\partial \varphi}.$$

The vector fields v and w form an orthogonal basis of the tangent plane at every point of the surface. The vector field w is always tangential to the lines of latitude, while v is tangential to the lines of longitude. As g_{22} does not depend on φ, we have that

$$\left\langle \frac{\partial^2 F}{\partial \varphi^2}, w \right\rangle = \frac{1}{2}\frac{\partial}{\partial \varphi}\left\langle \frac{\partial F}{\partial \varphi}, \frac{\partial F}{\partial \varphi} \right\rangle = \frac{1}{2}\frac{\partial g_{22}}{\partial \varphi} = 0.$$

Hence the tangential part of $\partial^2 F/\partial\varphi^2$ is proportional to v,

$$\nabla_w w = \alpha v,$$

for a suitable function α. We similarly argue that

$$\left\langle \frac{\partial^2 F}{\partial\varphi\partial t}, v \right\rangle = \frac{1}{2}\frac{\partial}{\partial\varphi}\left\langle \frac{\partial F}{\partial t}, \frac{\partial F}{\partial t} \right\rangle = \frac{1}{2}\frac{\partial g_{11}}{\partial\varphi} = 0$$

and hence

$$\nabla_v w = \nabla_w v = \beta w$$

for a suitable function β. We further have

$$\begin{aligned} 0 &= \frac{\partial g_{12}}{\partial\varphi} \\ &= \frac{\partial}{\partial\varphi}\left\langle \frac{\partial F}{\partial t}, \frac{\partial F}{\partial\varphi} \right\rangle \\ &= \langle \nabla_w v, w\rangle + \langle v, \nabla_w w\rangle \\ &= \langle \beta w, w\rangle + \langle v, \alpha v\rangle . \end{aligned}$$

Hence

$$\alpha \cdot |v|^2 = -\beta \cdot |w|^2.$$

For an arbitrary tangent vector $z = \gamma \cdot v + \delta \cdot w$ we have that

$$\begin{aligned} \langle \nabla_z w, z\rangle &= \gamma^2 \underbrace{\langle \nabla_v w, v\rangle}_{=0} + \gamma\delta\langle \underbrace{\nabla_w w}_{=\alpha v}, v\rangle + \gamma\delta\langle \underbrace{\nabla_v w}_{=\beta w}, w\rangle + \delta^2 \underbrace{\langle \nabla_w w, w\rangle}_{=0} \\ &= \gamma\delta\left(\alpha|v|^2 + \beta|w|^2\right) \\ &= 0. \end{aligned}$$

Hence a geodesic c satisfies

$$\frac{d}{dt}\langle w(c(t)), \dot c(t)\rangle = \underbrace{\langle \nabla_{\dot c(t)} w, \dot c(t)\rangle}_{=0} + \langle w(c(t)), \underbrace{\nabla_{\dot c(t)} \dot c(t)}_{=0}\rangle = 0.$$

Thus

$$\langle w(c(t)), \dot c(t)\rangle = \text{const} \cdot r(t) \cdot \cos(\theta(t))$$

is constant. □

Exercise 4.22 (a) Show that on surfaces of revolution the lines of longitude $t \mapsto F(t, \varphi_0)$ are length-minimising connecting curves of their end-points (and hence in particular geodesics after a reparametrisation by arc-length).

(b) What is the condition for a line of latitude $\varphi \mapsto F(t_0, \varphi)$ to be a geodesic?

Let us now return from surfaces of revolution to general regular surfaces with Riemannian metrics. We have seen that geodesics generalise the straight lines from plane geometry. Let us recall that the curvature of plane curves is defined as the deviation from a straight line. We can generalise this definition to curves in general (oriented) surfaces. We define them as the deviation from a geodesic.

Let S be an oriented regular surface with Riemannian metric g. Let $c : I \to S$ be a curve parametrised by arc-length. Let $n : I \to \mathbb{R}^3$ be the unit normal field along c that complements $\dot{c}$ to positively oriented orthonormal bases, i.e. for every $t \in I$ the pair $(\dot{c}(t), n(t))$ is a positively oriented orthonormal basis of $T_{c(t)}S$. Just like with plane curves differentiating the function $t \mapsto g_{c(t)}(\dot{c}(t), \dot{c}(t))$ shows that $(\nabla/dt)\dot{c}(t)$ is perpendicular to $\dot{c}(t)$. There is therefore a unique function $\kappa_g : I \to \mathbb{R}$ such that

$$\frac{\nabla}{dt}\dot{c}(t) = \kappa_g(t) \cdot n(t).$$

Definition 4.5.14 The function κ_g is called the ***geodesic curvature*** of c in S w.r.t. g.

It is clear from the definition that c is a geodesic if and only if $\kappa_g \equiv 0$. The geodesic curvature generalises the curvature of plane curves. If S is the x–y plane with the first fundamental form as a Riemannian metric, then κ_g is exactly the curvature of c, considered as a plane curve.

If we change the direction in which the curve is traversed, i.e. if we change c to $\tilde{c}(t) := c(-t)$, then κ_g changes its sign, $\tilde{\kappa}_g(t) = -\kappa_g(-t)$. Leaving curve c as it is and reversing the orientation of the surface S we change n to $-n$, and κ_g changes its sign, $\tilde{\kappa}_g(t) = -\kappa_g(t)$. It follows that for non-oriented surfaces at least the absolute value of the geodesic curvature is $|\kappa_g|$ is still well defined.

Exercise 4.23 Let S be an oriented regular surface with Riemannian metric g. Let $c : I \to S$ be a regular parametrised curve (not necessarily parametrised by arc-length). Let $n : I \to \mathbb{R}^3$ be the unit normal field along c that complements $\dot{c}$ to positively oriented orthogonal bases. Show that the geodesic curvature of c is given by

$$\kappa_g = \frac{g\left((\nabla/dt)\dot{c}, n\right)}{g(\dot{c}, \dot{c})}.$$

See also exercise 2.10.

Exercise 4.24 Let S be an oriented regular surface with Riemannian metric g and let $c : I \to S$ be a curve parametrised by arc-length, n as above. Derive the ***Frenet formulae***, which in matrix notation are given by

$$\left(\frac{\nabla}{dt}\dot{c}, \frac{\nabla}{dt}n\right) = (\dot{c}, n)\begin{pmatrix} 0 & -\kappa_g \\ \kappa_g & 0 \end{pmatrix}.$$

4.6 The exponential map

We now want to use geodesics to construct local parametrisations of surfaces that are particularly well suited to the geometry. Among other things this is important to make calculations easier to carry out; we have already seen in examples that the complexity of formulae can depend heavily on the coordinates used. As surfaces on the small scale are approximated by their tangent plane, we will try to translate the coordinates most commonly used on planes, i.e. Cartesian and polar coordinates, from the tangent plane to the surface itself. For this purpose we construct, using geodesics, geometrically natural maps from the tangent planes to the surface.

Let S be a regular surface with Riemannian metric g. Let $p \in S$ be a point. For a tangent vector $v \in T_pS$ we consider the (unique) geodesic $c : I \to S$ with $c(0) = p$, $\dot{c}(0) = v$ and maximal domain interval I. If c is still defined at time $t = 1$, i.e. if $1 \in I$, then we set

$$\exp_p(v) := c(1).$$

If $\exp_p$ is defined for $v \in T_pS$ and if $\delta \in [0, 1]$, then $\exp_p$ is also defined for $\delta v \in T_pS$. If c_v is the geodesic with $c_v(0) = p$ and $\dot{c}(0) = v$, $c_{\delta v}$ is analogously the geodesic with $c_{\delta v}(0) = p$ and $\dot{c}(0) = \delta v$, then $c_{\delta v}(t) = c_v(\delta t)$, see exercise 4.21. Hence $c_{\delta v}$ is defined at $t = 1$, since $c_v(0) = p$ is determined on the whole of $[0, 1]$. This argument shows that the domain $\mathcal{D}_p \subset T_pS$ of $\exp_p$ is a star-shaped subset of T_pS with respect to 0. The argument also shows that

$$c_v(t) = \exp_p(tv). \tag{4.12}$$

The domain is certainly not empty, since the constant geodesic $c(t) \equiv p$ is defined on the whole of $\mathbb{R}$, in particular for $t = 1$, and hence $0 \in \mathcal{D}_p$. The theorem about the dependence of solutions of ordinary differential equations on the initial values further ensures that $\mathcal{D}_p$ is an open subset of T_pS and that

$$\exp_p : \mathcal{D}_p \to S$$

is a smooth map.

Definition 4.6.1 The map $\exp_p : \mathcal{D}_p \to S$ is called the ***exponential map***.

Example 4.6.2 Let $S = \mathbb{R}^2 \times \{0\}$ be the x–y plane with the first fundamental form as Riemannian metric. Let $p \in S$ and $v \in T_pS = \mathbb{R}^2 \times \{0\}$. The geodesic c in S with $c(0) = p$ and $\dot{c}(0) = v$ is the straight line $c(t) = p + tv$. Hence $\mathcal{D}_p = T_pS = \mathbb{R}^2 \times \{0\}$ and

$$\exp_p(v) = p + v.$$

Example 4.6.3 Now let $S = \{(x, y, 0)^\top \in \mathbb{R}^3 \mid x^2 + y^2 < 1\}$ be the unit disc in the x–y plane with the first fundamental form as a Riemannian metric, as above. The geodesics are again segments of straight lines, but as they leave the disc after a finite amount of time, the domain of the (non-constant) geodesic is always a finite interval. For the exponential map we have the same formula as above:

$$\exp_p(v) = p + v,$$

but the domain of the exponential map is now

$$\mathcal{D}_p = \{v \in \mathbb{R}^3 \mid p + v \in S\} = S - p.$$

Example 4.6.4 Let $S = S^2$ be the sphere, again with the first fundamental form as a Riemannian metric. Let $p \in S$ and $v \in T_pS = p^\perp$. We write $v = \delta w$, where $w \in T_pS$ is a unit vector, $\|w\| = 1$ and $\delta = \|v\| \geq 0$. The geodesic c in S with $c(0) = p$ and $\dot{c}(0) = v$ is given by the great circle $c(t) = \cos(\delta t) \cdot p + \sin(\delta t) \cdot w$. Hence $\mathcal{D}_p = T_pS$ and

$$\exp_p(v) = \begin{cases} \cos(\|v\|) \cdot p + \sin(\|v\|) \cdot v/\|v\|, & v \neq 0, \\ p, & v = 0. \end{cases}$$

We want to use the exponential map to translate coordinates from the tangent plane to the surface. However, the example of the sphere shows that the exponential map need not be bijective. We therefore have to restrict it. The inverse function theorem will then tell us that the exponential map, restricted to a neighbourhood of 0, is a diffeomorphism. To be able to use the inverse function theorem, we calculate the differential of the exponential map at the point $0 \in \mathcal{D}_p$. We observe that $T_0\mathcal{D}_p = T_0T_pS = T_pS$.

Lemma 4.6.5 *The differential of the exponential map at the point* 0 *is the identity,*

$$d_0 \exp_p = \mathrm{Id} : T_pS \to T_pS.$$

Proof Let $v \in T_pS$. We know from (4.12) that the geodesic c with initial conditions $c(0) = p$ and $\dot{c}(0) = v$ is given by

$$c(t) = \exp_p(tv).$$

Further, $\tilde{c}(t) = tv$ is a curve in T_pS with $\tilde{c}(0) = 0$ and $\dot{\tilde{c}}(0) = v$. Hence

$$\begin{aligned} d_0 \exp_p(v) &= \frac{d}{dt} \exp_p(\tilde{c}(t))\Big|_{t=0} \\ &= \frac{d}{dt} \exp_p(tv)\Big|_{t=0} \\ &= \dot{c}(0) = v. \end{aligned}$$

□

By the inverse function theorem there exists a neighbourhood W of $0 \in \mathscr{D}_p$ such that $\exp_p|_W : W \to \exp_p(W) \subset S$ is a diffeomorphism. For a local parametrisation (U_1, F_1, V_1) of the tangent plane T_pS we obtain with the choices $U := F_1^{-1}(W)$, $F := \exp_p \circ F_1|_U$ and $V \subset \mathbb{R}^3$ open with $V \cap S = \exp_p(W)$ a local parametrisation (U, F, V) of S.

Example 4.6.6 Let S be an arbitrary regular surface with a Riemannian metric. Let $p \in S$ and let X_1, X_2 be an orthonormal basis of the tangent plane T_pS. We take the parametrisation in Cartesian coordinates for T_pS, i.e. $U_1 = \mathbb{R}^2$ and $F_1(u^1, u^2) = \sum_i u^i X_i$. The corresponding local parametrisation of S,

$$F(u^1, u^2) = \exp_p\Big(\sum_i u^i X_i\Big),$$

is called parametrisation in ***Riemann normal coordinates*** (at the point p).

Riemann normal coordinates have properties similar to those of the normal coordinates, as we constructed in theorem 3.6.15 for the first fundamental form.

Theorem 4.6.7 *Let S be a regular surface, let $p \in S$ and let F be a local parametrisation in Riemann normal coordinates at the point p. Then the corresponding component functions of the metric and the Christoffel symbols satisfy the following:*

(i) $F(0,0) = p$.
(ii) $g_{ij}(0,0) = \delta_{ij}, \qquad i,j = 1,2$.
(iii) $(\partial g_{ij}/\partial u^k)(0,0) = 0$ *and* $\Gamma_{ij}^k(0,0) = 0,\ i,j,k = 1,2$.

Proof Statement (i) is clear and statement (ii) means precisely that $d_0 \exp_p$ is a linear isometry. To prove statement (iii), we recall that the exponential map maps the straight lines through the origin to the geodesics through p. The map $t \mapsto tx$ describes a geodesic in Riemann normal coordinates for arbitrary

$x \in \mathbb{R}^2$. The geodesic equation (4.11) then becomes

$$\sum_{ij} \Gamma_{ij}^k(tx)x^i x^j = 0, \quad k = 1, 2.$$

For $t = 0$ in particular we therefore have

$$\sum_{ij} \Gamma_{ij}^k(0,0)x^i x^j = 0, \quad k = 1, 2,$$

for all x. As $\Gamma_{ij}^k(0,0)$ is symmetric in i and j for fixed k, we can polarise and obtain

$$\Gamma_{ij}^k(0,0) = 0$$

for all i, j, k. From (4.3)

$$\frac{\partial g_{jm}}{\partial u^i} = \sum_k \left(\Gamma_{ij}^k g_{km} + \Gamma_{im}^k g_{kj} \right),$$

it follows that

$$\frac{\partial g_{ij}}{\partial u^k}(0,0) = 0$$

for all i, j, k. □

Example 4.6.8 As above, let S be an arbitrary regular surface with a Riemannian metric, let $p \in S$ and let X_1, X_2 be an orthonormal basis of T_pS. This time we take polar coordinates for T_pS, $F_1(r,\varphi) = r \cdot (\cos(\varphi)X_1 + \sin(\varphi)X_2)$. The corresponding local parametrisation of S,

$$F(r,\varphi) = \exp_p \left(r \cdot (\cos(\varphi)X_1 + \sin(\varphi)X_2) \right),$$

is the parametrisation in ***geodesic polar coordinates*** (at the point p).

Theorem 4.6.9 (Gauss's lemma) *Let S be a regular surface with Riemannian metric g. Let $p \in S$ and let F be a local parametrisation in geodesic coordinates (r, φ).*

Then the Riemannian metric has the following form with respect to the local parametrisation:

$$(g_{ij}(r,\varphi))_{ij} = \begin{pmatrix} 1 & 0 \\ 0 & f(r,\varphi)^2 \end{pmatrix},$$

with a positive function f that satisfies

$$\lim_{r \to 0} f(r,\varphi) = 0, \qquad \lim_{r \to 0} \frac{\partial f}{\partial r}(r,\varphi) = 1.$$

Proof Let X_1, X_2 be the orthonormal basis of T_pS that we have used for the definition of the geodesic polar coordinates, $F(r,\varphi) = \exp_p(\tilde{F}(r,\varphi))$, where $\tilde{F}(r,\varphi) = r\cdot(\cos(\varphi)X_1 + \sin(\varphi)X_2)$. The curve $c(r) = F(r,\varphi_0)$ is for fixed $\varphi = \varphi_0$ by the definition of the exponential map the geodesic with $\dot{c}(0) = \cos(\varphi_0)X_1 + \sin(\varphi_0)X_2$. Geodesics are parametrised proportional to arc-length and $\dot{c}(0)$ is a unit vector, hence

$$g_{11}(r,\varphi_0) = g(\dot{c}(r),\dot{c}(r)) = 1.$$

Further

$$\begin{aligned}
\frac{\partial g_{12}}{\partial r}(r,\varphi_0) &= \frac{\partial}{\partial r} g\left(\frac{\partial F}{\partial r},\frac{\partial F}{\partial \varphi}\right)\bigg|_{(r,\varphi_0)} \\
&= g\left(\frac{\nabla}{\partial r}\frac{\partial F}{\partial r}(r,\varphi_0),\frac{\partial F}{\partial \varphi}(r,\varphi_0)\right) + g\left(\frac{\partial F}{\partial r}(r,\varphi_0),\frac{\nabla}{\partial r}\frac{\partial F}{\partial \varphi}(r,\varphi_0)\right) \\
&= g\left(\frac{\nabla}{dr}\dot{c}(r),\frac{\partial F}{\partial \varphi}(r,\varphi_0)\right) + g\left(\frac{\partial F}{\partial r}(r,\varphi_0),\frac{\nabla}{\partial r}\frac{\partial F}{\partial \varphi}(r,\varphi_0)\right) \\
&= 0 + g\left(\frac{\partial F}{\partial r}(r,\varphi_0),\frac{\nabla}{\partial \varphi}\frac{\partial F}{\partial r}(r,\varphi_0)\right) \\
&= \frac{1}{2}\frac{\partial}{\partial \varphi}\underbrace{g\left(\frac{\partial F}{\partial r},\frac{\partial F}{\partial r}\right)}_{\equiv 1}\Bigg|_{(r,\varphi_0)} \\
&= 0.
\end{aligned} \tag{4.13}$$

Hence the function g_{12} is constant in r for fixed $\varphi = \varphi_0$. As an abbreviation we set $Y_1 := \cos(\varphi_0)X_1 + \sin(\varphi_0)X_2$ and $Y_2 := -\sin(\varphi_0)X_1 + \cos(\varphi_0)X_2$. This is the orthonormal basis of T_pS that results from X_1, X_2 in a rotation through the angle φ_0. We have

$$\frac{\partial \tilde{F}}{\partial r}(r,\varphi_0) = Y_1, \qquad \frac{\partial \tilde{F}}{\partial \varphi}(r,\varphi_0) = r\cdot Y_2.$$

We calculate

$$\begin{aligned}
\lim_{r\to 0} g_{12}(r,\varphi_0) &= \lim_{r\to 0} g\left(\frac{\partial F}{\partial r}(r,\varphi_0),\frac{\partial F}{\partial \varphi}(r,\varphi_0)\right) \\
&= \lim_{r\to 0} g\left(d_{\tilde{F}(r,\varphi_0)}\exp_p\left(\frac{\partial \tilde{F}}{\partial r}(r,\varphi_0)\right), d_{\tilde{F}(r,\varphi_0)}\exp_p\left(\frac{\partial \tilde{F}}{\partial \varphi}(r,\varphi_0)\right)\right)
\end{aligned}$$

$$\begin{aligned}
&= \lim_{r\to 0} g\left(d_{\tilde{F}(r,\varphi_0)} \exp_p(Y_1), d_{\tilde{F}(r,\varphi_0)} \exp_p(r \cdot Y_2)\right) \\
&= g\left(d_0 \exp_p(Y_1), d_0 \exp_p(0)\right) \\
&= 0.
\end{aligned} \tag{4.14}$$

From (4.13) and (4.14) it follows that

$$g_{12} = g_{21} = 0.$$

Hence

$$(g_{ij})_{ij} = \begin{pmatrix} 1 & 0 \\ 0 & g_{22} \end{pmatrix}.$$

As the Riemannian metric is positive definite, we have $g_{22} > 0$ and we can write $g_{22} = f^2$. In a way similar to that used in (4.14) we calculate

$$\begin{aligned}
\lim_{r\to 0} g_{22}(r,\varphi_0) &= \lim_{r\to 0} g\left(d_{\tilde{F}(r,\varphi_0)} \exp_p\left(\frac{\partial \tilde{F}}{\partial \varphi}(r,\varphi_0)\right), d_{\tilde{F}(r,\varphi_0)} \exp_p\left(\frac{\partial \tilde{F}}{\partial \varphi}(r,\varphi_0)\right)\right) \\
&= \lim_{r\to 0} g\left(d_{\tilde{F}(r,\varphi_0)} \exp_p(r \cdot Y_2), d_{\tilde{F}(r,\varphi_0)} \exp_p(r \cdot Y_2)\right) \\
&= g\left(d_0 \exp_p(0), d_0 \exp_p(0)\right) \\
&= 0.
\end{aligned}$$

This proves that

$$\lim_{r\to 0} f(r,\varphi_0) = 0.$$

Further,

$$\begin{aligned}
\lim_{r\to 0} \frac{\partial f}{\partial r}(r,\varphi_0) &= \lim_{r\to 0} \frac{f(r,\varphi_0)}{r} \\
&= \lim_{r\to 0} \sqrt{\frac{g_{22}(r,\varphi_0)}{r^2}} \\
&= \sqrt{\lim_{r\to 0} \frac{g_{22}(r,\varphi_0)}{r^2}} \\
&= \sqrt{\lim_{r\to 0} g\left(d_{\tilde{F}(r,\varphi_0)} \exp_p(Y_2), d_{\tilde{F}(r,\varphi_0)} \exp_p(Y_2)\right)} \\
&= \sqrt{g\left(d_0 \exp_p(Y_2), d_0 \exp_p(Y_2)\right)} \\
&= \sqrt{g(Y_2, Y_2)} \\
&= 1.
\end{aligned}$$

□

Observe that polar coordinates are not defined for $r = 0$. The differential of the parametrisation F does not have full rank. The point p itself cannot be given in terms of geodesic coordinates at the point p. We cannot therefore evaluate the function f at $r = 0$, though we can consider the limit as r tends to 0. However, we can give p in terms of Riemann normal coordinates at p.

Lemma 4.6.10 *Keep the notation from theorem* 4.6.9. *Then the Gauss curvature satisfies*

$$K(F(r,\varphi)) = -\frac{1}{f(r,\varphi)}\frac{\partial^2 f}{\partial r^2}(r,\varphi).$$

Proof Theorem 4.6.9 tells us that in geodesic coordinates the Riemannian metric has the form

$$(g_{ij}(r,\varphi))_{ij} = \begin{pmatrix} 1 & 0 \\ 0 & f(r,\varphi)^2 \end{pmatrix}.$$

The inverse matrix is then

$$(g^{ij}(r,\varphi))_{ij} = \begin{pmatrix} 1 & 0 \\ 0 & f(r,\varphi)^{-2} \end{pmatrix}.$$

Using the formula

$$\Gamma^k_{ij} = \frac{1}{2}\sum_{m=1}^{2}\left(\frac{\partial g_{jm}}{\partial u^i} + \frac{\partial g_{im}}{\partial u^j} - \frac{\partial g_{ij}}{\partial u^m}\right) g^{mk}$$

from lemma 4.2.14, we easily calculate

$$\begin{aligned}
\Gamma^1_{11} &= \Gamma^2_{11} = \Gamma^1_{12} = \Gamma^1_{21} = 0, \\
\Gamma^2_{12} &= \Gamma^2_{21} = \frac{1}{f}\frac{\partial f}{\partial r}, \\
\Gamma^1_{22} &= -f \cdot \frac{\partial f}{\partial r}, \\
\Gamma^2_{22} &= \frac{1}{f}\frac{\partial f}{\partial \varphi}.
\end{aligned}$$

The vectors $\partial F/\partial r$ and $(1/f)(\partial F/\partial\varphi)$ form an orthonormal basis of the tangent plane. According to the *Theorema Egregium* (theorem 4.3.8) the Gauss curvature is given by

$$
\begin{aligned}
K &= g\left(R\left(\frac{\partial F}{\partial r}, \frac{1}{f}\frac{\partial F}{\partial \varphi}\right)\frac{1}{f}\frac{\partial F}{\partial \varphi}, \frac{\partial F}{\partial r}\right)\\
&= \frac{1}{f^2}R^1_{122}\\
&= \frac{1}{f^2}\left(\frac{\partial \Gamma^1_{22}}{\partial r} - \frac{\partial \Gamma^1_{21}}{\partial \varphi} + \Gamma^1_{11}\Gamma^1_{22} - \Gamma^1_{12}\Gamma^1_{21} + \Gamma^1_{21}\Gamma^2_{22} - \Gamma^1_{22}\Gamma^2_{21}\right)\\
&= \frac{1}{f^2}\left(-\left(\frac{\partial f}{\partial r}\right)^2 - f\cdot\frac{\partial^2 f}{\partial r^2} - 0 + 0 - 0 + 0 + f\cdot\frac{\partial f}{\partial r}\cdot\frac{1}{f}\cdot\frac{\partial f}{\partial r}\right)\\
&= -\frac{1}{f}\frac{\partial^2 f}{\partial r^2}.
\end{aligned}
$$

□

Remark In geodesic coordinates the function f carries the complete information about the Riemannian metric. We have seen that Gauss curvature can easily be calculated from f. Let us investigate which requirements on f correspond to the condition of constant Gauss curvature $K \equiv \kappa$. The condition

$$\kappa = -\frac{1}{f(r,\varphi)}\frac{\partial^2 f}{\partial r^2}(r,\varphi)$$

is equivalent to

$$\frac{\partial^2 f}{\partial r^2}(r,\varphi) = -\kappa\cdot f(r,\varphi).$$

This is an *ordinary* differential equation in $r \mapsto f(r,\varphi)$ if we keep φ fixed. This, together with the initial conditions $f(0,\varphi) = 0$ and $(\partial f/\partial r)(0,\varphi) = 1$, uniquely determines the solution, which is

$$f(r,\varphi) = \begin{cases} \frac{1}{\sqrt{\kappa}}\sin(\sqrt{\kappa}\cdot r), & \kappa > 0,\\ r, & \kappa = 0,\\ \frac{1}{\sqrt{-\kappa}}\sinh(\sqrt{-\kappa}\cdot r), & \kappa < 0.\end{cases}$$

This shows, in particular, that κ uniquely determines the metric in the coordinate neighbourhood.

Corollary 4.6.11 *If S_1 and S_2 are two regular surfaces with the same constant Gauss curvature κ, then S_1 and S_2 are locally isometric.*

We now see that the inhabitants of the surface from the introduction of this chapter about the inner geometry of surfaces cannot distinguish different surfaces with the same constant Gauss curvature if they can only take measurements in the close neighbourhood of a point. In particular, they cannot distinguish the plane, the cylinder and the cone.

For the proof of the Gauss–Bonnet formula in the last chapter we will need the following formula, which expresses the integral of Gauss curvature in the neighbourhood of a point. For a regular surface S with Riemannian metric g, a point $p \in S$ and $r > 0$ let

$$\bar{D}(p,r) := \{q \in S \mid \exists \text{ geodesic } c : [0,1] \to S \text{ with } c(0) = p, c(1) = q, L[c] \leq r\}.$$

Lemma 4.6.12 *Keep the notation from theorem 4.6.9. Let $r > 0$ be such that $\bar{D}(p,r)$ is (up to a null set) covered by the geodesic polar coordinate system. Then*

$$\int_{\bar{D}(p,r)} K\, dA = 2\pi - \int_{\varphi_0}^{\varphi_0+2\pi} \frac{\partial f}{\partial r}(r,\varphi)\, d\varphi.$$

Proof By the formula for the Riemannian metric from Gauss's lemma the surface element is given in geodesic polar coordinates by

$$dA = \sqrt{1 \cdot f^2}\, dr\, d\varphi = f\, dr\, d\varphi.$$

For the geodesic coordinate system to cover $\bar{D}(p,r)$ up to a null set, the domain of the local parametrisation must be of the form $U = (0,R) \times (\varphi_0, \varphi_0 + 2\pi)$ with $R \geq r$. We substitute the formula for Gauss curvature from lemma 4.6.10 and obtain

$$\begin{aligned}
\int_{\bar{D}(p,r)} K\, dA &= -\int_{\varphi_0}^{\varphi_0+2\pi} \int_0^r \frac{1}{f(r,\varphi)} \frac{\partial^2 f}{\partial r^2}(r,\varphi) f(r,\varphi)\, dr\, d\varphi \\
&= -\int_{\varphi_0}^{\varphi_0+2\pi} \int_0^r \frac{\partial^2 f}{\partial r^2}(r,\varphi)\, dr\, d\varphi \\
&= -\int_{\varphi_0}^{\varphi_0+2\pi} \left(\frac{\partial f}{\partial r}(r,\varphi) - 1 \right) d\varphi.
\end{aligned}$$

The last equality holds since $\lim_{r\to 0} (\partial f/\partial r)(r,\varphi) = 1$. □

Having looked at Riemann normal coordinates and geodesic polar coordinates, we now introduce coordinates that are particularly tailored to a given curve on the surface.

Lemma 4.6.13 *Let S be a regular surface with Riemannian metric g. Let $c : I \to S$ be a curve parametrised by arc-length, defined on an open interval I. Let $n : I \to \mathbb{R}^3$ be a vector field on S along c that has constant length 1 and $\langle \dot{c}, n \rangle \equiv 0$. Then there is for $t_0 \in I$ an $\varepsilon > 0$, such that*

$$F : (t_0 - \varepsilon, t_0 + \varepsilon) \times (-\varepsilon, \varepsilon) \to S, \quad F(t,s) := \exp_{c(t)}(sn(t)),$$

is a local parametrisation of S. Along c the Riemannian metric w.r.t. this parametrisation has the form

$$\left(g_{ij}(t,0)\right)_{ij} = \begin{pmatrix} 1 & 0 \\ 0 & 1 \end{pmatrix}.$$

Proof The partial derivatives of F at the point $(t,0)$ are given by

$$\frac{\partial F}{\partial t}(t,0) = \frac{d}{dt}\exp_{c(t)}(0) = \dot{c}(t)$$

and

$$\frac{\partial F}{\partial s}(t,0) = n(t)$$

and hence form an orthonormal basis of $T_{c(t)}S$. By the inverse function theorem, F is then after a suitable restriction a local parametrisation. Because of the orthonormality the statement about the $g_{ij}(t,0)$ is clear as well. □

Definition 4.6.14 The coordinates corresponding to such a parametrisation are called ***Fermi coordinates***.

Fermi coordinates will be useful in the proof of the divergence theorem (theorem 5.1.7).

Exercise 4.25 Show that we have for the Christoffel symbols in Fermi coordinates

$$\Gamma_{11}^{1}(t,0) = \Gamma_{12}^{2}(t,0) = \Gamma_{21}^{2}(t,0) = 0, \quad \Gamma_{22}^{1}(t,s) = \Gamma_{22}^{2}(t,s) = 0,$$

$$\Gamma_{11}^{2}(t,0) = \kappa_g(t), \quad \Gamma_{12}^{1}(t,0) = \Gamma_{21}^{1}(t,0) = -\kappa_g(t),$$

where κ_g is the geodesic curvature of c (w.r.t. the orientation of S making $(\dot{c}, n)$ positively oriented).

4.7 Parallel transport

If we have two points p and q in a plane $E \subset \mathbb{R}^3$, then we can identify tangent vectors from T_pE with those from T_qE, since $T_pE = T_qE = E$. If we now replace E by a general regular surface $S \subset \mathbb{R}^3$, then this is no longer straightforward, since in general $T_pS \neq T_qS$. In order still to be able to relate tangent planes at different points even on general surfaces we analyse the concept of "parallel" vectors in the plane a bit more carefully.

We can translate a vector $v_0 \in T_pE$ into the corresponding (the same) vector from T_qE in the following way: we choose a connecting smooth curve $c : [a,b] \to E$, $c(a) = p$, $c(b) = q$, and consider along c the constant vector field $v(t) \equiv v_0$ or, put differently, the vector field along c that satisfies $\dot{v} \equiv 0$. Then $v(b) \in T_q E$ is the vector translated to q. This approach can be used without complications on general regular surfaces by simply replacing the usual derivative of v by the covariant derivative.

Definition 4.7.1 Let S be a regular surface with Riemannian metric g. Let $c : I \to S$ be a smooth curve and $v : I \to \mathbb{R}^3$ be a vector field along c. Then v is ***parallel*** if

$$\frac{\nabla}{dt} v \equiv 0.$$

Remark Hence a smooth curve c is a geodesic if and only if its velocity field $\dot{c}$ is parallel.

We now investigate the property of parallelism w.r.t. a local parametrisation. For this purpose let (U, F, V) be a local parametrisation that contains the trace of the curve c, $c(I) \subset V$. Write $v(t) = \sum_k \xi^k(t)(\partial F/\partial u^k)(u(t))$, where $F \circ u = c$. By (4.2)

$$\frac{\nabla}{dt} v \equiv 0$$

is equivalent to

$$\dot{\xi}^k(t) + \sum_{i,j=1}^{2} \Gamma_{ij}^k(\tilde{c}(t))\xi^i(t)\dot{\tilde{c}}_j(t) \equiv 0 \tag{4.15}$$

for $k = 1, 2$. This is a system of *linear* ordinary differential equations of first order in the coefficients ξ^k. Because of the linearity of the system of equations we are in a situation which is better than the one we were in when we were dealing with the geodesic equations. Now solutions do exist not only locally, but on the whole interval I. Taking this into account, together with the fact that we are dealing with a system of first-order differential equations, so that it suffices to give an initial value (an initial derivative is not needed), we obtain through a reasoning similar to that in the proofs of theorems 4.5.11 and 4.5.12 the following theorem.

Theorem 4.7.2 (Existence and uniqueness of the parallel vector field) *Let S be a regular surface with Riemannian metric g. Let $c : I \to S$ be a smooth curve, $t_0 \in I$, $v_0 \in T_{c(t_0)}S$. Then there exists exactly one parallel vector field v along c with $v(t_0) = v_0$.*

Definition 4.7.3 Let S be a regular surface with Riemannian metric g. Let $c : [t_0, t_1] \to S$ be a smooth curve. The map $P_c : T_{c(t_0)}S \to T_{c(t_1)}S$ that maps v_0 to $v(t_1)$, where v is the unique parallel vector field along c with $v(t_0) = v_0$, is called ***parallel transport along*** c.

Proposition 4.7.4 *Let S be a regular surface with Riemannian metric g. Let $c : [t_0, t_1] \to S$ be a smooth curve. Then the following hold:*

(i) *If $v_0 \in T_{c(t_0)}S$, then the parallel vector field v along c with $v(t_0) = v_0$ is given by $v(t) = P_{c|[t_0,t]}(v_0)$.*
(ii) *The parallel transport $P_c : T_{c(t_0)}S \to T_{c(t_1)}S$ is a linear isometry.*
(iii) *Parallel transport is compatible with the reparametrisation of curve, i.e. if v is parallel along c and $\varphi : J \to [t_0, t]$ is a reparametrisation of c, then $v \circ \varphi$ is parallel along $c \circ \varphi$.*

Proof Statement (i) follows directly from the uniqueness of parallel transport. We have linearity of parallel transport because of the following: let v_0 and w_0 be from $T_{c(t_0)}S$, and let v and w be the parallel vector fields along c with $v(t_0) = v_0$ and $w(t_0) = w_0$. Let $\alpha, \beta \in \mathbb{R}$. Then the vector field $z(t) := \alpha \cdot v(t) + \beta \cdot w(t)$ is parallel to c, since

$$\frac{\nabla}{dt} z = \alpha \cdot \frac{\nabla}{dt} v + \beta \cdot \frac{\nabla}{dt} w = 0.$$

The field z satisfies the initial condition $z(t_0) = \alpha \cdot v_0 + \beta \cdot w_0$. Hence

$$P_c(\alpha \cdot v_0 + \beta \cdot w_0) = z(t_1) = \alpha \cdot v(t_1) + \beta \cdot w(t_1) = \alpha \cdot P_c(v_0) + \beta \cdot P_c(w_0).$$

This proves the linearity of P_c. For two parallel fields v and w along c we have

$$\begin{aligned} \frac{d}{dt} g(v, w) &= g\left(\frac{\nabla}{dt} v, w\right) + g\left(v, \frac{\nabla}{dt} w\right) \\ &= g(0, w) + g(v, 0) \\ &= 0. \end{aligned}$$

Thus $g_{c(t_0)}(v(t_0), w(t_0)) = g_{c(t_1)}(v(t_1), w(t_1))$, i.e.

$$g_{c(t_0)}(v_0, w_0) = g_{c(t_1)}(P_c(v_0), P_c(w_0)).$$

This means precisely that P_c is a linear *isometry*. Statement (ii) has therefore been proved.

Point (iii) follows from lemma 4.2.12:

$$\frac{\nabla}{dt}(v \circ \varphi) = \dot{\varphi} \cdot \left(\frac{\nabla}{dt} v \circ \varphi\right) = 0. \qquad \square$$

Exercise 4.26 Show that parallel transport is a concept of the inner geometry of the surface. More precisely, if $c : I \to S_1$ is a smooth curve, v a smooth vector field along c and $f : S_1 \to S_2$ a local isometry, then v is parallel along c

if and only if $df \circ v$ is parallel along $f \circ c$. Further conclude that the following diagram commutes for $I = [t_0, t_1]$, $c(t_0) = p$, $c(t_1) = q$:

$$\begin{array}{ccc} T_pS_1 & \xrightarrow{P_c} & T_qS_1 \\ \Big\downarrow d_pf & & \Big\downarrow d_qf \\ T_{f(p)}S_2 & \xrightarrow{P_{f\circ c}} & T_{f(q)}S_2 \end{array}$$

As the covariant and the usual derivative agree on the affine plane $E \subset S$, a vector field is parallel along a curve if and only if it is constant. In particular, the parallel transport $P_c : T_pE \to T_qE$ is independent of the curve c that connects the points p and q when we are dealing with the plane. This is not the case on general regular surfaces.

Example 4.7.5 Let S be the conical surface,

$$S = \left\{ (\xi, \eta, \zeta)^\top \;\middle|\; \xi^2 + \eta^2 = \tfrac{1}{3}\zeta^2, \zeta > 0 \right\}.$$

To investigate parallel vector fields along the curve $c : [0, \pi] \to S$, $c(t) = \frac{1}{2}(\sin(2t), \cos(2t), \sqrt{3})^\top$ in S, we consider the x–y plane $\tilde{S} = \mathbb{R}^2 \times \{0\}$. The map

$$f : \tilde{S} \to S, \quad f(x, y, 0) = \frac{1}{2\sqrt{x^2 + y^2}} \left(x^2 - y^2, 2xy, \sqrt{3}(x^2 + y^2) \right)^\top$$

is a local isometry, see exercise 4.2. The curve $\tilde{c} : [0, \pi] \to \tilde{S}$, $\tilde{c}(t) = (\cos(t), \sin(t), 0)^\top$, satisfies that $c = f \circ \tilde{c}$. In the plane the parallel vector fields are exactly the constant vector fields. The vector field $\tilde{v}(t) = (1, 0, 0)^\top$, for example, is parallel along $\tilde{c}$. It follows that $v := df \circ \tilde{v}$ is parallel along c in S. Using

$$\frac{\partial f}{\partial x} = \frac{1}{\sqrt{x^2 + y^2}} \begin{pmatrix} x - \frac{1}{2}\dfrac{(x^2 - y^2)\,x}{x^2 + y^2} \\ y - \dfrac{x^2 y}{x^2 + y^2} \\ \dfrac{\sqrt{3}}{2} x \end{pmatrix},$$

we calculate

$$\begin{aligned} v(t) &= d_{\tilde{c}(t)} f(\tilde{v}(t)) \\ &= d_{\tilde{c}(t)} f \begin{pmatrix} 1 \\ 0 \\ 0 \end{pmatrix} \\ &= \frac{\partial f}{\partial x}(\tilde{c}(t)) \end{aligned}$$

$$= \begin{pmatrix} \cos(t) - \frac{1}{2}\left(\cos(t)^2 - \sin(t)^2\right)\cos(t) \\ \sin(t) - \cos(t)^2 \sin(t) \\ \frac{\sqrt{3}}{2}\cos(t) \end{pmatrix}$$

$$= \begin{pmatrix} \cos(t)\left(1 - \frac{1}{2}\cos(2t)\right) \\ \sin(t)^3 \\ \frac{\sqrt{3}}{2}\cos(t) \end{pmatrix}.$$

In particular, $v(0) = (1/2, 0, \sqrt{3}/2)^\top$ and $v(\pi) = (-1/2, 0, -\sqrt{3}/2)^\top = -v(0)$. The parallel transport changes the tangent vector $v_0 = (1/2, 0, \sqrt{3}/2)^\top$ into its negative in one revolution around c. More generally, v_0 is mapped to itself for an even number k of revolutions, i.e. c, parametrised on $[0, k \cdot \pi]$, is mapped to itself by parallel translation, while it is mapped to $-v_0$ for uneven k.

Exercise 4.27 Illustrate this by drawing a constant vector field along half of a circular line on a sheet of paper, and then making a cone out of the half of the disc.

This example shows that parallel transport generally depends on the connecting curve on regular surfaces.

Exercise 4.28 Let S be a regular surface with Riemannian metric g. Let $c : I \to S$ be a constant curve, $c \equiv p$. A vector field along c is therefore a map $v : I \to T_pS$. Show that v is parallel if and only if v is constant.

4.8 Jacobi fields

We discussed the geometric meaning of Gauss curvature in section 3.6. Theorem 3.6.15 told us how the surface behaves near a point relative to the tangent plane at that point, in terms of the of the sign of the Gauss curvature at that point. But those investigations are only valid for the first fundamental form. In the case of a general Riemannian metric we want to understand what Gauss curvature tells us about the inner geometry of a surface. Jacobi fields are an important tool for this; they, as we will see below, contain information about the behaviour of neighbouring geodesics.

Definition 4.8.1 Let S be a regular surface with Riemannian metric g. Let c be a geodesic on S. A smooth vector field J on S along c is called a ***Jacobi field*** if

$$\frac{\nabla}{dt}\frac{\nabla}{dt}J = -R(J, \dot{c})\dot{c}.$$

This Jacobi equation is a linear ordinary differential equation of second order in J. We can therefore take arbitrary initial values $J(t_0)$ and $(\nabla/dt)J(t_0)$ in

$T_{c(t_0)}S$ for t_0 from the interval I on which c is defined, and obtain a unique solution of the equation, defined on the whole of I. As $T_{c(t_0)}S$ is two-dimensional the set of all Jacobi fields forms a four-dimensional vector space.

To what extent do Jacobi fields tell us something about the behaviour of neighbouring geodesics?

Definition 4.8.2 Let S be a regular surface with Riemannian metric g. Let $c_0 : I \to S$ be a geodesic on S. A ***geodesic variation*** of c_0 is a smooth map $c : (-\varepsilon, \varepsilon) \times I \to S$, $\varepsilon > 0$ such that $c(0, t) = c_0(t)$ and every curve $c_s : I \to S$, $c_s(t) := c(s, t)$ is a geodesic, $s \in (-\varepsilon, \varepsilon)$.

Proposition 4.8.3 *If c is a geodesic variation on S, then the corresponding variation field $J(t) := (\partial c/\partial s)(0, t)$ is a Jacobi field.*

Conversely, if J is a Jacobi field along a geodesic c_0, and if c_0 has a compact domain, then there exists a geodesic variation of c_0 with variation field J.

Proof (a) Let J be a variation field to a geodesic variation c. That every curve c_s is a geodesic means that

$$\frac{\nabla}{\partial t}\frac{\partial c}{\partial t} = 0.$$

We differentiate this equation with respect to s and obtain, using lemma 4.5.4, that

$$\begin{aligned} 0 &= \frac{\nabla}{\partial s}\frac{\nabla}{\partial t}\frac{\partial c}{\partial t} \\ &= \frac{\nabla}{\partial t}\frac{\nabla}{\partial s}\frac{\partial c}{\partial t} + R\left(\frac{\partial c}{\partial s}, \frac{\partial c}{\partial t}\right)\frac{\partial c}{\partial t} \\ &= \frac{\nabla}{\partial t}\frac{\nabla}{\partial t}\frac{\partial c}{\partial s} + R\left(\frac{\partial c}{\partial s}, \frac{\partial c}{\partial t}\right)\frac{\partial c}{\partial t}. \end{aligned}$$

In particular, setting $s = 0$ gives

$$0 = \frac{\nabla}{dt}\frac{\nabla}{dt}J + R(J, \dot{c}_0)\dot{c}_0,$$

the Jacobi equation.

(b) Conversely, let J be a Jacobi field along a geodesic c_0. Let t_0 be in the domain of c_0. We choose a curve $s \mapsto \varphi(s)$ with $\varphi(0) = c_0(t_0)$ and $\varphi'(0) = J(t_0)$, e.g.

$$\varphi(s) := \exp_{c_0(t_0)}(s \cdot J(t_0)).$$

Let $s \mapsto X(s)$ be the parallel vector field along φ with $X(0) = (\nabla/dt)J(t_0)$ and $s \mapsto Y(s)$ be the one with $Y(0) = \dot{c}_0(t_0)$. We define

$$c(s,t) := \exp_{\varphi(s)}\left((t-t_0)(sX(s)+Y(s))\right).$$

Then

$$c(0,t) = \exp_{c_0(t_0)}((t-t_0)\dot{c}_0(t_0)) = c_0(t)$$

and $t \mapsto c(s,t)$ is a geodesic for every s. Since the t-interval is compact, $c(s,t)$ is defined for all sufficiently small $|s|$ independent of t.

We have constructed a geodesic variation c of c_0. The variation field $\tilde{J}$ of c is again a Jacobi field by (a), and we have

$$\tilde{J}(t_0) = \frac{\partial c}{\partial s}(0,t_0) = \frac{d}{ds}\Big|_{s=0} \exp_{\varphi(s)}(0) = \varphi'(0) = J(t_0)$$

and

$$\begin{aligned}\frac{\nabla}{dt}\tilde{J}(t_0) &= \frac{\nabla}{\partial t}\frac{\partial c}{\partial s}(0,t_0) = \frac{\nabla}{\partial s}\frac{\partial c}{\partial t}(0,t_0) = \frac{\nabla}{ds}(sX(s)+Y(s)) \\ &= X(0) + \frac{\nabla Y}{ds}(0) = \frac{\nabla}{dt}J(t_0) + 0 = \frac{\nabla}{dt}J(t_0).\end{aligned}$$

Thus the Jacobi fields J and $\tilde{J}$ have the same initial values at $t = t_0$ and hence agree. It follows that J is the variation field of the geodesic variation c. □

Not all Jacobi fields are equally interesting. Jacobi fields are uninteresting if the corresponding geodesic variation originates from a simple reparametrisation of the geodesic c_0. As c_0 is a geodesic, the same is true for $t \mapsto c_0(\alpha t + \beta)$ with $\alpha, \beta \in \mathbb{R}$ arbitrary, see exercise 4.19. We can therefore define a geodesic variation of c_0 for $a, b \in \mathbb{R}$ via

$$c(s,t) := c_0(ast + t + bs).$$

Then the corresponding Jacobi field J has initial values

$$J(t_0) = \frac{\partial c}{\partial s}(0,t_0) = \frac{d}{ds}\Big|_{s=0} c_0(s(at_0+b)+t_0) = (at_0+b)\cdot \dot{c}_0(t_0)$$

and

$$\frac{\nabla}{dt}J(t_0) = \frac{\nabla}{dt}\Big|_{t=t_0}((at+b)\cdot \dot{c}_0(t)) = a \cdot \dot{c}_0(t_0).$$

We observe that for uninteresting Jacobi fields both $J(t_0)$ and $(\nabla/dt)J(t_0)$ are multiples of $\dot{c}_0(t_0)$, i.e. tangential to c_0. Indeed, we can give these Jacobi fields explicitly. Constraints due to the Jacobi equation and the initial values give that

$$J(t) = (at + b)\dot{c}_0(t).$$

These uninteresting Jacobi fields therefore form a two-dimensional subspace of the four-dimensional space of all Jacobi fields along c_0.

These Jacobi fields have not given us any information about the geometry of the surface. Let us now move on to the interesting Jacobi fields. Let $n(t_0) \in T_{c_0(t_0)}S$ be one of the two unit tangent vectors that are perpendicular to $\dot{c}_0(t_0)$. Parallel transport along c_0 then gives a unit normal field $n(t)$ along c_0. As c_0 is a geodesic, we know that $n(t)$ is perpendicular to $\dot{c}_0(t)$ for every t, not only for $t = t_0$. Now those Jacobi fields are interesting for which both $J(t_0)$ and $(\nabla/dt)J(t_0)$ are multiples of $n(t_0)$. We thus obtain another two-dimensional subspace of the vector space of Jacobi fields, complementary to the one of the uninteresting Jacobi fields. An arbitrary Jacobi field along c_0 can then be written uniquely as the sum of an interesting Jacobi field and an uninteresting one.

Let us now show that such an interesting Jacobi field is perpendicular to $\dot{c}_0(t)$ for all t, not only for $t = t_0$.

$$\begin{aligned} \left(\frac{d}{dt}\right)^2 g(J(t), \dot{c}_0(t)) &= g\left(\left(\frac{\nabla}{dt}\right)^2 J(t), \dot{c}_0(t)\right) \\ &= -g(R(J(t), \dot{c}_0(t))\dot{c}_0(t), \dot{c}_0(t)) \\ &= 0 \end{aligned}$$

because of lemma 4.3.10 (b). Hence the function $t \mapsto g(J(t), \dot{c}_0(t))$ is of the form $g(J(t), \dot{c}_0(t)) = \alpha t + \beta$ for certain $\alpha, \beta \in \mathbb{R}$. If this function and its first derivative vanish for $t = t_0$, then it must vanish for all t. The interesting Jacobi fields can therefore be written in the form

$$J(t) = \chi(t) \cdot n(t).$$

The Jacobi equation then translates to an ordinary differential equation in the coefficient function χ:

$$\ddot{\chi}(t) \cdot n(t) = \frac{\nabla}{dt}\frac{\nabla}{dt}J(t) = -R(J(t), \dot{c}(t))\dot{c}(t) = -\chi(t)R(n(t), \dot{c}(t))\dot{c}(t),$$

and hence

$$\ddot{\chi}(t) = -\chi(t)g(R(n(t), \dot{c}(t))\dot{c}(t), n(t)) = -\chi(t)K(c_0(t)).$$

We see that for the interesting Jacobi fields the Jacobi equation is equivalent to the following ordinary differential equation:

$$\ddot{\chi} = -\chi \cdot (K \circ c_0), \tag{4.16}$$

in which the Gauss curvature K plays a significant role. The solution to this differential equation cannot generally be found explicitly very easily. The interesting Jacobi fields give us actual information about the surface that tells us approximately how much neighbouring geodesics (from our geodesic variation) diverge from each other. Let us illustrate this with a simple but important example.

Example 4.8.4 If the Gauss curvature of a surface is constant, $K \equiv \kappa$, then (4.16) can be solved explicitly. Setting

$$\mathfrak{s}_\kappa(t) := \begin{cases} \sin(\sqrt{\kappa}t)/\sqrt{\kappa}, & \kappa > 0, \\ t, & \kappa = 0, \\ \sinh(\sqrt{|\kappa|}t)/\sqrt{|\kappa|}, & \kappa < 0, \end{cases} \qquad \mathfrak{c}_\kappa(t) := \begin{cases} \cos(\sqrt{\kappa}t), & \kappa > 0, \\ 1, & \kappa = 0, \\ \cosh(\sqrt{|\kappa|}t), & \kappa < 0, \end{cases}$$

we have

$$\mathfrak{c}_\kappa(t)^2 + \kappa\mathfrak{s}_\kappa(t)^2 = 1, \tag{4.17}$$

$$\dot{\mathfrak{s}}_\kappa = \mathfrak{c}_\kappa, \tag{4.18}$$

$$\dot{\mathfrak{c}}_\kappa = -\kappa\mathfrak{s}_\kappa. \tag{4.19}$$

In particular, the functions $\mathfrak{s}_\kappa$ and $\mathfrak{c}_\kappa$ solve the differential equation

$$\ddot{\chi} = -\kappa \cdot \chi$$

and are linearly independent. Hence every solution of this differential equation can be written as a linear combination of $\mathfrak{s}_\kappa$ and $\mathfrak{c}_\kappa$. The interesting Jacobi fields are thus of the form

$$J(t) = (\alpha\mathfrak{s}_\kappa(t) + \beta\mathfrak{c}_\kappa(t))n(t).$$

They behave very differently depending on the sign of κ. If $\kappa > 0$, then the length of J is periodic with period $2\pi/\kappa$. In particular, J is 0 again and again. For geodesics that diverge at a point this means that they will approach each other again after some time.

If $\kappa = 0$, then the length increases linearly. This is the familiar situation that we know from straight lines in the Euclidean plane.

In the case $\kappa < 0$ the length increases even more quickly, in fact exponentially. Summarising we can say that the more negative the curvature, the more the geodesics diverge.

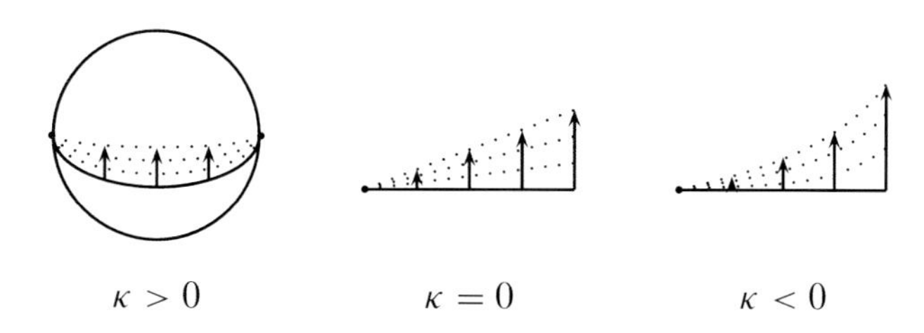

Exercise 4.29 Show that on the surface of revolution with parametrisation $F(t,\varphi) = (r(t)\cos(\varphi), r(t)\sin(\varphi), t)^\top$ the vector field $\partial F/\partial\varphi$ is a Jacobi field along the lines of longitude.

4.9 Spherical and hyperbolic geometry

We now want to investigate spherical geometry in more detail, in particular with regard to trigonometric rules analogous to those of Euclidean geometry, which we treated in the first chapter. At the same time we introduce a surface which is in a way the negatively curved analogue of the sphere, the hyperbolic plane. We can investigate its trigonometry simultaneously. The hyperbolic plane has great historic importance, since it satisfies all axioms of Euclidean geometry except for the parallel axiom. It also shows that the efforts to derive the parallel axiom from the other axioms, which had been going on for thousands of years, were inevitably in vain.

Let us construct the hyperbolic plane. For $\kappa \in \mathbb{R}$ we set

$$\widehat{\mathbb{M}}_\kappa := \left\{ (x,y,z)^\top \in \mathbb{R}^3 \,\middle|\, \kappa(x^2+y^2)+z^2 = 1 \right\}.$$

In the case $\kappa = 1$ this is simply the sphere $\widehat{\mathbb{M}}_1 = S^2$. In the case $\kappa = 0$, on the other hand, we obtain the union of two parallel planes through the points $(0,0,1)$ and $(0,0,-1)$. We are now mostly interested in the case $\kappa = -1$. The surface $\widehat{\mathbb{M}}_{-1}$ is the ***two-sheeted hyperboloid***, consisting of two sheets that are graphs of the functions $z = \pm\sqrt{1+x^2+y^2}$. Being graphs, they are, in particular, regular surfaces. Since we do not want to work with several sheets, we additionally define

$$\mathbb{M}_\kappa := \begin{cases} \widehat{\mathbb{M}}_\kappa, & \kappa > 0, \\ \{(x,y,z)^\top \in \widehat{\mathbb{M}}_\kappa \mid z > 0\}, & \kappa \le 0. \end{cases}$$

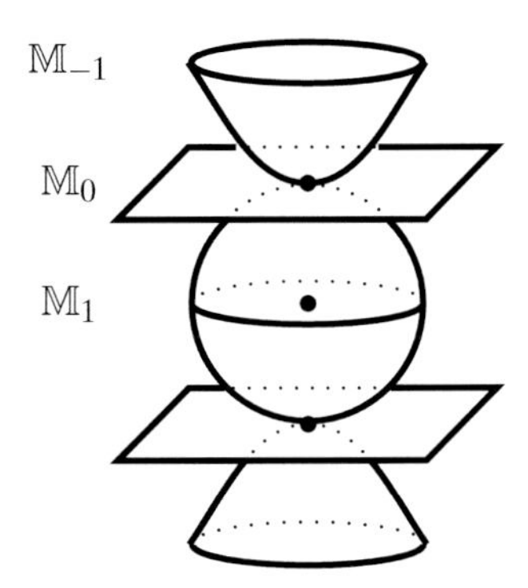

So we take only the upper plane or sheet. We already know that $\mathbb{M}_1$ has constant Gauss curvature $K \equiv 1$, while the plane has $K \equiv 0$, in each case with the first fundamental form as Riemannian metric. If we now take the first fundamental form on $\mathbb{M}_{-1}$, then it does not have curvature -1, which we would have liked.

Exercise 4.30 Show that $\mathbb{M}_{-1}$ with the first fundamental form as Riemannian metric has positive Gauss curvature.

Let us therefore take another Riemannian metric on $\mathbb{M}_{-1}$. Indeed, this example is so important that we have already introduced general Riemannian metrics because of this. For the construction of the Riemannian metric we define the following symmetric bilinear form on $\mathbb{R}^3$:

$$\left\langle \begin{pmatrix} x \\ y \\ z \end{pmatrix}, \begin{pmatrix} x' \\ y' \\ z' \end{pmatrix} \right\rangle_{\kappa} := xx' + yy' + \frac{zz'}{\kappa}.$$

In the case $\kappa = 1$, this is the usual Euclidean scalar product. If $\kappa = -1$, then the symmetric bilinear form is not degenerate, but indefinite. We call $\langle \cdot, \cdot \rangle_{-1}$ the ***Minkowski scalar product*** on $\mathbb{R}^3$. It plays an important role in Einstein's special relativity theory. We call the set of all vectors of "vanishing length" the ***light cone***

$$\left\{ X \in \mathbb{R}^3 \,\middle|\, \langle X, X \rangle_{-1} = 0 \right\} = \left\{ (x, y, z)^\top \in \mathbb{R}^3 \,\middle|\, z^2 = x^2 + y^2 \right\}.$$

We now define the Riemannian metric on $\mathbb{M}_\kappa$ by restricting the symmetric bilinear form $\langle \cdot, \cdot \rangle_\kappa$ for every point $p \in \mathbb{M}_\kappa$ to $T_p\mathbb{M}_\kappa$. The construction is thus similar to that of the first fundamental form, except that we have replaced the usual Euclidean scalar product on $\mathbb{R}^3$ by $\langle \cdot, \cdot \rangle_\kappa$.

In the case $\kappa = 0$, one could complain that we need to divide by 0 in the definition of $\langle \cdot, \cdot \rangle_0$, and that the definition is therefore not reasonable. But every tangent plane of $\mathbb{M}_0$ equals $\mathbb{R}^2 \times \{0\}$, i.e. the coefficient of the zz'-term does not play a role anyway. We could therefore without worry replace $\langle\cdot,\cdot\rangle_0$ by the usual Euclidean scalar product without changing the Riemannian metric induced on $\mathbb{M}_0$. It is simply the usual Euclidean metric on the plane.

Since for $\kappa < 0$ the bilinear form $\langle\cdot,\cdot\rangle_\kappa$ is no longer positive definite, it is a priori not clear that the restrictions to the tangent planes of $\mathbb{M}_\kappa$ are positive definite. They need to have this property if we want to obtain a Riemannian metric. To check that at least the restrictions of $\langle\cdot,\cdot\rangle_\kappa$ to the tangent planes are positive definite, we parametrise $\mathbb{M}_\kappa$ as follows:

$$F_\kappa : \mathbb{R}\times\mathbb{R} \to \mathbb{R}^3, \quad F_\kappa(r,\varphi) := \begin{pmatrix} \mathfrak{s}_\kappa(r)\cos(\varphi) \\ \mathfrak{s}_\kappa(r)\sin(\varphi) \\ \mathfrak{c}_\kappa(r) \end{pmatrix}.$$

Here $\mathfrak{c}_\kappa$ and $\mathfrak{s}_\kappa$ are the generalised sine and cosine functions defined in example 4.8.4. Using (4.17) it is easily seen that the image of F_κ is exactly $\mathbb{M}_\kappa$, more precisely

$$F_\kappa(\mathbb{R}\times\mathbb{R}) = F_\kappa([0,\pi/\sqrt{\kappa}]\times[0,2\pi)) = \mathbb{M}_\kappa$$

in the case $\kappa > 0$ and

$$F_\kappa(\mathbb{R}\times\mathbb{R}) = F_\kappa([0,\infty)\times[0,2\pi)) = \mathbb{M}_\kappa$$

in the case $\kappa \leq 0$. Using (4.18) and (4.19) we immediately get the partial derivatives of F_κ:

$$\frac{\partial F_\kappa}{\partial r}(r,\varphi) = \begin{pmatrix} \mathfrak{c}_\kappa(r)\cos(\varphi) \\ \mathfrak{c}_\kappa(r)\sin(\varphi) \\ -\kappa\mathfrak{s}_\kappa(r) \end{pmatrix}, \quad \frac{\partial F_\kappa}{\partial \varphi}(r,\varphi) = \begin{pmatrix} -\mathfrak{s}_\kappa(r)\sin(\varphi) \\ \mathfrak{s}_\kappa(r)\cos(\varphi) \\ 0 \end{pmatrix}.$$

The two partial derivatives are linearly independent for $r\in(0,\pi/\sqrt{\kappa})$ and $r\in(0,\infty)$. Restriction of F_κ to suitable domains therefore gives us local parametrisations cf. the $\mathbb{M}_\kappa$ except for the points $(0,0,1)^\top$ and, in the case $\kappa > 0$, $(0,0,-1)^\top$. At those two special points the tangent plane is exactly the x–y plane and a restriction of $\langle\cdot,\cdot\rangle_\kappa$ gives the usual Euclidean scalar product, and is thus positive definite. For the other points we calculate the coefficients g_{ij} of our restriction with respect to the local parametrisation F_κ, e.g.

$$\begin{aligned}\left\langle \frac{\partial F_\kappa}{\partial r}(r,\varphi), \frac{\partial F_\kappa}{\partial r}(r,\varphi)\right\rangle_\kappa &= \mathfrak{c}_\kappa(r)^2\cos(\varphi)^2 + \mathfrak{c}_\kappa(r)^2\sin(\varphi)^2 + \frac{1}{\kappa}\kappa^2\mathfrak{s}_\kappa(r)^2 \\ &= \mathfrak{c}_\kappa(r)^2 + \kappa\mathfrak{s}_\kappa(r)^2 = 1.\end{aligned}$$

The other components are obtained in a similar way:

$$\big(g_{ij}(r,\varphi)\big)_{ij} = \begin{pmatrix} 1 & 0 \\ 0 & \mathfrak{s}_\kappa(r)^2 \end{pmatrix}. \tag{4.20}$$

This matrix is positive definite for $r \in (0, \pi/\sqrt{\kappa})$ and $r \in (0, \infty)$. Therefore we really have a Riemannian metric on $\mathbb{M}_\kappa$.

Definition 4.9.1 The regular surface $\mathbb{M}_{-1}$ together with the Riemannian metric defined via the restriction of $\langle \cdot, \cdot \rangle_{-1}$, is called the ***hyperbolic plane***.

The special form of the metric in (4.20) shows that F_κ gave us a parametrisation in geodesic polar coordinates at the point $(0,0,1)^\top$. By lemma 4.6.10 the Gauss curvature is now given by

$$K = -\frac{\ddot{\mathfrak{s}}_\kappa(r)}{\mathfrak{s}_\kappa(r)} = \kappa.$$

In particular, the hyperbolic plane has constant Gauss curvature -1. Further, the curves $r \mapsto F_\kappa(r, \varphi)$ are the geodesics parametrised by arc-length that begin at $(0,0,1)^\top$. The traces of these geodesics have a particularly simple geometric characterisation. They are exactly the intersection of the surface $\mathbb{M}_\kappa$ with the plane through the origin, through $(0,0,1)^\top$ and through $(\cos(\varphi), \sin(\varphi), 0)^\top$. In the case of the sphere $\mathbb{M}_1$ this gives the great circles through the "north pole" $(0,0,1)^\top$. In the case of the plane we get the straight lines through this point. In the case of the hyperbolic plane we obtain hyperbolas. What do the other geodesics look like, those that do not pass through $(0,0,1)^\top$?

To understand this, we will use isometries to map the geodesics through $(0,0,1)^\top$ to other geodesics. To do so we need to obtain enough isometries. We investigate which linear maps on $\mathbb{R}^3$ become isometries after a restriction to $\mathbb{M}_\kappa$. For this purpose let $L \in \mathrm{GL}(3)$. Let us first consider the case $\kappa \neq 0$. If L preserves the symmetric bilinear form $\langle \cdot, \cdot \rangle_\kappa$, i.e.

$$\langle LX, LY \rangle_\kappa = \langle X, Y \rangle_\kappa$$

for all $X, Y \in \mathbb{R}^3$, then, in particular, $\widehat{\mathbb{M}}_\kappa$ is mapped to itself and for every $p \in \widehat{\mathbb{M}}_\kappa$ the map $d_pL = L|_{T_p\widehat{\mathbb{M}}_\kappa} : T_p\widehat{\mathbb{M}}_\kappa \to T_{Lp}\widehat{\mathbb{M}}_\kappa$ is a linear isometry. However, in the case $\kappa < 0$ it can happen that the two sheets of the hyperboloid are interchanged. To prevent this we additionally require that $(0,0,1)^\top$ is mapped back to the upper sheet, i.e. the third component of the image vector must be positive. Writing

$$L = \begin{pmatrix} L_{11} & L_{12} & L_{13} \\ L_{21} & L_{22} & L_{23} \\ L_{31} & L_{32} & L_{33} \end{pmatrix},$$

we see that $L_{33} > 0$. The other points of the upper sheet will for reasons of continuity be mapped to the upper sheet as well:

$$L(\mathbb{M}_\kappa) = \mathbb{M}_\kappa.$$

In the case $\kappa = 0$, the vectors of the form $(x, y, 1)^\top$ must be mapped to vectors whose third component is also equal to 1. This means precisely that $L_{31} = L_{32} = 0$ and $L_{33} = 1$. Further, the Euclidean map must be preserved in the x–y plane for the maps to be isometries. This means that

$$\begin{pmatrix} L_{11} & L_{12} \\ L_{21} & L_{22} \end{pmatrix} \in \mathrm{O}(2).$$

Summarising, we can say that we have already found rather large groups of isometries on $\mathbb{M}_\kappa$, i.e.

$$G_\kappa := \begin{cases} \{L \in \mathrm{GL}(3) \mid \langle LX, LY\rangle_\kappa = \langle X, Y\rangle_\kappa \text{ and } L_{33} > 0\}, & \kappa < 0, \\ \{L \in \mathrm{GL}(3) \mid L_{31} = L_{32} = 0, L_{33} = 1, (L_{ij})_{i,j=1,2} \in \mathrm{O}(2)\}, & \kappa = 0, \\ \{L \in \mathrm{GL}(3) \mid \langle LX, LY\rangle_\kappa = \langle X, Y\rangle_\kappa\}, & \kappa > 0. \end{cases}$$

For $\kappa = 1$ we have $G_1 = \mathrm{O}(3)$. In the case $\kappa = 0$, the group G_0 is isomorphic to the Euclidean group $\mathrm{E}(2)$, see definition 1.2.1. Indeed, $L \in G_0$ maps the vector $(x, y, 1)^\top$ to $(L_{11}x + L_{12}y + L_{13}, L_{21}x + L_{22}y + L_{23}, 1)^\top$, i.e. L acts like a Euclidean motion $F_{A,b}$ with $A = (L_{ij})_{i,j=1,2}$ and translational part $b = (L_{13}, L_{23})^\top$.

The elements of the group G_{-1} are called the ***time-preserving Lorentz transformations*** or ***orthochronous Lorentz transformations***, a terminology that comes from relativity theory.

For every κ, the group G_κ contains the rotations around the z-axis:

$$\begin{pmatrix} \cos(\varphi) & -\sin(\varphi) & 0 \\ \sin(\varphi) & \cos(\varphi) & 0 \\ 0 & 0 & 1 \end{pmatrix} \in G_\kappa.$$

Hence every point from $\mathbb{M}_\kappa$ can be moved into the x–z plane by an isometry, i.e. to a point of the form $(\mathfrak{s}_\kappa(r), 0, \mathfrak{c}_\kappa(r))^\top$, without moving the point $(0, 0, 1)^\top$ in this process.

For such a point $(\mathfrak{s}_\kappa(r), 0, \mathfrak{c}_\kappa(r))^\top$ we consider the linear map

$$L := \begin{pmatrix} -\mathfrak{c}_\kappa(r) & 0 & \mathfrak{s}_\kappa(r) \\ 0 & 1 & 0 \\ \kappa\mathfrak{s}_\kappa(r) & 0 & \mathfrak{c}_\kappa(r) \end{pmatrix} \in G_\kappa. \tag{4.21}$$

This L interchanges the two points $(\mathfrak{s}_\kappa(r), 0, \mathfrak{c}_\kappa(r))^\top$ and $(0, 0, 1)^\top$. Altogether we see that *every* point from $\mathbb{M}_\kappa$ can be mapped to the point $(0, 0, 1)^\top$ by a suitable isometry. We say that the isometry group *acts transitively* on $\mathbb{M}_\kappa$. Loosely one could say that $\mathbb{M}_\kappa$ with the Riemannian metric that we constructed is a surface with a very high symmetry.

To come back to our discussion of geodesics, for an arbitrary geodesic through a point $p \in \mathbb{M}_\kappa$ we can find an isometry $L \in G_\kappa$ which maps p to $(0, 0, 1)^\top$ and the geodesic to the intersection of $\mathbb{M}_\kappa$ with a two-dimensional subspace E of $\mathbb{R}^3$. As L is a linear map on $\mathbb{R}^3$, we know that $L^{-1}(E)$ is a two-dimensional subspace of $\mathbb{R}^3$ as well. Generally it no longer contains the point $(0,0,1)^\top$, but instead contains the point p. This is summarised in the following theorem:

Theorem 4.9.2 *The traces of geodesics of $\mathbb{M}_\kappa$ with the Riemannian metric defined by the restriction of $\langle \cdot, \cdot \rangle_\kappa$ are exactly the (non-empty) intersections of $\mathbb{M}_\kappa$ with the two-dimensional subspaces of $\mathbb{R}^3$.*

In the case $\kappa = 1$ these are the great circles on S^2, in the case $\kappa = 0$ they are the straight lines in the plane and in the case $\kappa = -1$ we obtain hyperbolas.

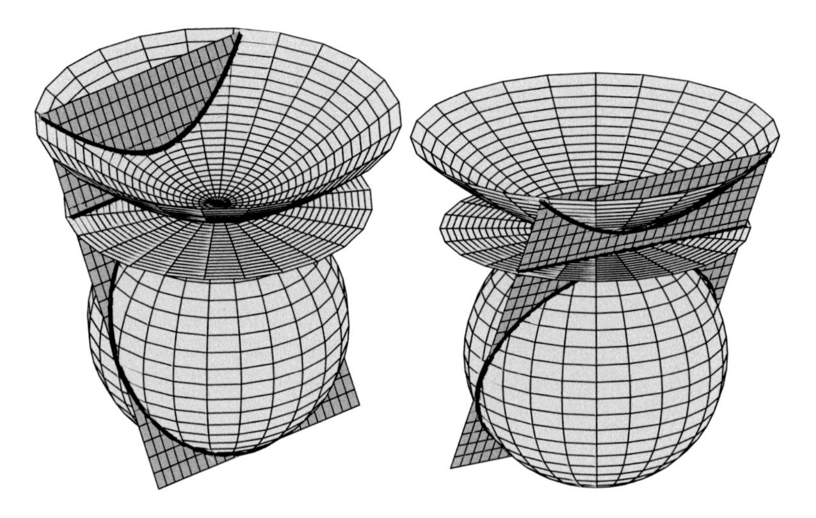

Exercise 4.31 Show that G_κ already contains all isometries of $\mathbb{M}_\kappa$.

Hint First use the exponential map to show that for $p \in \mathbb{M}_\kappa$ an isometry $f : \mathbb{M}_\kappa \to \mathbb{M}_\kappa$ is already uniquely determined by the image point $f(p)$ and the differential $d_pf : T_p\mathbb{M}_\kappa \to T_{f(p)}\mathbb{M}_\kappa$.

We now begin to derive the trigonometric theorems for $\mathbb{M}_\kappa$. For this purpose we consider a ***geodesic triangle*** on $\mathbb{M}_\kappa$, i.e. three points $A, B, C \in \mathbb{M}_\kappa$, which are connected by geodesics of length a, b, c. We will denote the interior angles at the vertices by α, β, γ. We will in the following always assume that the geodesic triangles are not degenerate, i.e. the lengths of the sides are positive and in the case $\kappa > 0$ smaller than $\pi/\sqrt{\kappa}$ and the angles are in $(0, \pi)$.

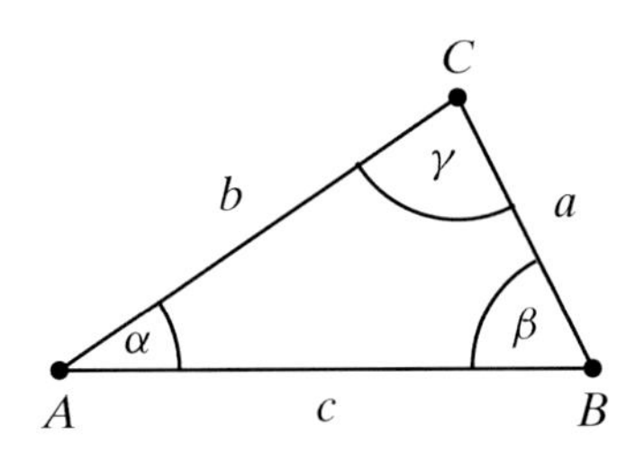

We want to find relations between the angles and the lengths of sides. As isometries do not change angles or the lengths of sides, we can first use an isometry to map the point A to $(0, 0, 1)^\top$ and then rotate B into the x–z plane. Without loss of generality we therefore have

$$A = \begin{pmatrix} 0 \\ 0 \\ 1 \end{pmatrix}, \quad B = \begin{pmatrix} \mathfrak{s}_\kappa(c) \\ 0 \\ \mathfrak{c}_\kappa(c) \end{pmatrix}, \quad C = \begin{pmatrix} \mathfrak{s}_\kappa(b)\cos(\alpha) \\ \mathfrak{s}_\kappa(b)\sin(\alpha) \\ \mathfrak{c}_\kappa(b) \end{pmatrix}.$$

We apply isometry L from (4.21), which interchanges points A and B, with $r = c$ to the entire triangle.

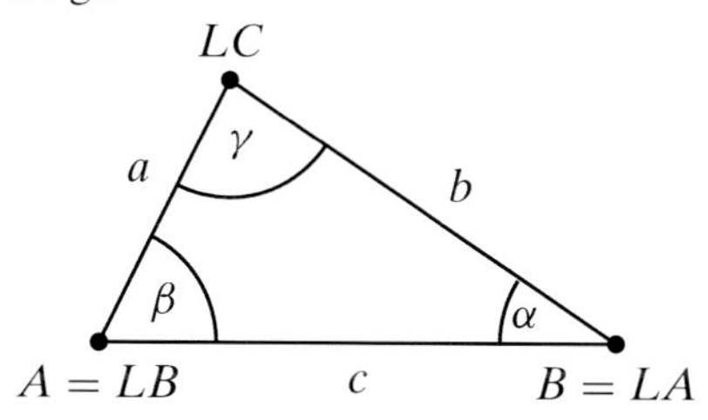

As the side from LC to LB has length a and the angle at A is β, we have

$$LC = \begin{pmatrix} \mathfrak{s}_\kappa(a)\cos(\beta) \\ \mathfrak{s}_\kappa(a)\sin(\beta) \\ \mathfrak{c}_\kappa(a) \end{pmatrix}. \tag{4.22}$$

On the other hand, we calculate

$$\begin{aligned} LC &= \begin{pmatrix} -\mathfrak{c}_\kappa(c) & 0 & \mathfrak{s}_\kappa(c) \\ 0 & 1 & 0 \\ \kappa\mathfrak{s}_\kappa(c) & 0 & \mathfrak{c}_\kappa(c) \end{pmatrix} \cdot \begin{pmatrix} \mathfrak{s}_\kappa(b)\cos(\alpha) \\ \mathfrak{s}_\kappa(b)\sin(\alpha) \\ \mathfrak{c}_\kappa(b) \end{pmatrix} \\ &= \begin{pmatrix} -\mathfrak{c}_\kappa(c)\mathfrak{s}_\kappa(b)\cos(\alpha) + \mathfrak{s}_\kappa(c)\mathfrak{c}_\kappa(b) \\ \mathfrak{s}_\kappa(b)\sin(\alpha) \\ \kappa\mathfrak{s}_\kappa(c)\mathfrak{s}_\kappa(b)\cos(\alpha) + \mathfrak{c}_\kappa(c)\mathfrak{c}_\kappa(b) \end{pmatrix}. \end{aligned} \tag{4.23}$$

We compare the components in (4.22) and (4.23) and obtain

$$\mathfrak{s}_\kappa(a)\cos(\beta) = -\mathfrak{c}_\kappa(c)\mathfrak{s}_\kappa(b)\cos(\alpha) + \mathfrak{s}_\kappa(c)\mathfrak{c}_\kappa(b), \tag{4.24}$$

$$\mathfrak{s}_\kappa(a)\sin(\beta) = \mathfrak{s}_\kappa(b)\sin(\alpha), \tag{4.25}$$

$$\mathfrak{c}_\kappa(a) = \kappa\mathfrak{s}_\kappa(c)\mathfrak{s}_\kappa(b)\cos(\alpha) + \mathfrak{c}_\kappa(b)\mathfrak{c}_\kappa(c). \tag{4.26}$$

Equation 4.25 gives the *sine rule*

$$\frac{\mathfrak{s}_\kappa(a)}{\sin(\alpha)} = \frac{\mathfrak{s}_\kappa(b)}{\sin(\beta)},$$

and (4.26) is called the *cosine rule for sides*. Multiplying (4.24) by $\cos(\alpha)$ and (4.25) by $\sin(\alpha)$ and subtracting the resulting equations one from the other gives

$$\begin{aligned}\mathfrak{s}_\kappa(a)\,(\cos(\alpha)\cos(\beta)-\sin(\alpha)\sin(\beta))\\ = -\mathfrak{c}_\kappa(c)\mathfrak{s}_\kappa(b)\cos(\alpha)^2+\mathfrak{s}_\kappa(c)\mathfrak{c}_\kappa(b)\cos(\alpha)-\mathfrak{s}_\kappa(b)\sin(\alpha)^2.\end{aligned} \tag{4.27}$$

Interchanging the roles of b and c in (4.24) (i.e. considering triangle ABC instead of ACB) we obtain

$$\mathfrak{s}_\kappa(a)\cos(\gamma) = -\mathfrak{c}_\kappa(b)\mathfrak{s}_\kappa(c)\cos(\alpha)+\mathfrak{s}_\kappa(b)\mathfrak{c}_\kappa(c). \tag{4.28}$$

We substitute (4.28) into (4.27) and finally use the sine rule:

$$\begin{aligned}&\mathfrak{s}_\kappa(a)\,(\cos(\alpha)\cos(\beta)-\sin(\alpha)\sin(\beta))\\ &\quad= -\mathfrak{c}_\kappa(c)\mathfrak{s}_\kappa(b)\cos(\alpha)^2-\mathfrak{s}_\kappa(a)\cos(\gamma)+\mathfrak{s}_\kappa(b)\mathfrak{c}_\kappa(c)-\mathfrak{s}_\kappa(b)\sin(\alpha)^2\\ &\quad= \mathfrak{s}_\kappa(b)\mathfrak{c}_\kappa(c)\sin(\alpha)^2-\mathfrak{s}_\kappa(a)\cos(\gamma)-\mathfrak{s}_\kappa(b)\sin(\alpha)^2\\ &\quad= \mathfrak{s}_\kappa(a)\mathfrak{c}_\kappa(c)\sin(\alpha)\sin(\beta)-\mathfrak{s}_\kappa(a)\cos(\gamma)-\mathfrak{s}_\kappa(a)\sin(\alpha)\sin(\beta).\end{aligned}$$

We divide this equation by $\mathfrak{s}_\kappa(a)$, add $\sin(\alpha)\sin(\beta)$ on both sides and obtain

$$\cos(\alpha)\cos(\beta) = \mathfrak{c}_\kappa(c)\sin(\alpha)\sin(\beta)-\cos(\gamma).$$

This is the *cosine rule for angles*. We summarise these rules in the following theorem:

Theorem 4.9.3 (Spherical and hyperbolic trigonometry) *Let ABC be a geodesic triangle with side-lengths a, b, c and interior angles α, β, γ on $\mathbb{M}_\kappa$. Then the following hold:*

(i) *the sine rule:*

$$\frac{\mathfrak{s}_\kappa(a)}{\sin(\alpha)} = \frac{\mathfrak{s}_\kappa(b)}{\sin(\beta)} = \frac{\mathfrak{s}_\kappa(c)}{\sin(\gamma)};$$

(ii) *the cosine rule for sides:*

$$\begin{aligned}\mathfrak{c}_\kappa(a) &= \mathfrak{c}_\kappa(b)\mathfrak{c}_\kappa(c)+\kappa\mathfrak{s}_\kappa(b)\mathfrak{s}_\kappa(c)\cos(\alpha),\\ \mathfrak{c}_\kappa(b) &= \mathfrak{c}_\kappa(a)\mathfrak{c}_\kappa(c)+\kappa\mathfrak{s}_\kappa(a)\mathfrak{s}_\kappa(c)\cos(\beta),\\ \mathfrak{c}_\kappa(c) &= \mathfrak{c}_\kappa(a)\mathfrak{c}_\kappa(b)+\kappa\mathfrak{s}_\kappa(a)\mathfrak{s}_\kappa(b)\cos(\gamma);\end{aligned}$$

(iii) *the cosine rule for angles:*

$$\cos(\alpha) = \mathfrak{c}_\kappa(a)\sin(\beta)\sin(\gamma) - \cos(\beta)\cos(\gamma),$$
$$\cos(\beta) = \mathfrak{c}_\kappa(b)\sin(\alpha)\sin(\gamma) - \cos(\alpha)\cos(\gamma),$$
$$\cos(\gamma) = \mathfrak{c}_\kappa(c)\sin(\alpha)\sin(\beta) - \cos(\alpha)\cos(\beta).$$

This theorem contains three trigonometric rules for both spherical ($\kappa = 1$) and hyperbolic geometry ($\kappa = -1$). The theorem holds for all $\kappa \in \mathbb{R}$, in particular for $\kappa = 0$. Indeed, the sine rule of Euclidean geometry (theorem 1.2.6) is included in theorem 4.9.3. The cosine rule for sides on the other hand is trivial in the Euclidean case. The cosine rule for angles says that in the case $\kappa = 0$

$$\cos(\gamma) = \sin(\alpha)\sin(\beta) - \cos(\alpha)\cos(\beta) = -\cos(\alpha+\beta) = \cos(\pi - \alpha - \beta),$$

i.e. we obtain theorem 1.2.7 about the sum of interior angles in a Euclidean triangle. In the hyperbolic case $\kappa = -1$ or more generally if $\kappa < 0$, we have for $c > 0$ that $\mathfrak{c}_\kappa(c) > 1$, hence

$$\cos(\gamma) > \sin(\alpha)\sin(\beta) - \cos(\alpha)\cos(\beta) = -\cos(\alpha+\beta) = \cos(\pi - \alpha - \beta)$$

and thus

$$\alpha + \beta + \gamma < \pi.$$

Hyperbolic triangles are therefore “narrow” compared to Euclidean ones. Analogously we see that in the spherical case ($\kappa > 0$)

$$\alpha + \beta + \gamma > \pi.$$

Spherical triangles are “fat”.

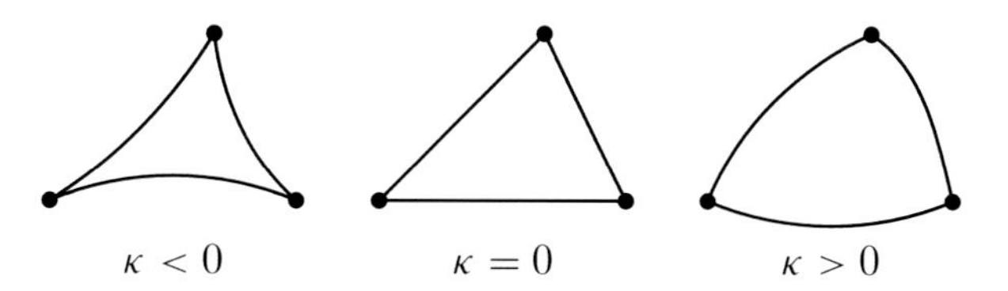

Exercise 4.32 Use the sine rule for geodesic triangles on $\mathbb{M}_\kappa$ to derive the *height formula*

$$\mathfrak{s}_\kappa(h_c) = \mathfrak{s}_\kappa(b)\sin(\alpha) = \mathfrak{s}_\kappa(a)\sin(\beta).$$

h_c is the height of the triangle with over side $\overline{AB}$.

Exercise 4.33 Derive the cosine rule (theorem 1.2.4) in the Euclidean case $\kappa = 0$ from the sine rule and the cosine rule for angles.

4.10 Cartography

Cartography is the science of geographical maps. Mathematically speaking, we want to study charts of S^2. For simplicity, we will ignore all complications arising from the fact that the earth is better described by a certain rotational ellipsoid than by a sphere. Ideally, one would like to find charts which preserve lengths, angles, and areas up to a scale factor. However, this is not possible since such a chart would be an isometry between a portion of S^2 and a portion of $\mathbb{R}^2$. By Gauss's *Theorema Egregium* no such isometry exists because S^2 has Gauss curvature $K \equiv 1$ while the Euclidean plane has $K \equiv 0$. In the literature on cartography one nevertheless finds statements of the extent that certain maps are length-preserving. This is rather misleading because it means that *some* lengths are preserved, not all.

As we will see it is possible to construct charts that are angle-preserving or area-preserving. A chart is area-preserving if the area element of the sphere expressed in the coordinates u^1, u^2 coincides with the area element of the Euclidean metric of the plane, $dA = du^1\, du^2$. How can we see whether or not a chart is angle-preserving?

Lemma 4.10.1 *Let g and g' be two Euclidean metrics on $\mathbb{R}^n$, $n \geq 2$. Then g and g' define the same angles if and only if there exists a number $c > 0$ such that*

$$g' = c \cdot g.$$

Proof If $g' = c \cdot g$, then for any non-zero vectors $X, Y \in \mathbb{R}^n$ we have

$$\frac{g(X,Y)}{\sqrt{g(X,X)}\sqrt{g(Y,Y)}} = \frac{g'(X,Y)}{\sqrt{g'(X,X)}\sqrt{g'(Y,Y)}},$$

because the factor c cancels. Thus the angles between X and Y are the same for g and for g'.

Conversely, let g and g' define the same angles. Fix a non-zero vector $Y \in \mathbb{R}^n$. For any $X \in \mathbb{R}^n$, not a multiple of Y, consider the triangle with vertices 0, Y and X. Denote the angle at the corner X by η and the one at the corner Y by ξ. It is important that the angles are the same for g and for g'. By the sine rule we have

$$\frac{\|X\|_g}{\|Y\|_g} = \frac{\sin(\xi)}{\sin(\eta)} = \frac{\|X\|_{g'}}{\|Y\|_{g'}},$$

hence

$$g'(X,X) = c \cdot g(X,X)$$

with $c = g'(Y,Y)/g(Y,Y)$. Since the set of vectors X which are not a multiple of Y is dense in $\mathbb{R}^n$ this relation holds for *all* $X \in \mathbb{R}^n$ by continuity. Polarisation implies

$$g'(X,Z) = c \cdot g(X,Z)$$

for all $X, Z \in \mathbb{R}^n$. □

This justifies the following.

Definition 4.10.2 Let S and S' be regular surfaces equipped with Riemannian metrics g and g' respectively. A local diffeomorphism $\Phi : S \to S'$ is called ***angle-preserving*** or ***conformal*** if there exists a positive function $c : S \to \mathbb{R}$ such that

$$\Phi^* g' = c \cdot g.$$

By the previous lemma this means that the angle between $X, Y \in T_pS$ is the same as the angle between $d_p\Phi(X), d_p\Phi(Y) \in T_{\Phi(p)}S$. Note that the *conformal factor* c will, in general, depend on $p \in S$. Since g and Φ^*g' are smooth, c is a smooth function on S.

A local parametrisation (U, F, V) of a regular surface S with Riemannian metric g will be called conformal if it is conformal as a map $F : U \to V$, where U carries the Euclidean metric. In other words, there exists a positive function $c : U \to \mathbb{R}$ such that

$$(g_{ij}(u^1, u^2)) = c(u^1, u^2) \cdot \begin{pmatrix} 1 & 0 \\ 0 & 1 \end{pmatrix}.$$

The most common way to specify a location on the surface of the earth is to give its longitude λ and latitude φ. For example, Greenwich (England) has longitude $\lambda = 0$ and latitude $\varphi = 51°28'44'' \approx 0.286 \cdot \pi$. Mathematically, this means that we use the parametrisation introduced in example 3.3.3:

$$F : \left(-\frac{\pi}{2}, \frac{\pi}{2}\right) \times (0, 2\pi) \to \mathbb{R}^3,$$

$$F(\varphi, \lambda) = \begin{pmatrix} \cos(\varphi) \cdot \cos(\lambda) \\ \cos(\varphi) \cdot \sin(\lambda) \\ \sin(\varphi) \end{pmatrix}.$$

The computation of the first fundamental form gave us

$$\left(g_{ij}(\varphi, \lambda)\right)_{ij} = \begin{pmatrix} 1 & 0 \\ 0 & \cos^2(\varphi) \end{pmatrix}$$

and hence

$$dA = \cos(\varphi)\, d\varphi\, d\lambda.$$

Thus longitude and latitude do not give rise to an area- or angle-preserving map.

Let us instead look at so-called *azimuthal* charts of the sphere. They are obtained by projecting the sphere onto one of its tangent planes. For simplicity, we will always use the tangent plane to the "north pole" $e_3 = (0,0,1)^\top$, but any other point would do equally well. The simplest such chart is the *orthographic projection* or *parallel projection*.

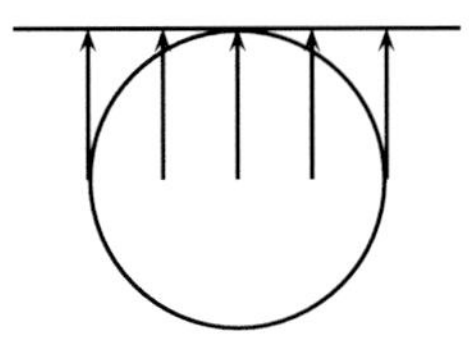

Here we have $U = \{(x,y)^\top \mid x^2 + y^2 < 1\}$, $V = \{(x,y,z)^\top \in \mathbb{R}^3 \mid z > 0\}$ and $F(x,y) = (x,y,\sqrt{1-x^2-y^2})^\top$, cf. example 3.1.5. Note that the local parametrisation F is the inverse of the projection. One computes

$$(g_{ij}(x,y)) = \frac{1}{1-x^2-y^2}\begin{pmatrix} 1-y^2 & xy \\ xy & 1-x^2 \end{pmatrix}$$

and hence

$$dA = \sqrt{\det(g_{ij}(x,y))}\, dx\, dy = (1-x^2-y^2)^{-1/2}\, dx\, dy.$$

Thus the orthographic projection is neither area- nor angle-preserving. It shows the earth as it is seen from outer space.

The next azimuthal chart is the *gnomonic projection* or *central projection*.

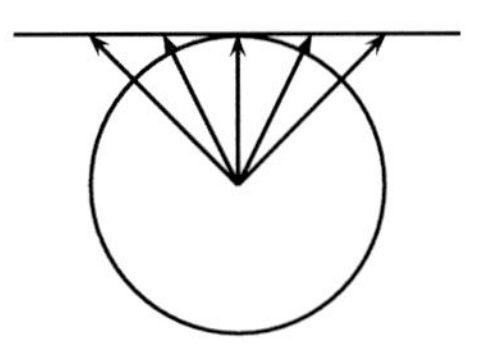

Here we have $U = \mathbb{R}^2$, $V = \{(x,y,z)^\top \in \mathbb{R}^3 \mid z > 0\}$, and $F(x,y) = (1/\sqrt{1+x^2+y^2})(x,y,1)^\top$. One computes

$$(g_{ij}(x,y)) = \frac{1}{(1+x^2+y^2)^2}\begin{pmatrix} 1+y^2 & -xy \\ -xy & 1+x^2 \end{pmatrix}$$

and

$$dA = (1+x^2+y^2)^{-3/2}\, dx\, dy.$$

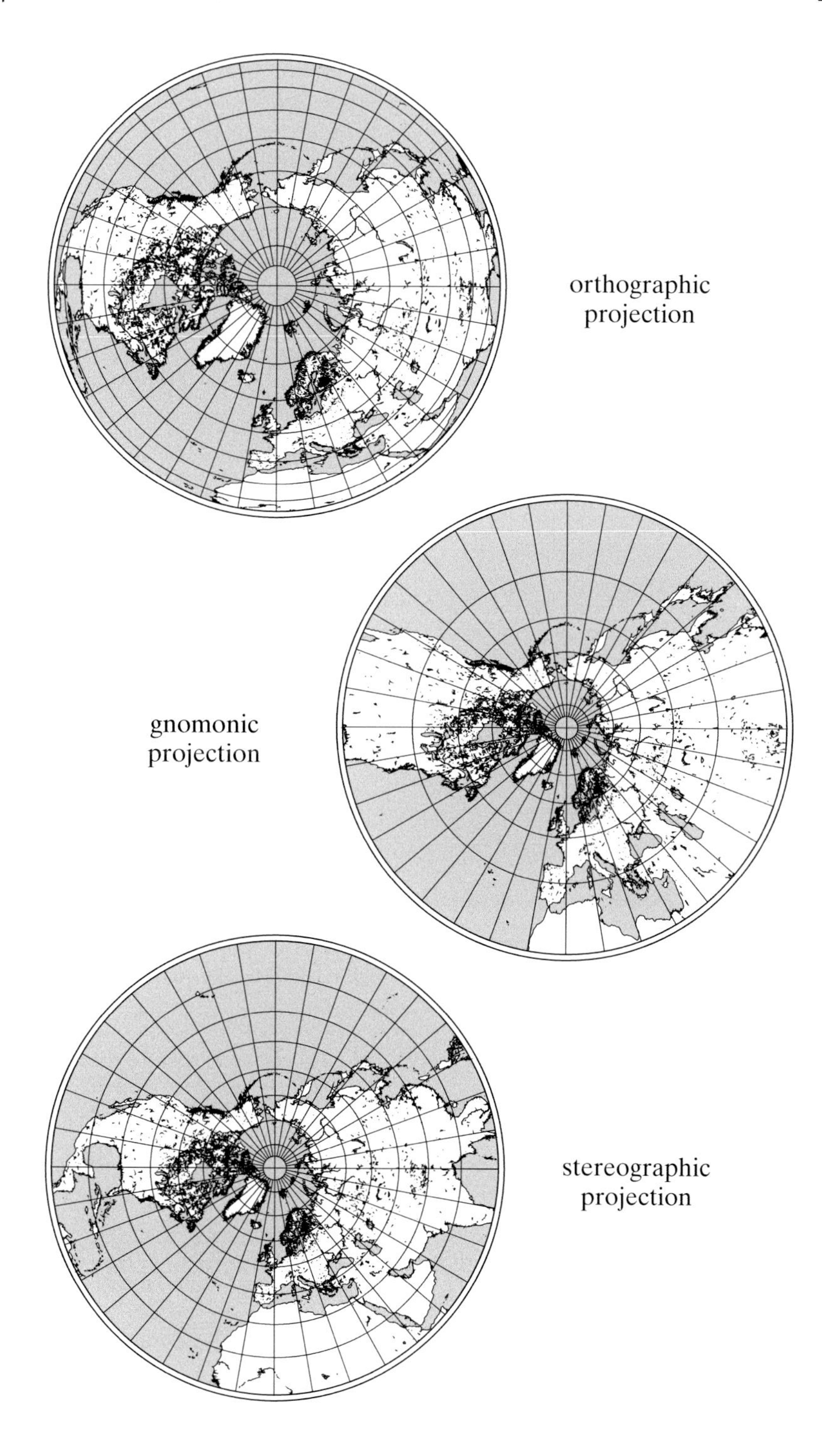
orthographic projection
gnomonic projection
stereographic projection

Hence this chart is also neither area- nor angle-preserving but it has the pleasing property that the geodesics on the sphere, the great circles, are represented in the chart by straight lines. This is because a great circle is the intersection of S^2 with a plane E through the origin which is projected to the intersection of E with the tangent plane, a straight line. Thus this type of map is well suited for finding the shortest route from one point to another.

If we change the projection point from the centre of the sphere to the south pole we obtain the *stereographic projection*.

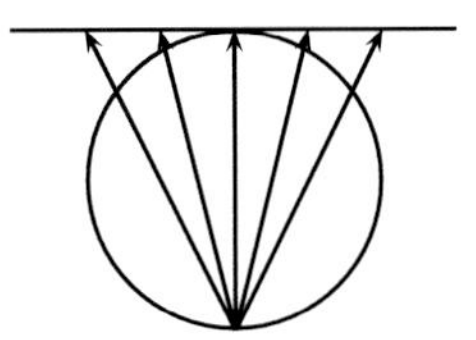

In this case

$$U = \mathbb{R}^2,\ V = \mathbb{R}^3 \setminus \left\{(0,0,-1)^\top\right\},\ F(x,y) = \frac{1}{4+x^2+y^2}(4x, 4y, 4-x^2-y^2)^\top,$$

$$(g_{ij}(x,y)) = \frac{16}{(4+x^2+y^2)^2}\begin{pmatrix}1 & 0\\ 0 & 1\end{pmatrix}$$

and

$$dA = \frac{16}{(4+x^2+y^2)^2}\,dx\,dy.$$

Thus the stereographic projection is conformal but not area-preserving.

All three projections considered so far satisfy $(g_{ij}(0,0)) = \begin{pmatrix}1 & 0\\ 0 & 1\end{pmatrix}$. Thus the differential of the local parametrisation is an isometry at $(0,0,)^\top$. The distortion of the map is small near the north pole (or whatever point has been chosen as the projection centre).

To get a map with small distortion in a whole neighbourhood of the equator (or any other great circle) one wraps a cylinder around the equator, projects the sphere onto the cylinder and then unwraps it.

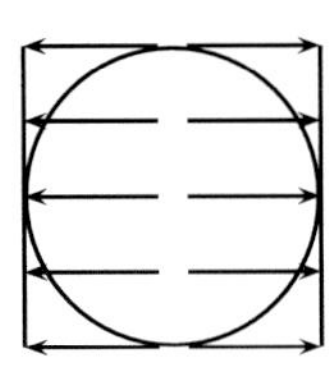

We then have $U = (0, 2\pi) \times (-1, 1)$, $V = \mathbb{R}^3 \setminus \left\{(x,0,z)^\top \mid x > 0\right\}$, and

$$F(\varphi, h) = \begin{pmatrix}\cos(\varphi)\cdot\sqrt{1-h^2}\\ \sin(\varphi)\cdot\sqrt{1-h^2}\\ h\end{pmatrix}.$$

One computes

$$(g_{ij}(\varphi, h)) = \begin{pmatrix} 1 - h^2 & 0 \\ 0 & \dfrac{1}{1 - h^2} \end{pmatrix}.$$

Thus this map is not conformal but $dA = d\varphi\, dh$, hence it is area-preserving.

Let us modify this example. We introduce a new parameter x and let the height h be a function of x yet to be determined, i.e.

$$F(\varphi, x) = \begin{pmatrix} \cos(\varphi) \cdot \sqrt{1 - h(x)^2} \\ \sin(\varphi) \cdot \sqrt{1 - h(x)^2} \\ h(x) \end{pmatrix}.$$

Then we get

$$(g_{ij}(\varphi, x)) = \begin{pmatrix} 1 - h(x)^2 & 0 \\ 0 & \dfrac{h'(x)^2}{1 - h(x)^2} \end{pmatrix}.$$

We see that we need $h' \neq 0$. We may and will assume that $h' > 0$ because otherwise we may simply replace $h(x)$ by $h(-x)$. In order to get a conformal chart we need

$$1 - h(x)^2 = \frac{h'(x)^2}{1 - h(x)^2},$$

i.e.

$$h'(x) = 1 - h(x)^2.$$

The solution to this ordinary differential equation with initial condition $h(0) = 0$ is

$$h(x) = \tanh(x).$$

The resulting geographical map is known as *Mercator's projection* named after the Flemish geographer Gheert Cremer (1512–1594), who latinised his name to Gerardus Mercator.

Exercise 4.34 What can you say about the solutions to this ordinary differential equation with other initial values $h(0)$?

There are many more geographical maps. For example, in order to get a good map representation of a neighbourhood of a circle of latitude other than the equator, one can wrap a cone around the sphere along a small circle and then unwrap it. The interested reader should consult a book on cartography such as [28].

Exercise 4.35 Show that there are no geographical maps that are at the same time conformal and area-preserving.

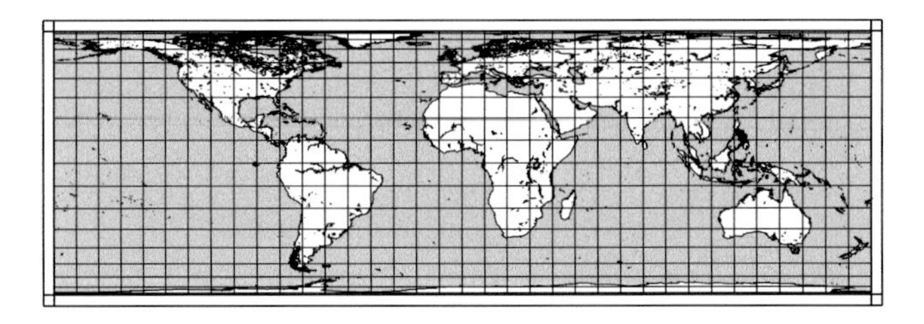

cylindrical equal-area projection

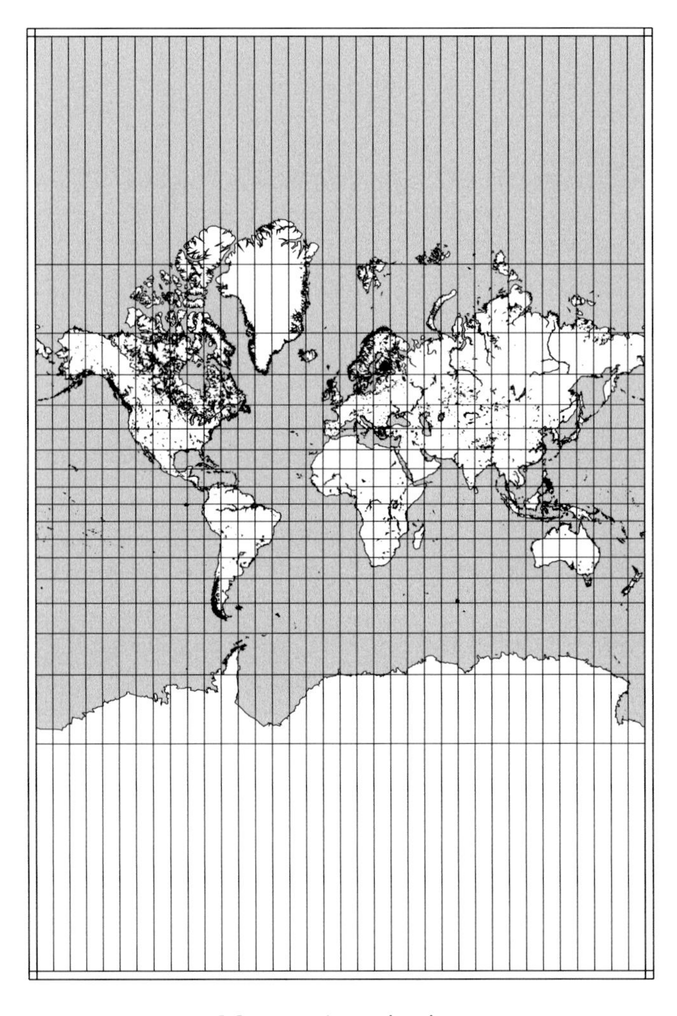

Mercator's projection

4.11 Further models of hyperbolic geometry

We now study charts of the hyperbolic plane. In contrast to the spherical case it is possible to find charts covering the whole hyperbolic plane at once. We will call such charts ***models of hyperbolic geometry***. If (U, F, V) is such a chart with $\mathbb{M}_{-1} \subset V$, then F is a diffeomorphism from U to $\mathbb{M}_{-1}$ and all investigations of hyperbolic geometry can be performed on U using the functions $g_{ij} : U \to \mathbb{R}$ describing the hyperbolic Riemannian metric in this chart. Since the hyperbolic metric was obtained by restricting the Minkowski scalar product of $\mathbb{R}^3$ instead of the Euclidean scalar product, we have

$$g_{ij}\left(u^1, u^2\right) = \left\langle \frac{\partial F}{\partial u^i}\left(u^1, u^2\right), \frac{\partial F}{\partial u^j}\left(u^1, u^2\right) \right\rangle_{-1}.$$

The first chart is obtained by central projection with respect to the origin of the unit disc in $T_{e_1}\mathbb{M}_{-1}$ to $\mathbb{M}_{-1}$.

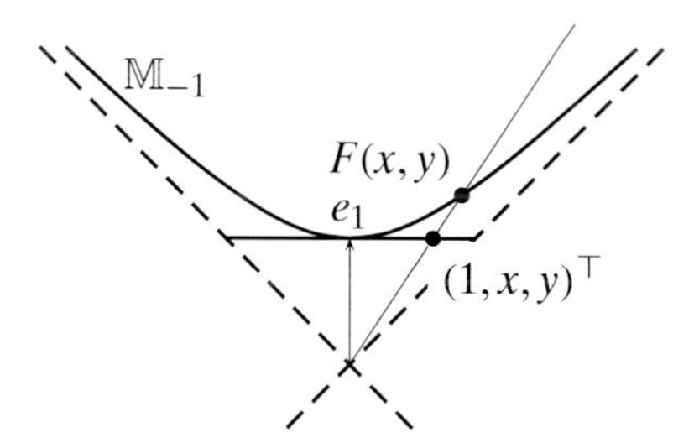

Expressed as formulae, this becomes, $U = \left\{(x, y)^\top \in \mathbb{R}^2 \mid x^2 + y^2 < 1\right\}$ and

$$F(x, y) = \frac{1}{\sqrt{1 - x^2 - y^2}} \begin{pmatrix} 1 \\ x \\ y \end{pmatrix}. \tag{4.29}$$

This chart is known as the ***Klein model***, as the ***Cayley–Klein model***, as the ***projective model***, and also as the ***Beltrami–Klein model*** of hyperbolic geometry. We compute

$$\frac{\partial F}{\partial x} = \left(1 - x^2 - y^2\right)^{-3/2} \begin{pmatrix} x \\ 1 - y^2 \\ xy \end{pmatrix}, \qquad \frac{\partial F}{\partial y} = \left(1 - x^2 - y^2\right)^{-3/2} \begin{pmatrix} y \\ xy \\ 1 - x^2 \end{pmatrix}$$

and hence

$$g_{11}(x, y) = \left\langle \frac{\partial F}{\partial x}, \frac{\partial F}{\partial x} \right\rangle_{-1}$$

$$= \left(1 - x^2 - y^2\right)^{-3} \cdot \left(-x^2 + \left(1 - y^2\right)^2 + (xy)^2\right)$$

$$= \frac{1 - y^2}{\left(1 - x^2 - y^2\right)^2}$$

and similarly for the other coefficients for the metric. We obtain

$$(g_{ij}(x, y)) = \frac{1}{\left(1 - x^2 - y^2\right)^2} \begin{pmatrix} 1 - y^2 & xy \\ xy & 1 - x^2 \end{pmatrix}.$$

Exercise 4.36 Show that the hyperbolic plane has infinite area.

The metric does not look particularly simple in the Klein model but this model has one very pleasing feature: geodesics in $\mathbb{M}_{-1}$ correspond to straight line segments in U. This is clear from the construction of F. The geodesics in $\mathbb{M}_{-1}$ are the intersections of $\mathbb{M}_{-1}$ with planes E containing the origin. Under the central projection this is mapped to the intersection of E with the unit disc in $T_{e_1}\mathbb{M}_{-1}$. The intersection of the two planes E and $T_{e_1}\mathbb{M}_{-1}$ is a straight line.

This allows a simple discussion of the parallel axiom. Keeping the terminology and the notation from the first chapter we now take the hyperbolic plane $\mathbb{M}_{-1}$ as the set of points, the set of "straight lines" will be the set of traces of all geodesics,

$$\mathcal{P} := \mathbb{M}_{-1},$$

$$\mathcal{G} := \{\mathbb{M}_{-1} \cap E \mid E \subset \mathbb{R}^3 \text{ is a two-dimensional subspace}\} - \{\emptyset\}.$$

The incidence relation is the usual set-theoretical inclusion. We immediately see that the incidence axioms I_1–I_4 are satisfied. Unlike in spherical geometry, in which geodesics are great circles, there is an obvious useful definition for when one point lies between two others on the hyperbolas in hyperbolic geometry. The validity of the ordering axioms A_1–A_5 is then easy to see. For the definition of congruence of lengths and angles we use isometries as in the discussion of the Cartesian model of Euclidean geometry. This time we use the group G_{-1} of isometries on $\mathbb{M}_{-1}$ instead of the Euclidean group. It is then not difficult to show the validity of the congruence axioms. By theorem 1.1.9 we already know that parallels therefore exist. The completeness axioms can be derived in a way similar to the one used for the Cartesian model.

Only the parallel axiom is not valid, since parallels are not unique. This is easily seen in the Klein model. Given a line segment L in U there is an infinity of line segments not intersecting L but having one point in common.

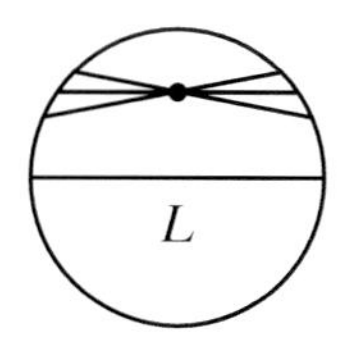

Hyperbolic geometry shows that the parallel axiom does not follow from the other axioms of Euclidean geometry.

Note that the Klein model is the hyperbolic analogue of the gnomonic chart of the sphere. Also compare their metric coefficients. Now we move on to the hyperbolic analogue of the stereographic projection in spherical geometry. In principle, we only have to change the projection point from the origin to $-e_1$ as in spherical geometry. To get slightly simpler formulae we do not project to $T_{e_1}\mathbb{M}_{-1}$ but to the parallel plane spanned by e_2 and e_3. The difference is only a stretching by a factor of 2.

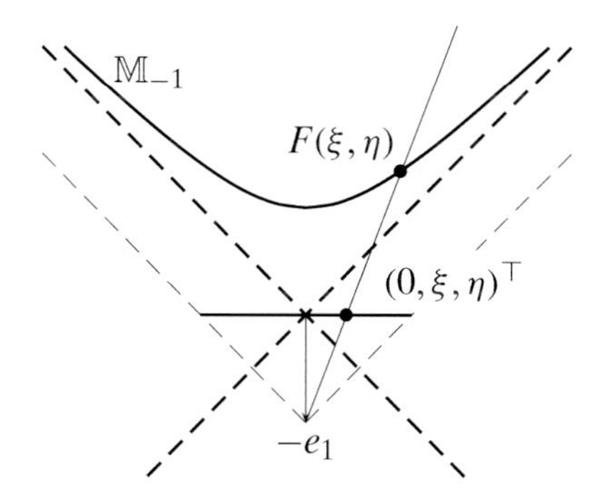

Set up in this way U is the unit disc as for the Klein model and one easily computes that

$$F(\xi, \eta) = \frac{1}{1 - \xi^2 - \eta^2} \begin{pmatrix} 1 + \xi^2 + \eta^2 \\ 2\xi \\ 2\eta \end{pmatrix}.$$

A calculation similar to the one for the Klein model yields

$$(g_{ij}(\xi, \eta)) = \frac{4}{(1 - \xi^2 - \eta^2)^2} \begin{pmatrix} 1 & 0 \\ 0 & 1 \end{pmatrix}.$$

Thus this model of hyperbolic geometry is conformal just like the stereographic projection. So it comes at no surprise that this model is known as the ***conformal disc model*** and also as the ***Poincaré disc model***. It was popularised by the Dutch artist Maurits Cornelis Escher (1898–1972) in his famous woodcuts such as "Circle Limit I" from 1958.[1]

[1] M. C. Escher's "Circle Limit I" © 2009 The M. C. Escher Company – Holland. All rights reserved. www.mcescher.com

The straight line segments through the centre of the disc correspond to the geodesics of $\mathbb{M}_{-1}$ given by intersections with those planes that contain both the origin and e_1. The other geodesics correspond to circular arcs in the disc meeting its boundary perpendicularly. Escher's woodcut displays some of them. The black fish are all congruent with respect to the hyperbolic metric. In particular, their hyperbolic areas are all the same and similarly for the white fish.

One may wonder what this picture would look like in the Klein model. The circular arcs must be replaced by straight line segments. Indeed, Escher's Circle Limit I transformed to the Klein model looks like this:

There is another interesting and important model. We construct it as follows. We start with the Klein model and do a parallel projection to the upper hemisphere, i.e. the point $(x,y)^\top$ is mapped to $(\sqrt{1-x^2-y^2},x,y)^\top$. This is the transformation that gave rise to the orthographic chart in cartography. The straight line segments in the disc are mapped to circular arcs meeting the equator of the hemisphere at a right angle.

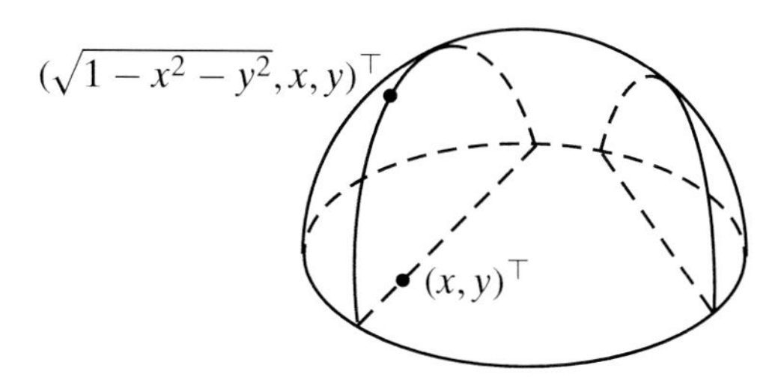

Now we choose a point on the equator of the hemisphere, say e_2, and perform a stereographic projection with respect to e_2 onto the tangent plane of the sphere at the antipodal point $-e_2$. This maps the upper hemisphere onto the upper half-plane. The spherical arcs are mapped to spherical arcs in the half-plane except for those arcs in the hemisphere that contain e_2. These are projected onto straight lines. Both the spherical arcs and the straight lines meet the boundary of the half-plane at a right angle.

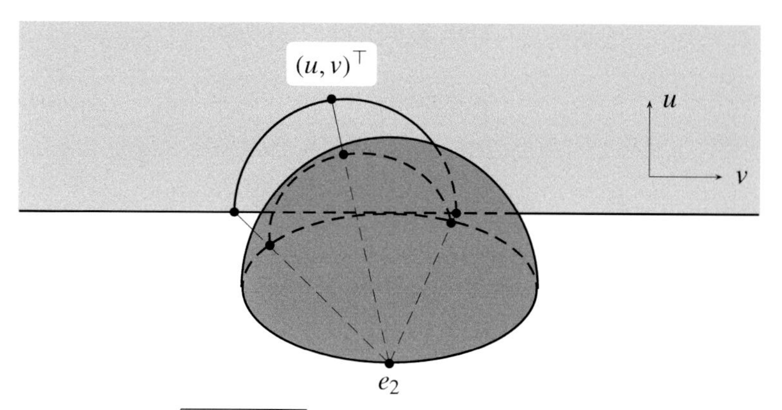

The image of $(\sqrt{1-x^2-y^2},x,y)^\top$ under this stereographic projection is easily computed to be

$$\begin{pmatrix} u \\ v \end{pmatrix} = \frac{1}{1-x}\begin{pmatrix} 2\sqrt{1-x^2-y^2} \\ 2y \end{pmatrix}.$$

Solving for $(x,y)^\top$ yields

$$\begin{pmatrix} x \\ y \end{pmatrix} = \frac{1}{u^2+v^2+4}\begin{pmatrix} u^2+v^2-4 \\ 4v \end{pmatrix}.$$

Plugging this into the parametrisation (4.29) of the Klein model gives us the ***Poincaré half-plane model***

$$F(u,v) = \frac{1}{4u}\begin{pmatrix} u^2+v^2+4 \\ u^2+v^2-4 \\ 4v \end{pmatrix}$$

with $U = \{(u,v)^\top \mid u > 0\}$. As before one computes

$$(g_{ij}(u,v)) = \frac{1}{u^2}\begin{pmatrix} 1 & 0 \\ 0 & 1 \end{pmatrix}.$$

Hence the Poincaré half-plane model is also conformal. Escher's woodcut transformed to this model now looks like this:

Exercise 4.37 The grey region Δ in the Poincaré half-plane model shown in the figure below is a geodesic triangle with vertices at infinity. Compute its area.

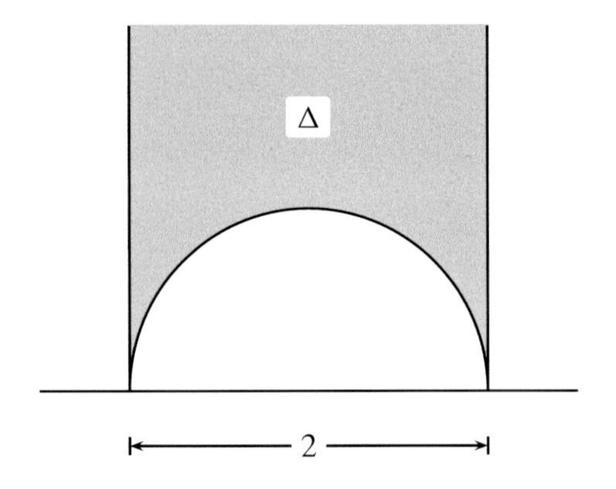

5 Geometry and analysis

Surfaces with boundary are introduced. The divergence theorem of Gauss is derived and used to show that the total Gauss curvature of a compact regular surface does not depend on the Riemannian metric.

5.1 The divergence theorem

In this section we want to derive a two-dimensional analogue of the fundamental theorem of calculus. In this theorem the integral of a derivative over a one-dimensional interval is identified with the difference of the values at the end-points. This term in the values at the end-points can be considered as the integral of the function over the (zero-dimensional) boundary of the interval. The divergence theorem expresses the integral of a derivative of a vector field as a one-dimensional line integral. To make all this precise we first need the notion of a surface with boundary.

Definition 5.1.1 A ***surface with boundary*** is a closed subset S of a regular surface $S_{\text{reg}} \subset \mathbb{R}^3$ such that for every point $p \in S$ there exists a local parametrisation $F : U \to S_{\text{reg}}$ of S_{reg} with $p \in F(U)$, such that either

- $F(U) \subset S$ (then p is called an ***interior point*** of S) or
- $F^{-1}(p) = (x, 0)^\top$ for an $x \in \mathbb{R}$ and $F^{-1}(S) = \{(x, y)^\top \in U \mid y \geq 0\}$ (then p is called a ***boundary point*** of S).

The set of all interior points is the ***interior*** of S, the set of all boundary points is the ***boundary*** ∂S.

Example 5.1.2 The closed circular disc $S = \{(x, y, 0)^\top \mid x^2 + y^2 \leq 1\}$ is a surface with boundary. We can take the x–y plane as the regular surface $S_{\text{reg}} = \mathbb{R}^2 \times \{0\}$. The points $(x, y, 0)^\top$ with $x^2 + y^2 < 1$ are interior points, since for $r := \sqrt{1 - (x^2 + y^2)} > 0$ the set $U := \{(\xi, \eta)^\top \in \mathbb{R}^2 \mid (\xi - x)^2 + (\eta - y)^2 < r^2\}$

with the function $F(\xi, \eta) = (\xi, \eta, 0)^\top$ is a local parametrisation of S_{reg}, whose image is fully contained in S.

The points $(x, y, 0)^\top$ with $x^2 + y^2 = 1$, on the other hand, are boundary points. For example, a local parametrisation of S_{reg} for the point $(0, -1, 0)^\top$ is given by

$$F : \left(-\frac{1}{2}, \frac{1}{2}\right) \times (-\infty, 1) \to \mathbb{R}^3, \quad F(\xi, \eta) = \left(\xi, \eta - \sqrt{1 - \xi^2}, 0\right)^\top,$$

with $F(0, 0) = (0, -1, 0)^\top$ and $F(\xi, \eta) \in S$ if and only if $\eta \geq 0$. For the other points $(x, y, 0)^\top$ with $x^2 + y^2 = 1$ we can obtain such a parametrisation by composing F with a suitable rotation.

Exercise 5.1 Show that the upper hemisphere $S = \{(x, y, z)^\top \in \mathbb{R}^3 | \, x^2 + y^2 + z^2 = 1, \; z \geq 0\}$ is a surface with boundary.

Exercise 5.2 Let $c : \mathbb{R} \to \mathbb{R}^3$ be a simple closed space curve with period L. Let S_{reg} be the ruled surface given by the parametrisation $F(t, s) = c(t) + s v(t)$, $v(t + L) = \pm v(t)$, $s \in (-1, 1)$.

(a) Show that $S := F\left(\mathbb{R} \times \left[-\frac{1}{2}, \frac{1}{2}\right]\right)$ is a surface with boundary.
(b) Show that the following statements are equivalent:
  (i) S_{reg} is orientable;
  (ii) $v(t + L) = +v(t)$ for all t;
  (iii) the boundary of S is the disjoint union of *two* space curves.
(c) What is the boundary of S in the non-orientable case, e.g. for the Möbius strip, see example 3.8.4?

Regular surfaces S are also surfaces with boundaries; they are exactly those surfaces with boundary whose boundary is empty, $\partial S = \emptyset$.

If p is a boundary point of a surface with boundary, then the boundary near p can be parametrised by a regular curve. If F is a neighbourhood of the local parametrisation of S_{reg} at p as in definition 5.1.1, then $c(t) := F(t, 0)$ is the afore-mentioned regular curve. For $c(t_0) = p$ we have $\dot{c}(t_0) \in T_p S_{\text{reg}}$. There are therefore exactly two unit vectors $\pm\nu(p) \in T_p S_{\text{reg}}$ that are perpendicular to $\dot{c}(t_0)$. They are called ***unit normal vectors to the boundary*** of S. We have $(d_u F)^{-1}(\dot{c}(t_0)) = (1, 0)^\top$ for $u = F^{-1}(p)$ and therefore $(d_u F)^{-1}(\pm\nu(p))$ cannot have a vanishing y-component as well: $\langle (d_u F)^{-1}(\pm\nu(p)), (0, 1)^\top \rangle \neq 0$. If we choose $\nu(p)$ such that $\langle (d_u F)^{-1}(\nu(p)), (0, 1)^\top \rangle < 0$, then $\nu(p)$ is called the ***outer unit normal vector to the boundary*** at the point p, while $-\nu(p)$ is the ***inner unit normal vector***.

Exercise 5.3 Let S be surface with boundary, $p \in \partial S$. Show that there is a regular curve $c : [0, \varepsilon) \to S \subset \mathbb{R}^3$ with $\dot{c}(0) = -\nu(p)$ for the inner unit normal vector $-\nu(p)$, while the analogue for the outer unit normal vector does not exist.

We will also have to integrate functions over the boundary of a surface with boundary, and hence make the following definition.

Definition 5.1.3 Let S be a surface with boundary, let $f : \partial S \to \mathbb{R}$ be a smooth function with compact support. We write $\partial S \cap \operatorname{supp} f = C_1 \dot{\cup} \cdots \dot{\cup} C_n$ as a disjoint union, where every C_j is a part of the boundary that can be parametrised by a regular curve. We choose parametrisations by arc-length $c_j : I_j \to \mathbb{R}^3$ with $c_j(I_j) = C_j$ and define the ***line integral*** as

$$\int_{\partial S} f ds := \sum_{j=1}^{n} \int_{I_j} f \circ c_j(t) dt.$$

Using lemma 2.1.14 it is easy to see that the definition is independent of the choices made.

Exercise 5.4 Show that if $c_j : I_j \to C_j$ are regular parametrisations of C_j (not necessarily by arc-length), then

$$\int_{\partial S} f ds = \sum_{j=1}^{n} \int_{I_j} f(c_j(t)) \cdot \|\dot{c}_j(t)\| \, dt.$$

We have seen in example 4.2.2 that differentiation of a function f on a regular surface yields a vector field $\operatorname{grad} f$, the gradient of f, on the surface. The statements from examples 4.2.2 and 4.2.3 remain valid if we endow the regular surface with an arbitrary Riemannian metric instead of the first fundamental form. However, for a given f the gradient also depends on the choice of the Riemannian metric; this is not the case for the differential $d_p f$.

We will now see how we can obtain a function from a vector field by differentiating. For this purpose let S be a regular surface with Riemannian metric g. Let X be a differentiable vector field on S. For $p \in S$ we can regard the covariant derivative of X as an endomorphism on T_pS,

$$\nabla_{.} X : T_pS \to T_pS, \quad Y_p \mapsto \nabla_{Y_p} X.$$

Definition 5.1.4 The trace of the above-mentioned endomorphism is called the ***divergence*** of X at the point p,

$$\operatorname{div} X(p) := \operatorname{Trace}(Y_p \mapsto \nabla_{Y_p} X).$$

Lemma 5.1.5 *If we write the vector field X with respect to a local parametrisation F as $X = \sum_i \xi^i \partial F / \partial u^i$, then the divergence is given by*

$$\operatorname{div} X = \sum_j \Big(\frac{\partial \xi^j}{\partial u^j} + \sum_i \Gamma^j_{ij} \xi^i \Big) = \frac{1}{\sqrt{\det(g_{k\ell})}} \sum_j \frac{\partial}{\partial u^j} \Big(\sqrt{\det(g_{k\ell})} \xi^j \Big).$$

Proof We find the matrix representation of the endomorphism $Y_p \mapsto \nabla_{Y_p} X$ with respect to the basis $\partial F/\partial u^1, \partial F/\partial u^2$.

$$\begin{aligned}
\nabla_{\frac{\partial F}{\partial u^j}} X &= \nabla_{\frac{\partial F}{\partial u^j}} \Big(\sum_i \xi^i \frac{\partial F}{\partial u^i} \Big) \\
&= \sum_i \Big(\frac{\partial \xi^i}{\partial u^j} \frac{\partial F}{\partial u^i} + \xi^i \sum_k \Gamma^k_{ij} \frac{\partial F}{\partial u^k} \Big) \\
&= \sum_k \Big(\frac{\partial \xi^k}{\partial u^j} + \sum_i \xi^i \Gamma^k_{ij} \Big) \frac{\partial F}{\partial u^k}.
\end{aligned}$$

The endomorphism $Y_p \mapsto \nabla_{Y_p} X$ therefore has the following matrix representation w.r.t. this basis:

$$\Big(\frac{\partial \xi^k}{\partial u^j} + \sum_i \xi^i \Gamma^k_{ij} \Big)_{jk}.$$

The trace is then

$$\operatorname{div} X = \sum_j \Big(\frac{\partial \xi^j}{\partial u^j} + \sum_i \Gamma^j_{ij} \xi^i \Big).$$

This proves the first part of the claim.

As

$$\begin{aligned}
&\frac{1}{\sqrt{\det(g_{k\ell})}} \sum_j \frac{\partial}{\partial u^j} \Big(\sqrt{\det(g_{k\ell})} \xi^j \Big) \\
&= \frac{1}{\sqrt{\det(g_{k\ell})}} \sum_j \frac{\partial \sqrt{\det(g_{k\ell})}}{\partial u^j} \cdot \xi^j + \frac{1}{\sqrt{\det(g_{k\ell})}} \sum_j \sqrt{\det(g_{k\ell})} \cdot \frac{\partial \xi^j}{\partial u^j} \\
&= \sum_j \frac{\partial \xi^j}{\partial u^j} + \sum_j \frac{\partial}{\partial u^j} \Big(\ln \Big(\sqrt{\det(g_{k\ell})} \Big) \Big) \xi^j \\
&= \sum_j \frac{\partial \xi^j}{\partial u^j} + \frac{1}{2} \sum_j \frac{\partial}{\partial u^j} \left(\ln \left(\det(g_{k\ell}) \right) \right) \xi^j,
\end{aligned}$$

the second part follows from the first if we can show that

$$\frac{1}{2} \frac{\partial}{\partial u^j} \left(\ln \left(\det(g_{k\ell}) \right) \right) = \sum_i \Gamma^i_{ji}. \tag{5.1}$$

Using lemma 4.2.14 we calculate

$$\sum_i \Gamma^i_{ji} = \frac{1}{2}\sum_{ik}\left(\frac{\partial g_{jk}}{\partial u^i} + \frac{\partial g_{ik}}{\partial u^j} - \frac{\partial g_{ji}}{\partial u^k}\right) g^{ik} \tag{5.2}$$

$$= \frac{1}{2}\sum_{ik}\left(\frac{\partial g_{ik}}{\partial u^j}\right) g^{ik}, \tag{5.3}$$

since the first and the third sum in (5.2) cancel. Writing g for the matrix $(g_{ij})_{ij}$ we can write (5.3) in a more compact way as

$$\sum_i \Gamma^i_{ji} = \frac{1}{2}\,\mathrm{Trace}\left(g^{-1}\frac{\partial g}{\partial u^j}\right).$$

Equation (5.1) follows by lemma 5.1.6. □

Lemma 5.1.6 (Derivative of the determinant) *Let $t \mapsto g(t)$ be a differentiable curve of invertible real $n \times n$ matrices. Then*

$$\frac{d}{dt}\ln\det g = \mathrm{Trace}\left(g^{-1}\frac{d}{dt}g\right).$$

Proof We first prove the equation for $t = t_0$ if $g(t_0) = \mathrm{Id}$. Then the claim is simply

$$\frac{d}{dt}\det g(t_0) = \mathrm{Trace}\left(\frac{dg}{dt}(t_0)\right). \tag{5.4}$$

As commonly known (see [17, p. 171, theorem 7.2]), the determinant is given by

$$\det g = \sum_\sigma \mathrm{sign}(\sigma) g_{1\sigma(1)}\cdots g_{n\sigma(n)},$$

where the sum is taken over all permutations $\sigma : \{1,\ldots,n\} \to \{1,\ldots,n\}$. It follows that

$$\frac{d}{dt}\det g(t_0)$$
$$= \sum_\sigma \mathrm{sign}(\sigma)\left(\frac{dg_{1\sigma(1)}}{dt}(t_0)\cdots g_{n\sigma(n)}(t_0) + \cdots + g_{1\sigma(1)}(t_0)\cdots\frac{dg_{n\sigma(n)}}{dt}(t_0)\right).$$

As $g_{ij}(t_0) = 0$ for $i \neq j$, the only permutation that can contribute a non-vanishing summand is the trivial permutation $\sigma = \mathrm{Id}$:

$$\begin{aligned}\frac{d}{dt}\det g(t_0) &= \frac{dg_{11}}{dt}(t_0)\cdots g_{nn}(t_0)+\cdots+g_{11}(t_0)\cdots\frac{dg_{nn}}{dt}(t_0)\\ &= \frac{dg_{11}}{dt}(t_0)\cdot 1\cdots 1+\cdots+1\cdots 1\cdot\frac{dg_{nn}}{dt}(t_0)\\ &= \operatorname{Trace}\left(\frac{dg}{dt}(t_0)\right).\end{aligned}$$

In the case $g(t_0)=\mathrm{Id}$, we set $h(t) := g(t_0)^{-1}g(t)$. Then $h(t_0)=\mathrm{Id}$ and an application to h of the part already proved gives

$$\begin{aligned}\det(g(t_0))^{-1}\frac{d}{dt}\det g(t_0) = \frac{d}{dt}\det h(t_0) &= \operatorname{Trace}\left(\frac{dh}{dt}(t_0)\right)\\ &= \operatorname{Trace}\left(g(t_0))^{-1}\frac{dg}{dt}(t_0)\right),\end{aligned}$$

which proves the lemma. □

Let us now move on to the main result of this section.

Theorem 5.1.7 (Gauss's divergence theorem) *Let S_{reg} be a regular surface with Riemannian metric g. Let X be a continuously differentiable vector field with compact support on S_{reg}. Let $S \subset S_{\mathrm{reg}}$ be a surface with boundary. Let ν be the outer unit normal field of S. Then*

$$\int_S \operatorname{div} X\, dA = \int_{\partial S} g(X,\nu)\, ds.$$

Proof For every point in $\operatorname{supp} X \cap S$ there is a local parametrisation of S_{reg} as in definition 5.1.1. As $\operatorname{supp} X \cap S$ is compact, we can cover $\operatorname{supp} X \cap S$ with finitely many neighbourhoods from such parametrisations. In a way similar to the one used in the proof of theorem 3.8.8 we choose smooth functions $\rho_j : \mathbb{R}^3 \to \mathbb{R}$ with $0 \le \rho_j \le 1$ and $\sum_j \rho_j \equiv 1$ in a neighbourhood of $\operatorname{supp} X \cap S$, such that every support $\operatorname{supp} \rho_j$ is contained in a neighbourhood as described above. If we now write $X_j := \rho_j \cdot X$, then $X = \sum_j X_j$ on S and every X_j has its support in a neighbourhood as in definition 5.1.1. By the linearity of the integral and of the divergence it suffices to prove the claim for the X_j. There is thus no loss of generality if we assume that $\operatorname{supp} X \cap S$ is contained in such a neighbourhood.

Let us therefore suppose that $\operatorname{supp} X \cap S$ is contained in such a neighbourhood. We first consider the case that the neighbourhood meets the boundary of S, i.e. that it is of the second type in definition 5.1.1. The local parametrisation $F : U \to \mathbb{R}^3$ therefore has the property $F^{-1}(S) = \{(u^1,u^2)^\top \in U \mid u^2 \ge 0\}$ and $\operatorname{supp} X \cap S \subset F(U)$. Without loss of generality we further assume that

$$F : U = (-a, a) \times (-a, a) \to S$$

are Fermi coordinates along c, cf. lemma 4.6.13, where c is a parametrisation by arc-length of $\partial S \cap F(U)$. Along the boundary of S we then have

$$\left(g_{ij}(u^1, 0)\right)_{ij} = \begin{pmatrix} 1 & 0 \\ 0 & 1 \end{pmatrix} \tag{5.5}$$

and

$$\nu(F(u^1, 0)) = -\frac{\partial F}{\partial u^2}(u^1, 0). \tag{5.6}$$

In particular $\sqrt{\det(g_{k\ell})(u^1, 0)} = 1$. We now find the formula of the divergence theorem w.r.t. this parametrisation, using lemma 5.1.5. For $X = \sum_j \xi^j \partial F / \partial u^j$ we obtain

$$\begin{aligned} &\int_S \operatorname{div} X \, dA \\ &\quad = \int_0^a \int_{-a}^a \frac{1}{\sqrt{\det(g_{k\ell})}} \sum_j \frac{\partial}{\partial u^j} \left(\sqrt{\det(g_{k\ell})} \xi^j\right) \sqrt{\det(g_{k\ell})} du^1 du^2 \\ &\quad = \sum_j \int_0^a \int_{-a}^a \frac{\partial}{\partial u^j} \left(\sqrt{\det(g_{k\ell})} \xi^j\right) du^1 du^2. \end{aligned} \tag{5.7}$$

The summands with $j = 1$ are first integrated over u^1 and we obtain

$$\begin{aligned} &\int_{-a}^a \frac{\partial}{\partial u^1} \left(\sqrt{\det(g_{k\ell})} \xi^1\right) du^1 \\ &\quad = \sqrt{\det(g_{k\ell})(a, u^2)} \xi^1(-a, u^2) - \sqrt{\det(g_{k\ell})(-a, u^2)} \xi^1(-a, u^2) \\ &\quad = 0, \end{aligned}$$

since $\xi^1(-a, u^2) = \xi^1(a, u^2) = 0$ by the assumption about the support of X. The summands with $j = 2$ are first integrated over u^2 and we obtain

$$\begin{aligned} &\int_0^a \frac{\partial}{\partial u^2} \left(\sqrt{\det(g_{k\ell})} \xi^2\right) du^2 \\ &\quad = \sqrt{\det(g_{k\ell})(u^1, a)} \xi^2(u^1, a) - \sqrt{\det(g_{k\ell})(u^1, 0)} \xi^2(u^1, 0) \\ &\quad = -\xi^2(u^1, 0). \end{aligned}$$

Equation (5.7) then simplifies to

$$\int_S \operatorname{div} X \, dA = -\int_{-a}^a \xi^2(u^1, 0) du^1.$$

Evaluation of the line integral gives

$$\int_{\partial S} g(X, \nu) ds = \int_{-a}^{a} (-\xi^2(u^1, 0)) du^1$$

by (5.5) and (5.6).

If the neighbourhood does not meet the boundary, then the above arguments show that

$$\int_S \mathrm{div} X \, dA = 0.$$

This proves the divergence theorem. □

The boundary term in the divergence theorem measures the proportion of the vector field that points out of the surface with boundary S. If we imagine the vector field X as the current density of a liquid that flows on the regular surface S_{reg}, then the boundary term gives the net amount of liquid that flows out of S. In particular, if the field is divergence-free, i.e. $\mathrm{div} X \equiv 0$, then the amount of liquid in S remains unchanged. Arguments of this type are important in the derivations of many conservation laws in physics.

The divergence theorem now also gives us a graphic interpretation of the divergence of a vector field. If p is a point on the regular surface, then we take the circular disc $\bar{D}(p, r)$ with centre p and radius r as the regular surface with boundary, see lemma 4.6.12. The mean of the divergence on this disc is by the divergence theorem

$$\frac{\int_{\bar{D}(p,r)} \mathrm{div} X \, dA}{A[\bar{D}(p, r)]} = \frac{\int_{\partial \bar{D}(p,r)} g(X, \nu) ds}{A[\bar{D}(p, r)]}.$$

Passing to the limit $r \searrow 0$ gives

$$\mathrm{div} X(p) = \lim_{r \searrow 0} \frac{\int_{\partial \bar{D}(p,r)} g(X, \nu) ds}{A[\bar{D}(p, r)]}.$$

The divergence at p therefore tells us how far the vector field at the point p points out of the disc. If the divergence at p is positive, then we can imagine p as a source for the vector field, if it is negative then p is a sink.

Exercise 5.5 Sketch the following vector fields in the x–y plane near the origin and calculate their divergence (w.r.t. the usual Euclidean metric) at the origin:

(a) $X(x, y) = (x, y, 0)^\top$;
(b) $X(x, y) = (-x, -y, 0)^\top$;
(c) $X(x, y) = (1, 0, 0)^\top$;
(d) $X(x, y) = (-y, x, 0)^\top$.

Let S be a regular surface with Riemannian metric g. For a twice continuously differentiable function $f : S \to \mathbb{R}$ we set

$$\Delta f : S \to \mathbb{R}, \quad \Delta f := \operatorname{div}\operatorname{grad} f.$$

Definition 5.1.8 We call Δ the ***Laplace operator***. A function that satisfies

$$\Delta f \equiv 0$$

is called ***harmonic***.

Example 5.1.9 Let $S \subset \mathbb{R}^3$ be a minimal surface with the first fundamental form as Riemannian metric. Let $\ell : \mathbb{R}^3 \to \mathbb{R}$ be a linear function, e.g. one of the three Cartesian coordinate functions. We then argue that

$$f := \ell|_S : S \to \mathbb{R}$$

is harmonic. As ℓ is linear, there exists a vector $Z \in \mathbb{R}^3$ such that

$$\ell(X) = \langle X, Z\rangle$$

for all $X \in \mathbb{R}^3$. The gradient of f at the point p is given by the projection of Z on T_pS:

$$\operatorname{grad} f(p) = Z - \langle Z, N(p)\rangle\, N(p),$$

where N is one of the two unit normal fields on S that are defined at least near p. This formula for the gradient follows from the fact that the gradient at p is the unique *tangent* vector which satisfies that

$$\big\langle \operatorname{grad} f(p), X\big\rangle = \partial_X f = \langle X, Z\rangle$$

for all $X \in T_pS$. The covariant derivative of $\operatorname{grad} f$ is therefore given by

$$\begin{aligned}
\nabla_X \operatorname{grad} f &= d\operatorname{grad} f(X) - \big\langle d\operatorname{grad} f(X), N\big\rangle N \\
&= -\langle Z, dN(X)\rangle\, N - \langle Z, N\rangle\, dN(X) \\
&\quad + \langle \langle Z, dN(X)\rangle\, N, N\rangle\, N + \langle \langle Z, N(X)\rangle\, dN(X), N\rangle\, N \\
&= \langle Z, W(X)\rangle\, N + \langle Z, N\rangle\, W(X) - \langle Z, W(X)\rangle\, N + 0 \\
&= \langle Z, N\rangle\, W(X).
\end{aligned}$$

We equate the traces of the left and the right hand sides of the equation to obtain

$$\Delta f = \operatorname{div}\operatorname{grad} f = \langle Z, N\rangle \operatorname{Trace} W = 2\,\langle Z, N\rangle\, H = 0.$$

We can, of course, apply the divergence theorem in the special case $\partial S = \emptyset$. The boundary integral then vanishes and we obtain the following corollary.

Corollary 5.1.10 *Let S be a compact regular surface with Riemannian metric g. Then every continuously differentiable vector field X and every twice continuously differentiable function f on S satisfy*

$$\int_S \mathrm{div} X \, dA = 0 = \int_S \Delta f \, dA.$$

Exercise 5.6 Show that differentiable functions f and differentiable vector fields X satisfy

$$\mathrm{div}(fX) = f \mathrm{div} X + g(\mathrm{grad} f, X).$$

Exercise 5.7 Let S be a compact surface with boundary and let $f_1, f_2 : S \to \mathbb{R}$ be functions which are sufficiently often differentiable. Prove the ***Green formulae***:

(a)

$$\int_S \Delta f_1 \cdot f_2 \, dA = -\int_S g(\mathrm{grad} f_1, \mathrm{grad} f_2) \, dA + \int_{\partial S} \partial_\nu f_1 \cdot f_2 \, ds;$$

(b) if $\partial S = \emptyset$:

$$\int_S \Delta f_1 \cdot f_2 \, dA = \int_S f_1 \cdot \Delta f_2 \, dA.$$

Exercise 5.8 Let S be a compact regular surface (without boundary). Let S be connected in the sense that any two points of S can be connected by a smooth curve in S. Show that the constant functions are the only harmonic functions on S.

Definition 5.1.11 Let S be a regular surface. A ***symmetric* $(2,0)$ *tensor field*** on S is a map that assigns to every point $p \in S$ a symmetric bilinear form b_p on T_pS, so that w.r.t. local parametrisations $F : U \to S$ the functions

$$b_{ij} : U \to \mathbb{R}, \quad b_{ij}(u) := b_{F(u)}\left(\frac{\partial F}{\partial u^i}(u), \frac{\partial F}{\partial u^j}(u)\right),$$

are always smooth.

Riemannian metrics are exactly those symmetric $(2,0)$ tensor fields that are positive definite at every point $p \in S$.

Now let S be a regular surface with a Riemannian metric g and another symmetric bilinear $(2,0)$ tensor field b.

Definition 5.1.12 The ***trace*** of b is the function $\operatorname{Trace} b : S \to \mathbb{R}$, which, w.r.t. a local parametrisation F, is given by

$$(\operatorname{Trace} b) \circ F = \sum_{ij} g^{ij} b_{ij}.$$

The ***divergence*** of b is the vector field $\operatorname{div} b$ on S, which, w.r.t. a local parametrisation, is given by

$$(\operatorname{div} b)^\ell = \sum_{ijk} g^{k\ell} g^{ij} \left(\frac{\partial b_{jk}}{\partial u^i} - \sum_\alpha \left(\Gamma_{ij}^\alpha b_{\alpha k} + \Gamma_{ik}^\alpha b_{\alpha j} \right) \right).$$

Remark The trace and the divergence of symmetric $(2, 0)$ tensor fields can, of course, also be characterised without the use of local parametrisations. If we fix $p \in S$, then there is for the symmetric bilinear form b_p on T_pS exactly one endomorphism $B_p : T_pS \to T_pS$ with $b_p(X, Y) = g_p(B_p(X), Y)$ for all $X, Y \in T_pS$ that is self-adjoint w.r.t. g_p. In local coordinates this endomorphism has the matrix coefficients $B_j^i = \sum_k g^{ik} b_{kj}$ and we have

$$(\operatorname{Trace} b)(p) = \operatorname{Trace}(B_p).$$

The divergence of symmetric $(2, 0)$ tensor fields, like the divergence of vector fields, can be defined as a trace of a suitably chosen covariant derivative of such tensor fields. However, as we do not need this, we will omit the details.

5.2 Variation of the metric

We will spend the rest of the chapter discussing how geometric quantities, e.g. the surface element or the Gauss curvature, change if we distort the Riemannian metric.

Definition 5.2.1 Let S be a regular surface and let $I \subset \mathbb{R}$ be an interval. A ***one-parameter family of Riemannian metrics*** on S is a map that assigns every $t \in I$ and every $p \in S$ a Euclidean scalar product $g_{t,p}$ on T_pS, such that for every local parametrisation (U, F, V) of S the maps

$$I \times U \to \mathbb{R}, \quad (t, u^1, u^2) \mapsto g_{ij}(t, u^1, u^2) := g_{t,F(u)} \left(\frac{\partial F}{\partial u^i}(u), \frac{\partial F}{\partial u^j}(u) \right)$$

are smooth.

Put briefly, for every fixed $t \in I$ there is a Riemannian metric $p \mapsto g_{t,p}$ such that everything depends smoothly on t.

Example 5.2.2 If there are two metrics g_0 and g_1 defined on a regular surface S, then

$$g_{t,p} := (1-t)g_{0,p} + tg_{1,p}, \quad t \in [0,1]$$

defines a one-parameter family of Riemannian metrics, which represents a transition from g_0 to g_1.

Example 5.2.3 If g is a Riemannian metric on S, then

$$g_{t,p} := tg_p, \quad t \in (0,\infty)$$

defines a one-parameter family of Riemannian metrics, which are simply scalings of the original metric.

Given a one-parameter family of Riemannian metrics with $t_0 \in I$ we can find its Taylor expansion w.r.t. t about $t = t_0$ and write

$$g_{ij}(t,u^1,u^2) = g_{ij}(u^1,u^2) + (t-t_0)\cdot \dot{g}_{ij}(u^1,u^2) + \mathrm{O}((t-t_0)^2).$$

Analogously we define $\dot{g}^{ij}$, $\dot{\Gamma}^k_{ij}$, $\dot{R}^\ell_{ijk}$, $\dot{K}$, $\dot{dA}$, and so forth. This definition of $\dot{g}_{ij}$ determines a $(2,0)$ tensor field $\dot{g}$, the derivative of this one-parameter family w.r.t. t at the point $t = t_0$. The derivatives of all quantities that only depend on the inner geometry of the surface can be found using the derivative of the metric. We will now investigate this in more detail.

Lemma 5.2.4 *The following are implications of the definitions made above:*

(a)

$$\dot{g}^{jk} = -\sum_{i\ell} g^{ij}\dot{g}_{i\ell}g^{\ell k}.$$

(b)

$$\dot{\Gamma}^k_{ij} = \frac{1}{2}\sum_{\alpha} g^{k\alpha}\left(\frac{\partial \dot{g}_{j\alpha}}{\partial u^i} + \frac{\partial \dot{g}_{i\alpha}}{\partial u^j} - \frac{\partial \dot{g}_{ij}}{\partial u^\alpha}\right) - \sum_{\beta,\ell}\Gamma^\beta_{ij}g^{\ell k}\dot{g}_{\beta\ell}.$$

Proof In the proof we will use the abbreviation $\partial_i := \partial/\partial u^i$ and the *Einstein summation*, which allows a more compact notation and is popular in literature in physics. The convention consists of omitting the sigma and regarding an expression as a sum if an index appears twice, once at the top and once at the bottom. The formula from lemma 4.2.14 for the Christoffel symbols then becomes

$$\Gamma^k_{ij} = \tfrac{1}{2}g^{k\alpha}(\partial_i g_{j\alpha} + \partial_j g_{i\alpha} - \partial_\alpha g_{ij}). \tag{5.8}$$

Note that the summation indices, here α, can be renamed at any time. We will make extensive use of this.

For the proof of (a) we differentiate the (constant) Kronecker symbol and obtain

$$0 = \dot{\delta}_i^k = (g_{ij}g^{jk})^\cdot = \dot{g}_{ij}g^{jk} + g_{ij}\dot{g}^{jk}.$$

It follows that

$$g_{ij}\dot{g}^{jk} = -\dot{g}_{ij}g^{jk}$$

and hence claim (a).

Differentiating (5.8) with respect to parameter t gives

$$\begin{aligned}
\dot{\Gamma}_{ij}^k &= \tfrac{1}{2}\dot{g}^{k\alpha}(\partial_i g_{j\alpha} + \partial_j g_{i\alpha} - \partial_\alpha g_{ij}) + \tfrac{1}{2}g^{k\alpha}(\partial_i \dot{g}_{j\alpha} + \partial_j \dot{g}_{i\alpha} - \partial_\alpha \dot{g}_{ij}) \\
&= -\tfrac{1}{2}g^{\beta\alpha}\dot{g}_{\beta\ell}g^{\ell k}(\partial_i g_{j\alpha} + \partial_j g_{i\alpha} - \partial_\alpha g_{ij}) + \tfrac{1}{2}g^{k\alpha}(\partial_i \dot{g}_{j\alpha} + \partial_j \dot{g}_{i\alpha} - \partial_\alpha \dot{g}_{ij}) \\
&= -\Gamma_{ij}^\beta g^{\ell k}\dot{g}_{\beta\ell} + \tfrac{1}{2}g^{k\alpha}(\partial_i \dot{g}_{j\alpha} + \partial_j \dot{g}_{i\alpha} - \partial_\alpha \dot{g}_{ij}).
\end{aligned}$$

The last equality is obtained by substituting (5.8) again. □

Lemma 5.2.5 *The variation of the surface element is given by*

$$\dot{dA} = \tfrac{1}{2}\,\mathrm{Trace}(\dot{g})dA.$$

Proof The claim essentially follows from lemma 5.1.6:

$$\begin{aligned}
\sqrt{\det(g_{ij})}^{\,\cdot} &= \frac{\det(g_{ij})^\cdot}{2\sqrt{\det(g_{ij})}} \\
&= \frac{g^{k\ell}\dot{g}_{k\ell}\det(g_{ij})}{2\sqrt{\det(g_{ij})}} \\
&= \frac{1}{2}\,\mathrm{Trace}(\dot{g})\sqrt{\det(g_{ij})}.
\end{aligned}$$

□

We therefore already know how the area changes if the Riemannian metric on a compact regular surface is deformed:

$$\left.\frac{d}{dt}\right|_{t=t_0} A[S, g_t] = \frac{1}{2}\int_S \mathrm{Trace}(\dot{g})dA.$$

Exercise 5.9 Use this formula to derive theorem 3.8.8 again.

Hint Write the normal field Φ (locally) in the form $\Phi = f \cdot N$, where N is a unit normal field, and show for the resulting one-parameter family of first fundamental forms that

$$\dot{g} = -2f \cdot II.$$

Lemma 5.2.6 *The variation of the Gauss curvature is given by*

$$2\dot{K} = \mathrm{div}(\mathrm{div}(\dot{g})) - \Delta(\mathrm{Trace}(\dot{g})) - K \cdot \mathrm{Trace}(\dot{g}).$$

Proof To prove the claim for a given point $p \in S$, we choose Riemann normal coordinates (w.r.t g_{t_0}) at p. This has the advantage that the first derivatives of the metric coefficients and the Christoffel symbols vanish at the point under consideration, which simplifies the calculation significantly. Lemma 4.3.5 and lemma 5.2.4 applied at point p give

$$\begin{aligned}
\dot{R}^{\ell}_{ijk} &= \partial_i \dot{\Gamma}^{\ell}_{kj} - \partial_j \dot{\Gamma}^{\ell}_{ki} + \dot{\Gamma}^{\ell}_{\alpha i}\Gamma^{\alpha}_{kj} + \Gamma^{\ell}_{\alpha i}\dot{\Gamma}^{\alpha}_{kj} - \dot{\Gamma}^{\ell}_{\alpha j}\Gamma^{\alpha}_{ki} - \Gamma^{\ell}_{\alpha j}\dot{\Gamma}^{\alpha}_{ki} \\
&= \partial_i \dot{\Gamma}^{\ell}_{kj} - \partial_j \dot{\Gamma}^{\ell}_{ki} \\
&= \tfrac{1}{2} g^{\alpha\ell} \partial_i \left(\partial_k \dot{g}_{j\alpha} + \partial_j \dot{g}_{k\alpha} - \partial_\alpha \dot{g}_{kj} \right) - \partial_i \Gamma^{\beta}_{kj} g^{\alpha\ell} \dot{g}_{\alpha\beta} \\
&\quad - \tfrac{1}{2} g^{\alpha\ell} \partial_j \left(\partial_k \dot{g}_{i\alpha} + \partial_i \dot{g}_{k\alpha} - \partial_\alpha \dot{g}_{ki} \right) + \partial_j \Gamma^{\beta}_{ki} g^{\alpha\ell} \dot{g}_{\alpha\beta} \\
&= \tfrac{1}{2} g^{\alpha\ell} \left(\partial_i \partial_k \dot{g}_{j\alpha} - \partial_i \partial_\alpha \dot{g}_{kj} - \partial_j \partial_k \dot{g}_{i\alpha} + \partial_j \partial_\alpha \dot{g}_{ki} \right) - R^{\beta}_{ijk} g^{\alpha\ell} \dot{g}_{\alpha\beta}.
\end{aligned}$$

We substitute this and lemma 4.3.11 into the derivative of the formula

$$K = \tfrac{1}{2} g^{jk} R^{i}_{ijk}$$

and obtain

$$\begin{aligned}
2\dot{K} &= g^{jk} \dot{R}^{i}_{ijk} + \dot{g}^{jk} R^{i}_{ijk} \\
&= \tfrac{1}{2} g^{jk} g^{\alpha i} \left(\partial_i \partial_k \dot{g}_{j\alpha} - \partial_i \partial_\alpha \dot{g}_{kj} - \partial_j \partial_k \dot{g}_{i\alpha} + \partial_j \partial_\alpha \dot{g}_{ki} \right) \\
&\quad - g^{jk} R^{\beta}_{ijk} g^{\alpha i} \dot{g}_{\alpha\beta} - g^{\alpha j} \dot{g}_{\alpha\ell} g^{\ell k} R^{i}_{ijk} \\
&= g^{jk} g^{\alpha i} \left(\partial_i \partial_k \dot{g}_{j\alpha} - \partial_i \partial_\alpha \dot{g}_{kj} \right) \\
&\quad - g^{jk} K \left(g_{jk} \delta^{\beta}_{i} - g_{ik} \delta^{\beta}_{j} \right) g^{\alpha i} \dot{g}_{\alpha\beta} - g^{\alpha j} \dot{g}_{\alpha\ell} g^{\ell k} K \left(g_{jk} \delta^{i}_{i} - g_{ik} \delta^{i}_{j} \right) \\
&= -2K \cdot \mathrm{Trace}(\dot{g}) + g^{jk} g^{\alpha i} \left(\partial_i \partial_k \dot{g}_{j\alpha} - \partial_i \partial_\alpha \dot{g}_{kj} \right),
\end{aligned} \tag{5.9}$$

where we used $\delta^{i}_{i} = 2 = g^{jk} g_{jk}$ and renamed parameters, e.g. $g^{jk} g^{\alpha i} \partial_j \partial_\alpha \dot{g}_{ki} = g^{jk} g^{\alpha i} \partial_k \partial_i \dot{g}_{j\alpha}$. Lemma 5.1.5 and the definition of the divergence of a (2,0) tensor field give

$$\begin{aligned}
\mathrm{div}(\mathrm{div}(\dot{g})) &= \partial_\ell (\mathrm{div}(\dot{g}))^{\ell} \\
&= \partial_\ell \left(g^{k\ell} g^{ij} \left(\partial_i \dot{g}_{jk} - \Gamma^{\alpha}_{ij} \dot{g}_{\alpha k} - \Gamma^{\alpha}_{ik} \dot{g}_{j\alpha} \right) \right) \\
&= g^{k\ell} g^{ij} \left(\partial_\ell \partial_i \dot{g}_{jk} - \partial_\ell \Gamma^{\alpha}_{ij} \dot{g}_{\alpha k} - \partial_\ell \Gamma^{\alpha}_{ik} \dot{g}_{j\alpha} \right).
\end{aligned} \tag{5.10}$$

Equation (4.3) implies that

$$g^{ij}\partial_i\partial_j g^{k\ell} = -g^{ij}g^{k\alpha}\partial_i\Gamma^{\ell}_{\alpha j} - g^{ij}g^{\ell\alpha}\partial_i\Gamma^{k}_{\alpha j}$$

and hence

$$\begin{aligned}\Delta(\mathrm{Trace}(\dot g)) &= g^{ij}\partial_i\partial_j(g^{k\ell}\dot g_{k\ell}) \\ &= g^{ij}(\partial_i\partial_j g^{k\ell})\dot g_{k\ell} + g^{ij}g^{k\ell}\partial_i\partial_j\dot g_{k\ell} \\ &= -\left(g^{ij}g^{k\alpha}\partial_i\Gamma^{\ell}_{\alpha j} + g^{ij}g^{\ell\alpha}\partial_i\Gamma^{k}_{\alpha j}\right)\dot g_{k\ell} \\ &\quad + g^{ij}g^{k\ell}\partial_i\partial_j\dot g_{k\ell}. \end{aligned} \tag{5.11}$$

Subtracting (5.11) from (5.10) yields

$$\begin{aligned}\mathrm{div}(\mathrm{div}(\dot g)) - \Delta(\mathrm{Trace}(\dot g)) &= g^{k\ell}g^{ij}\Big(\partial_\ell\partial_i\dot g_{jk} - \partial_i\partial_j\dot g_{k\ell} - \partial_\ell\Gamma^{\alpha}_{ij}\dot g_{\alpha k} \\ &\qquad -\partial_\ell\Gamma^{\alpha}_{ik}\dot g_{j\alpha} + \partial_i\Gamma^{\alpha}_{\ell j}\dot g_{k\alpha} + \partial_i\Gamma^{\alpha}_{kj}\dot g_{\alpha\ell}\Big) \\ &= g^{k\ell}g^{ij}\left(\partial_\ell\partial_i\dot g_{jk} - \partial_i\partial_j\dot g_{k\ell} - \partial_\ell\Gamma^{\alpha}_{ij}\dot g_{\alpha k} + \partial_i\Gamma^{\alpha}_{\ell j}\dot g_{k\alpha}\right) \\ &= g^{k\ell}g^{ij}\left(\partial_\ell\partial_i\dot g_{jk} - \partial_i\partial_j\dot g_{k\ell} + R^{\alpha}_{i\ell j}\dot g_{k\alpha}\right) \\ &= g^{k\ell}g^{ij}(\partial_\ell\partial_i\dot g_{jk} - \partial_i\partial_j\dot g_{k\ell}) - K\cdot\mathrm{Trace}(\dot g).\end{aligned}$$

Comparing this to (5.9) concludes the proof. □

The variation formulae proved up to now imply a remarkable theorem.

Theorem 5.2.7 *Let S be a compact regular surface. Then the number*

$$\int_S K\, dA$$

is independent of the Riemannian metric.

This theorem is remarkable because the Gauss curvature K as a function on S depends significantly on the choice of the Riemannian metric. For example, we studied two very different Riemannian metrics on the torus. On the one hand, we know the torus as a tubular surface from example 3.8.18 with its first fundamental form as the Riemannian metric. In this case the Gauss curvature is not constant. On the other hand, we met a Riemannian metric with Gauss curvature $K \equiv 0$ in example 4.4.2. Theorem 5.2.7 tells us without any calculation that also the tubular surface must satisfy $\int_S K\, dA = 0$. Compare this with exercise 3.32.

Proof We see from lemma 5.2.5 and lemma 5.2.6 that every one-parameter family of Riemannian metrics g_t with Gauss curvature K_t and surface element dA_t satisfies

$$\begin{aligned} \frac{d}{dt}\int_S K_t\, dA_t &= \int_S (\dot{K}_t\, dA_t + K_t\, d\dot{A}_t) \\ &= \int_S \left(\frac{1}{2}\mathrm{div}(\mathrm{div}(\dot{g})) - \frac{1}{2}\Delta(\mathrm{Trace}(\dot{g}))\right) dA_t = 0 \end{aligned}$$

by corollary 5.1.10. If only two Riemannian metrics g_0 and g_1 on S are given, then as in example 5.2.2 we consider the one-parameter family $g_t := (1-t)g_0 + tg_1$ and observe that

$$\int_S K_1\, dA_1 - \int_S K_0\, dA_0 = \int_0^1 \frac{d}{dt}\int_S K_t\, dA_t\, dt = 0. \qquad \square$$

Corollary 5.2.8 *Let S_1 and S_2 be compact regular surfaces with Riemannian metrics. If S_1 and S_2 are diffeomorphic, then*

$$\int_{S_1} K\, dA = \int_{S_2} K\, dA.$$

Proof The equation certainly holds if the metric on S_1 is the pulled-back metric of S_2, since the two metrics would then be isomorphic. Otherwise theorem 5.2.7 tells us that $\int_{S_1} K\, dA$ w.r.t. the pulled-back metric and w.r.t. given metric agree. $\square$

Example 5.2.9 The sphere and the torus *cannot* be diffeomorphic, since $\int_{\mathrm{torus}} K\, dA = 0$, while $\int_{S^2} K\, dA = 4\pi$.

If one wants to show that two given surfaces, e.g. the sphere and the ellipsoid, are diffeomorphic, then one will try to construct a diffeomorphism. But how can one show that two surfaces are not diffeomorphic? The number $\int_S K\, dA$ gives us the possibility to distinguish surfaces that are not diffeomorphic, e.g. the sphere and the torus. In the next chapter we will learn to understand this number better by showing how we can find it by cutting the surface into triangles and then counting vertices, edges and triangles. This will be the subject of the Gauss–Bonnet theorem.

6 Geometry and topology

Differential geometry is a fine, quantitative geometry, in which relationships between lengths and angles are important. Topology, by contrast, is of a much coarser and more qualitative nature. Here only those quantities that are preserved under distortions are studied. In order to obtain a topological description of the total Gauss curvature, we triangulate the surfaces, i.e. we cut them into triangles. The theorem of Gauss–Bonnet now tells us that we can determine the total curvature by counting vertices, edges and triangles.

6.1 Polyhedra

In the last sections of this book we want to study global properties of surfaces. For example, we want be able to decide whether two given surfaces are homeomorphic or not. For this purpose we will cut the surfaces into triangles, which will then allow us to use combinatorial methods. We make the following definition.

Definition 6.1.1 A ***polyhedron*** X is a finite set of triangles $\Delta_j \subset \mathbb{R}^n$,

$$X = \{\Delta_1, \ldots, \Delta_k\},$$

with the following property: any two of those triangles intersect

- not at all, or
- exactly at one vertex, or
- exactly at one edge.

The union of those triangles,

$$|X| := \cup_{j=1}^{k} \Delta_j,$$

is called the ***geometric realisation*** of X.

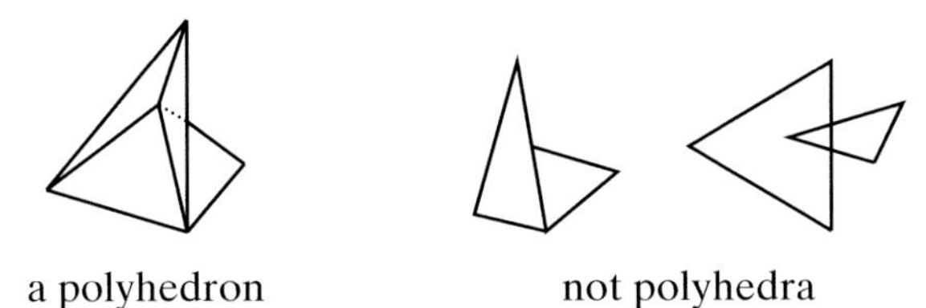

a polyhedron not polyhedra

We use the following notation:

$$\begin{aligned} e(X) &= \text{number of vertices of } X, \\ k(X) &= \text{number of edges of } X, \\ f(X) &= \text{number of triangles } X. \end{aligned}$$

Definition 6.1.2 The number

$$\chi(X) := e(X) - k(X) + f(X)$$

is called the ***Euler–Poincaré characteristic*** of X.

Example 6.1.3 Let us take a ***tetrahedron*** in $\mathbb{R}^3$ as X:

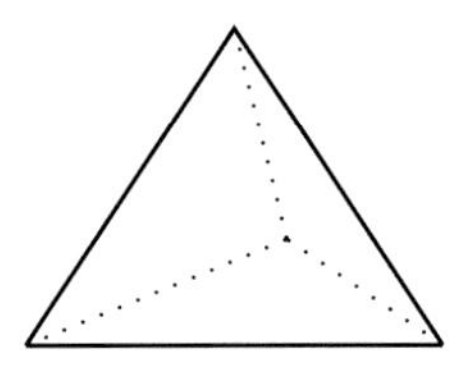

Counting gives

$$\begin{aligned} e(X) &= 4, \\ k(X) &= 6, \\ f(X) &= 4 \end{aligned}$$

and hence

$$\chi(X) = 2.$$

Exercise 6.1 Show that the geometric realisation of the tetrahedron is homeomorphic to S^2.

Remark If we refine the polyhedron X by adding a vertex inside a triangle and obtain a new polyhedron X', then we see that

$$e(X') = e(X) + 1,$$
$$k(X') = k(X) + 3,$$
$$f(X') = f(X) + 2.$$

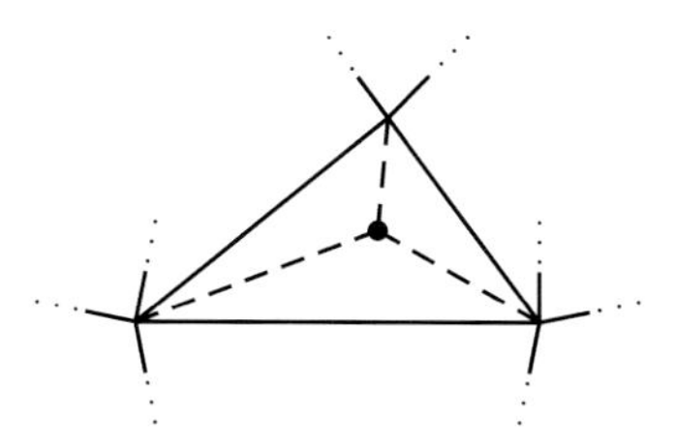

Hence the Euler–Poincaré characteristic is not changed at all by such refinements:

$$\chi(X') = \chi(X).$$

The polyhedra X and X' obviously have the same geometric realisation: $|X| = |X'|$. Thus, if we look for quantities that only depend on the geometric realisation of a polyhedron, then $e(X)$, $k(X)$ and $f(X)$ are unsuitable, but the Euler–Poincaré characteristic $\chi(X)$ is a good candidate.

Definition 6.1.4 Let X be a polyhedron and let v be a vertex of X. Let $\alpha_1, \ldots, \alpha_\ell$ be the interior angles at vertex v in all triangles of X of which v is a vertex. Then

$$\operatorname{def}(v) := 2\pi - \sum_{j=1}^{\ell} \alpha_j$$

is called the ***angle defect*** of X at v.

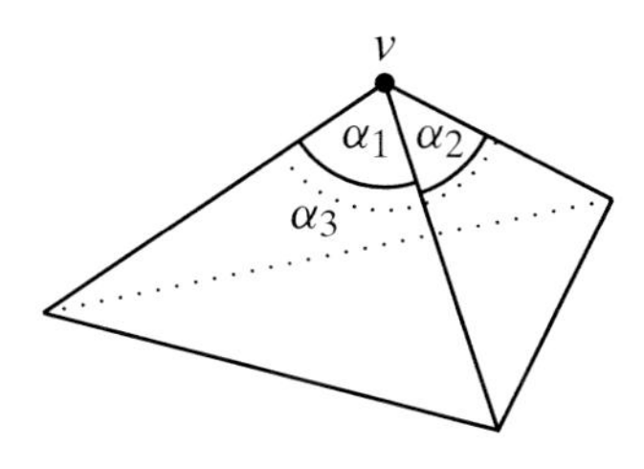

The angle defect at v vanishes precisely when the interior angles at v add up to 2π, i.e. to 360 degrees.

Theorem 6.1.5 (the Gauss–Bonnet theorem for polyhedra) *Let X be a polyhedron such that exactly two triangles meet at every edge. Then*

$$\sum_v \operatorname{def}(v) = 2\pi\chi(X),$$

where the sum is taken over all vertices v of X.

The total angle defect therefore agrees with the Euler–Poincaré characteristic up to a factor of 2π.

Proof Since exactly two triangles meet at every edge by assumption and every triangle has exactly three edges, we have

$$2k(X) = 3f(X).$$

It follows that

$$\begin{aligned}\chi(X) &= f(X) - k(X) + e(X)\\ &= f(X) - \tfrac{3}{2}f(X) + e(X)\\ &= -\tfrac{1}{2}f(X) + e(X).\end{aligned} \tag{6.1}$$

If $\Delta_1, \ldots, \Delta_{f(X)}$ are the triangles of X, then we obtain

$$\begin{aligned}\sum_v \operatorname{def}(v) &= 2\pi \cdot e(X) - \sum_{j=1}^{f(X)} \sum_{\substack{v \text{ vertex}\\ \text{of } \Delta_j}} \text{interior angles of } \Delta_j \text{ at } v\\ &= 2\pi \cdot e(X) - \sum_{j=1}^{f(X)} \pi\\ &= 2\pi \cdot e(X) - \pi \cdot f(X).\end{aligned} \tag{6.2}$$

In the second line we used that the sum of the interior angles of a Euclidean triangle is exactly π. The claim follows from (6.1) and (6.2). □

Exercise 6.2 Verify the Gauss–Bonnet theorem for polyhedra in the example of a tetrahedron with equilateral triangles.

6.2 Triangulations

We want to decompose surfaces into triangles to then investigate them with combinatorial tools. Such a decomposition is called a triangulation.

Definition 6.2.1 Let $S \subset \mathbb{R}^3$ be a compact regular surface. A ***triangulation*** of S is a pair (X, Φ) consisting of a polyhedron X and a homeomorphism

$$\Phi : |X| \to S,$$

such that for every triangle Δ of X the restriction $\Phi|_\Delta : \Delta \to \Phi(\Delta)$ is a diffeomorphism. More precisely,

> for every triangle Δ of X, there exists an open neighbourhood U of Δ in the plane spanned by Δ, and there is an open neighbourhood V of $\Phi(\Delta)$ in S and a diffeomorphism $\varphi : U \to V$ such that $\varphi|_\Delta = \Phi|_\Delta$.

The diffeomorphic images $\Phi(\Delta)$ of the triangles of X decompose the surface S into "curved triangles".

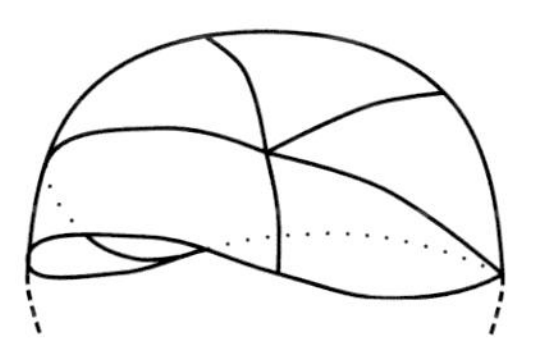

Remark A polyhedron X that belongs to a triangulation of a compact regular surface always has the property required in theorem 6.1.5, which states that exactly two triangles meet at every edge. This is the case because a regular surface is locally divided into exactly two parts by a regular curve.

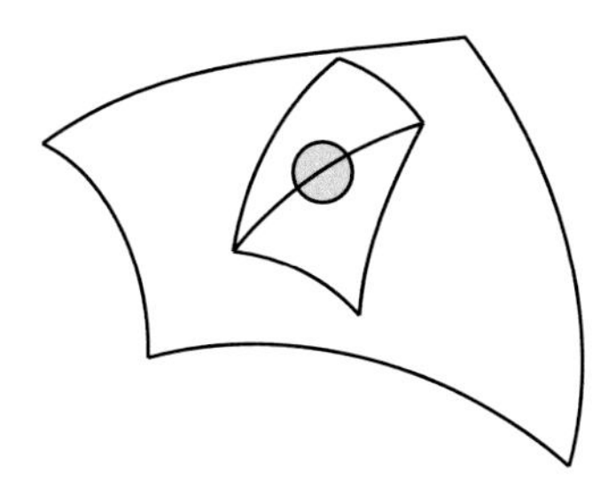

Definition 6.2.2 Let S be a compact regular surface and let (X, Φ) be a triangulation of S. Then

$$\chi(S) := \chi(X)$$

is called the ***Euler–Poincaré characteristic*** of S.

Exercise 6.3 Give a triangulation of S^2 and calculate the Euler–Poincaré characteristic of S^2.

Definition 6.2.2 still needs some justification. It is a priori possible that different triangulations of the same surface lead to different Euler–Poincaré characteristics. However, a corollary of the Gauss–Bonnet theorem will tell us that this is not the case, see corollary 6.3.3.

It also needs to be shown that a compact regular surface can always be triangulated, i.e. has a triangulation. We will prove this using a series of lemmas. The reader willing to believe in the existence of triangulations may continue reading with theorem 6.2.17 on page 257.

For a compact subset $S \subset \mathbb{R}^n$ and a point $q \in \mathbb{R}^n$ we define the distance from q to S by

$$\operatorname{dist}(q, S) := \min\{\|q - p\| \mid p \in S\}.$$

By the compactness of S the minimum is attained. We have $\operatorname{dist}(q, S) = 0$ if and only if $q \in S$. We further define the ***ρ-neighbourhood*** of S by

$$U_\rho(S) := \{q \in \mathbb{R}^n \mid \operatorname{dist}(q, S) < \rho\}.$$

Lemma 6.2.3 *Let S be a compact orientable regular surface with unit normal field N. Then there exists a $\rho > 0$ such that for every point $q \in U_\rho(S)$ there is exactly one point $\mathscr{P}(q) \in S$ with*

$$\operatorname{dist}(q, S) = \|\mathscr{P}(q) - q\|.$$

Furthermore, the map

$$U_\rho(S) \to S \times (-\rho, \rho), \quad q \mapsto (\mathscr{P}(q), \pm\operatorname{dist}(q, S)),$$

is a diffeomorphism, where the sign in front of $\operatorname{dist}(q, S)$ *depends on whether q lies on the side of S to which N points or on the other side. The inverse map is given by*

$$\mathscr{E} : S \times (-\rho, \rho) \to U_\rho(S), \quad \mathscr{E}(p, t) = p + t \cdot N(p).$$

Proof Let us first find the differential of $\mathscr{E}$. If F is a local parametrisation of S, then

$$\frac{\partial}{\partial u^j}\mathscr{E}(F(u), t) = \frac{\partial F}{\partial u^j}(u) + t\frac{\partial N}{\partial u^j}(u) = (\mathrm{Id} - tW)\left(\frac{\partial F}{\partial u^j}\right) \in T_{F(u)}S$$

and

$$\frac{\partial}{\partial t}\mathscr{E}(F(u), t) = N(F(u)).$$

As S is compact, the absolute values of the eigenvalues of W, the principal curvatures, are bounded by a constant $C > 0$. The map $\mathrm{Id} - tW$ is therefore invertible for $|t| < \rho_1 := 1/C$ and hence $(\partial/\partial u^1)\mathscr{E}(F(u),t)$, $(\partial/\partial u^2)\mathscr{E}(F(u),t)$ and $(\partial/\partial t)\mathscr{E}(F(u),t)$ are linearly independent. We see that $\mathscr{E}$ defined on $S \times (-\rho_1, \rho_1)$ is a local diffeomorphism.

$\mathscr{E} : S \times (-\rho, \rho) \to U_\rho(S)$ is surjective for every $\rho \in (0, \rho_1)$, since if $p \in S$ is the point nearest to $q \in U_\rho(S)$, then by corollary 3.6.18 the vector $p - q$ is perpendicular to T_pS and hence $q = \mathscr{E}(p, \pm\|p - q\| \cdot N(p))$.

It remains to show that for sufficiently small $\rho \in (0, \rho_1)$ the map $\mathscr{E} : S \times (-\rho, \rho) \to U_\rho(S)$ is also injective. If this were not the case, then there would be sequences (p_i, t_i) and (p_i', t_i') with $\mathscr{E}(p_i, t_i) = \mathscr{E}(p_i', t_i')$, $t_i, t_i' \to 0$ and $p_i \neq p_i'$. By the compactness of S we can after passing to a subsequence assume that the two sequences of points converge, $p_i \to p$ and $p_i' \to p'$ for $i \to \infty$. By

$$\|p_i - p_i'\| \leq \|p_i - \mathscr{E}(p_i, t_i)\| + \|p_i' - \mathscr{E}(p_i', t_i)\| \leq |t_i| + |t_i'| \longrightarrow 0$$

we have $p = p'$. But as $\mathscr{E}$ is a diffeomorphism in a neighbourhood of $(p, 0)$, and thus in particular injective, it follows from $\mathscr{E}(p_i, t_i) = \mathscr{E}(p_i', t_i')$ that $p_i = p_i'$ for a sufficiently large i, which contradicts the assumption. □

Example 6.2.4 In the case of the sphere $S = S^2$ the value $\rho = 1$ is the maximal ρ that satisfies the condition in lemma 6.2.3, since the origin has the same distance from all points on S.

To obtain a triangulation of a regular surface S, we will construct the corresponding polyhedron in such a way that all vertices lie on the surface. The triangles, and hence the entire geometric realisation $|X|$ of the polyhedron, will still be contained in $U_\rho(S)$. In particular, $\mathscr{P} : |X| \to S$ is a continuous map, restricted to each triangle even a smooth one. Nevertheless, we must proceed very carefully positioning the triangles, since otherwise even the restriction of $\mathscr{P}$ to a single triangle might not be injective. Very narrow triangles might be aligned perpendicularly to the surface.

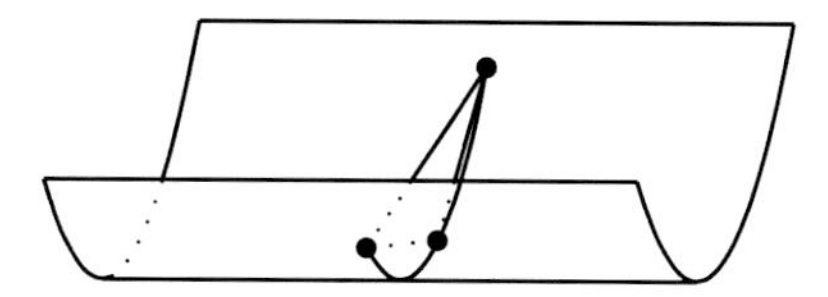

This effect can occur with arbitrarily small triangles, although one might naively expect that a sufficiently fine partition into triangles will provide a triangulation. We must therefore also ensure that the triangles are not too narrow.

Definition 6.2.5 Let $S \subset \mathbb{R}^3$ and let $\varepsilon > 0$. A subset $A \subset S$ is called ***ε-separated*** if for any two distinct points $p, q \in A$

$$\|p - q\| \geq \varepsilon.$$

An ε-separated subset $A \subset S$ is called an ***ε-configuration*** on S if it is a maximal ε-separated subset of S, i.e. if we have equality $A = A'$ for every ε-separated subset $A' \subset S$ with $A \subset A'$.

The condition $\|p - q\| \geq \varepsilon$ is equivalent to the condition that the open balls of radius $\varepsilon/2$ with centres p and q are disjoint. We can therefore imagine an ε-separated subset of S as a collection of pairwise disjoint open $\varepsilon/2$-balls with centres on S.

Lemma 6.2.6 *If S is contained in a ball of radius R, then every ε-separated subset of $A \subset S$ has at most*

$$\left(\frac{2R + \varepsilon}{\varepsilon}\right)^3$$

many elements. In particular, S has an ε-configuration with finitely many elements.

Proof Let $S \subset B(q, R)$. As the balls of radius $\varepsilon/2$ around the points from A are pairwise disjoint and since

$$\bigcup_{p \in A} B(p, \varepsilon/2) \subset U_{\varepsilon/2}(S) \subset B(q, R + \varepsilon/2),$$

the volumes satisfy

$$\sum_{p \in A} \frac{4\pi}{3}(\varepsilon/2)^3 \leq \frac{4\pi}{3}(R + \varepsilon/2)^3$$

and hence

$$|A| \leq \left(\frac{2R + \varepsilon}{\varepsilon}\right)^3. \tag{6.3}$$

An arbitrary ε-separated subset is extended by adding points and making sure that the new set of points is still ε-separated, until no more points can be added in this way. This process stops after finitely many steps by (6.3). We have then obtained an ε-configuration. □

Lemma 6.2.7 *Let $S \subset \mathbb{R}^3$, let $\varepsilon > 0$, let A be an ε-configuration on S. Then the ε-balls with centres in A cover S entirely:*

$$S \subset \bigcup_{p \in A} B(p, \varepsilon).$$

Proof Suppose that there is a $q \in S$ with $q \notin \bigcup_{p \in A} B(p, \varepsilon)$. Then $\|q - p\| \geq \varepsilon$ for all $p \in A$ and $A' := A \cup \{q\}$ is ε-separated as well. This contradicts the maximality of A. □

Definition 6.2.8 Let $A \subset S \subset \mathbb{R}^3$ be an ε-configuration. The ***star*** of $p \in A$ is given by

$$\mathrm{St}(p) := \{x \in \mathbb{R}^3 | \ \|x - p\| \leq \|x - q\| \text{ for all } q \in A\}.$$

Points $p_1, \ldots, p_k \in A$ are called ***common neighbours*** if

$$S \cap \bigcap_{j=1}^{k} \mathrm{St}(p_j) \neq \emptyset.$$

If the ε-configuration consists of two points p_1 and p_2 only, then $\mathrm{St}(p_1)$ and $\mathrm{St}(p_2)$ are the closed half-spaces in which $\mathbb{R}^3$ is divided by the plane that is perpendicular to $p_1 - p_2$ and goes through the midpoint $\frac{1}{2}(p_1 + p_2)$. In the general case, $\mathrm{St}(p)$ is the intersection of finitely many closed half-spaces.

Lemma 6.2.9 *Let $S \subset \mathbb{R}^3$, let $\varepsilon > 0$ and let A be an ε-configuration on S. Then all $p \in A$ satisfy*

$$\mathrm{St}(p) \cap S \subset B(p, \varepsilon).$$

For any two distinct neighbouring points $p, p' \in A$ we have

$$\varepsilon \leq \|p - p'\| \leq 2\varepsilon.$$

Proof Let $q \in S$ with $\|q - p\| \geq \varepsilon$. As the ε-balls with centres in A cover S entirely by lemma 6.2.7, there is a $p' \in A$ with $\|q - p'\| < \varepsilon$. Hence $\|q - p'\| < \|q - p\|$ and thus $q \notin \mathrm{St}(p)$. This proves $\mathrm{St}(p) \cap S \subset B(p, \varepsilon)$.

The lower bound for $\|p - p'\|$ holds for all distinct points from A. As p and p' are neighbours, the set

$$B(p, \varepsilon) \cap B(p', \varepsilon) \supset \mathrm{St}(p) \cap \mathrm{St}(p') \cap S$$

cannot be empty. It follows that $\|p - p'\| \leq 2\varepsilon$. □

Exercise 6.4 Compare volumes as in the proof of lemma 6.2.6 to show that there can never be more than 27 common neighbours.

We now use the following notation: for three non-collinear points $p_1, p_2, p_3 \in \mathbb{R}^3$, let $E(p_1, p_2, p_3) \subset \mathbb{R}^3$ be the plane spanned by those points and let $\Delta(p_1, p_2, p_3) \subset E(p_1, p_2, p_3)$ be the triangle with those vertices.

From now on $S \subset \mathbb{R}^3$ denotes a compact regular surface. Let $\varepsilon > 0$ and let A be an ε-configuration on S. The polyhedron we construct will only contain triangles $\Delta(p_1, p_2, p_3)$ for which the points p_1, p_2 and p_3 are common neighbours. We therefore investigate such triangles in a little more detail. To show that such triangles cannot become arbitrarily narrow, we consider their height $h(p_1, p_2, p_3)$ over the side connecting p_1 and p_3.

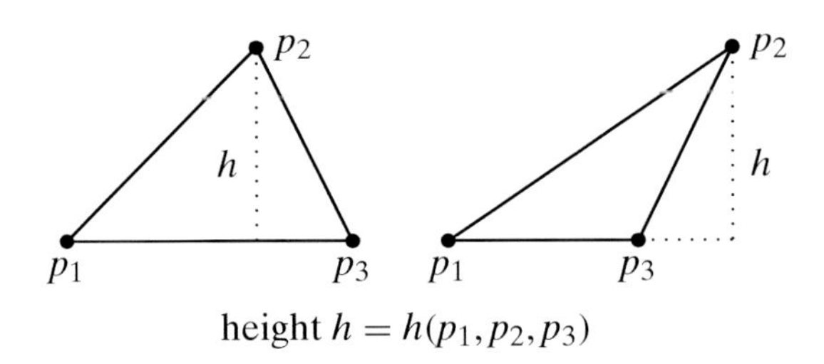

height $h = h(p_1, p_2, p_3)$

Lemma 6.2.10 *There is a universal constant $C_H > 0$ such that we have*

$$h(p_1, p_2, p_3) \geq C_H \cdot \|p_1 - p_3\|$$

for every set $S \subset \mathbb{R}^3$, for every $\varepsilon > 0$, for every ε-configuration and for every triangle $\Delta(p_1, p_2, p_3)$ whose vertices are common neighbours. In particular,

$$h(p_1, p_2, p_3) \geq C_H \cdot \varepsilon.$$

Proof As p_1, p_2 and p_3 are common neighbours, there exists $q \in S \cap \mathrm{St}(p_1) \cap \mathrm{St}(p_2) \cap \mathrm{St}(p_3)$. In particular, lemma 6.2.9 implies that

$$\|p_1 - q\| = \|p_2 - q\| = \|p_3 - q\| =: r \leq \varepsilon.$$

Likewise, because of lemma 6.2.9 it follows for $i \neq j$ that

$$\|p_i - p_j\| \geq \varepsilon \geq r.$$

There is no loss of generality if we suppose that $q = (0,0,0)^\top$. Further, the ratio $h(p_1, p_2, p_3)/\|p_1 - p_3\|$ remains unchanged if all points p_1, p_2, p_3 are stretched by a factor of r. Hence

$$C_H := \min \left\{ \frac{h(p_1, p_2, p_3)}{\|p_1 - p_3\|} \;\middle|\; p_1, p_2, p_3 \in S^2, \quad \|p_i - p_j\| \geq 1 \right\} > 0$$

is the desired constant. The minimum is attained and is positive, since the function

$$(p_1, p_2, p_3) \mapsto \frac{h(p_1, p_2, p_3)}{\|p_1 - p_3\|}$$

is continuous and positive on the compact set

$$\left\{(p_1,p_2,p_3) \in S^2 \times S^2 \times S^2 \,\middle|\, \|p_i - p_j\| \geq 1\right\}. \qquad \square$$

Lemma 6.2.11 *Let $S \subset \mathbb{R}^3$ be a compact orientable regular surface. Then there exists a positive ρ_1, smaller than the ρ from lemma 6.2.3, such that for every ε-configuration A on S with $\varepsilon \in (0,\rho_1]$ and for every triangle $\Delta = \Delta(p_1,p_2,p_3)$ whose vertices are common neighbours the restriction of $\mathscr{P}$ to a small neighbourhood $U(\Delta)$ of Δ in the plane $E(p_1,p_2,p_3)$ is a diffeomorphism*

$$\mathscr{P}|_{U(\Delta)} : U(\Delta) \to \mathscr{P}(U(\Delta)) \subset S.$$

Proof (a) We first show that for a sufficiently small ε the restriction of $\mathscr{P}$ to a neighbourhood of Δ in $E := E(p_1,p_2,p_3)$ is injective. If this were not the case, then there would be a sequence of ε_i-configurations A_i on S with $\varepsilon_i \searrow 0$ and common neighbours $p_{1,i}, p_{2,i}, p_{3,i}$ w.r.t. A_i, such that $\mathscr{P}$ is not injective on the ε_i-neighbourhood of $\Delta(p_{1,i},p_{2,i},p_{3,i})$ in $E(p_{1,i},p_{2,i},p_{3,i})$. Let $x_i \neq y_i \in E(p_{1,i},p_{2,i},p_{3,i})$ be in this ε_i-neighbourhood, with $\mathscr{P}(x_i) = \mathscr{P}(y_i) =: q_i \in S$. Then both x_i and y_i lie on the straight line through q_i that is parallel to $N(q_i)$. Hence $x_i - y_i$ is parallel to $N(q_i)$. Parallel transport and possibly re-enumerating the vertices gives a point r_i on the side connecting $p_{1,i}$ and $p_{3,i}$ such that $p_{2,i} - r_i$ is parallel to $N(q_i)$.

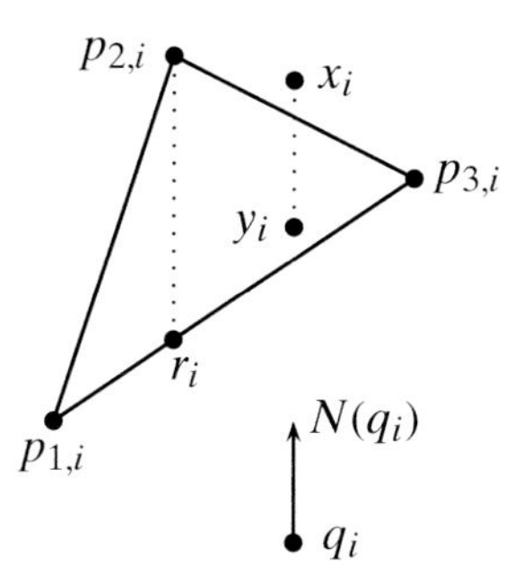

The following sequences converge by the compactness of S after passing to a subsequence:

$$p_{1,i} \to p_1, \quad p_{2,i} \to p_2, \quad p_{3,i} \to p_3, \quad q_i \to q.$$

From $\|p_{j,i} - p_{k,i}\| \leq 2\varepsilon_i \to 0$ and $\|y_i - q_i\| \to 0$ it follows that

$$p_1 = p_2 = p_3 = q.$$

By lemma 6.2.9 and lemma 6.2.10 we have that

$$C_H \cdot \varepsilon_i \le h(p_{1,i}, p_{2,i}, p_{3,i}) \le \|p_{2,i} - r_i\| \le 2\varepsilon_i$$

and thus

$$C_H \le \frac{\|p_{2,i} - r_i\|}{\varepsilon_i} \le 2.$$

By the compactness of the set of all vectors X with $C_H \le \|X\| \le 2$, passing to a subsequence gives convergence

$$\frac{p_{2,i} - r_i}{\varepsilon_i} \to Z$$

with $C_H \le \|Z\| \le 2$. As $p_{2,i} - r_i$ is parallel to $N(q_i)$ and $N(q_i) \to N(q)$ converges, Z must be a non-trivial multiple of $N(q)$. On the other hand the vectors $(p_{1,i} - p_{2,i})/\varepsilon_i$ and $(p_{3,i} - p_{2,i})/\varepsilon_i$ converge to the vectors $X, Y \in T_qS$ after passing to a subsequence, for similar reasons. Those vectors X and Y form a basis of T_qS, since the triangle $\Delta(X, 0, Y)$ in T_qS has height $\ge C_H$ and is therefore non-degenerate. As $p_{2,i} - r_i$ lies in the linear hull of $p_{1,i} - p_{2,i}$ and $p_{3,i} - p_{2,i}$, Z must be a linear combination of X and Y and hence lies in T_qS. But the vector S cannot lie in T_qS and at the same time be a non-trivial multiple of $N(q)$, contradiction.

(b) We now know that for a sufficiently small ε the restriction of $\mathscr{P}$ to a small neighbourhood $U(\Delta)$ is a smooth and bijective map $U(\Delta) \to \mathscr{P}(U(\Delta))$. To show that it is a diffeomorphism, we still need to verify that the differential

$$d_x\mathscr{P} : E(p_1, p_2, p_3) \to T_{\mathscr{P}(x)}S$$

is invertible for all $x \in U(\Delta)$. Suppose that this were not the case for very small $\varepsilon > 0$. Then we obtain in a way similar to the proof of (a) a subsequence of ε_i-configurations A_i on S with $\varepsilon_i \searrow 0$, common neighbours $p_{1,i}, p_{2,i}, p_{3,i}$ w.r.t. A_i, further $X_i \in E(p_{1,i}, p_{2,i}, p_{3,i})$ with $\|X_i\| = 1$ and $d_{x_i}\mathscr{P}(X_i) = (0, 0, 0)^\top$, where x_i lies in the ε_i-neighbourhood of $\Delta(p_{1,i}, p_{2,i}, p_{3,i})$ in $E(p_{1,i}, p_{2,i}, p_{3,i})$. We set $q_i := \mathscr{P}(x_i)$. As in (a) we can assume that

$$p_{1,i}, p_{2,i}, p_{3,i}, x_i, q_i \to q \in S$$

and

$$X_i \to X \in T_qS$$

converge, $\|X\| = 1$. The condition $d_{x_i}\mathscr{P}(X_i) = (0, 0, 0)^\top$ is by lemma 6.2.3 equivalent to

$$X_i = t_i N(q_i)$$

for a suitable t_i. But $N(q_i) \to N(q)$, so X must be perpendicular to T_qS, contradiction. □

We now know that for every ε-configuration with ε sufficiently small and every triangle whose vertices are common neighbours the restriction of $\mathscr{P}$ to this triangle is a diffeomorphism. We want to piece our triangulation together from such triangles. However, we must proceed carefully, since we cannot simply take *all* those triangles, as the example with four common neighbours already shows.

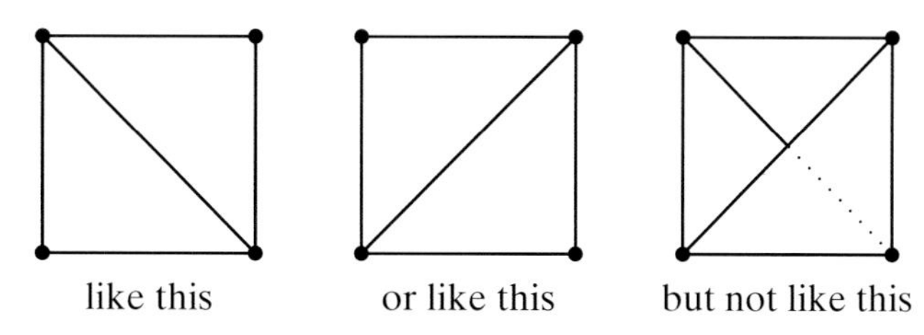

We therefore need to think about which triangles of this type we want to use for the construction of a triangulation.

Lemma 6.2.12 *Let $S \subset \mathbb{R}^3$ be a compact orientable regular surface. Then there exists a positive ρ_2 such that for every ε-configuration A on S with $\varepsilon \in (0, \rho_2]$ and for any three common neighbours p_1, p_2, p_3 the set*

$$\mathrm{St}(p_1) \cap \mathrm{St}(p_2) \cap \mathrm{St}(p_3) \cap S$$

has exactly one element.

Proof The condition that p_1, p_2 and p_3 are common neighbours says precisely that the set $\mathrm{St}(p_1) \cap \mathrm{St}(p_2) \cap \mathrm{St}(p_3) \cap S$ is not empty. We therefore need to show that if ε is sufficiently small, then the set $\mathrm{St}(p_1) \cap \mathrm{St}(p_2) \cap \mathrm{St}(p_3) \cap S$ does not have two distinct elements.

Suppose that there were a sequence of ε_i-configurations A_i on S with $\varepsilon_i \searrow 0$, common neighbours $p_{1,i}, p_{2,i}, p_{3,i}$ w.r.t A_i and for each i two distinct points $x_i, y_i \in \mathrm{St}(p_1) \cap \mathrm{St}(p_2) \cap \mathrm{St}(p_3) \cap S$. Then $x_i - y_i$ would be perpendicular to $E(p_{1,i}, p_{2,i}, p_{3,i})$.

As in the preceding proofs, we can again assume that

$$p_{1,i}, p_{2,i}, p_{3,i}, x_i, y_i \to q \in S,$$

$$\frac{p_{1,i} - p_{2,i}}{\varepsilon_i} \to X, \quad \frac{p_{3,i} - p_{2,i}}{\varepsilon_i} \to Y,$$

where X and Y form a basis of T_qS, and

$$\frac{x_i - y_i}{\|x_i - y_i\|} \to Z$$

for $i \to \infty$. The vector Z is a unit normal vector on T_qS since $x_i - y_i$ is perpendicular to $E(p_{1,i}, p_{2,i}, p_{3,i})$. On the other hand, Z must be a tangent vector because $x_i, y_i \in S$, contradiction. □

Corollary 6.2.13 *Let $S \subset \mathbb{R}^3$ be a compact orientable regular surface with an ε-configuration A with $\varepsilon \in (0, \rho_2]$ and ρ_2 as in lemma 6.2.12. Let $M_1, M_2 \subset A$ be two sets, each consisting of common neighbours.*

If M_1 and M_2 have more than two points in common, then $M_1 \cup M_2$ consists of common neighbours as well.

Proof Let p_1, p_2, p_3 be three common points of M_1 and M_2. Because of

$$1 = |\mathrm{St}(p_1) \cap \mathrm{St}(p_2) \cap \mathrm{St}(p_3) \cap S| \geq \left| \bigcap_{p \in M_1} \mathrm{St}(p) \cap S \right| > 0$$

we have

$$\mathrm{St}(p_1) \cap \mathrm{St}(p_2) \cap \mathrm{St}(p_3) \cap S = \bigcap_{p \in M_1} \mathrm{St}(p) \cap S.$$

Analogously,

$$\mathrm{St}(p_1) \cap \mathrm{St}(p_2) \cap \mathrm{St}(p_3) \cap S = \bigcap_{p \in M_2} \mathrm{St}(p) \cap S.$$

Hence

$$\bigcap_{p \in M_1 \cup M_2} \mathrm{St}(p) \cap S = \bigcap_{p \in M_1} \mathrm{St}(p) \cap \bigcap_{p \in M_2} \mathrm{St}(p) \cap S = \mathrm{St}(p_1) \cap \mathrm{St}(p_2) \cap \mathrm{St}(p_3) \cap S$$

is not empty. Thus $M_1 \cup M_2$ also consists of common neighbours. □

Lemma 6.2.14 *Let $D \subset \mathbb{R}^2$ be an open disc. Let $f : D \to \mathbb{R}$ be a smooth function.*

Let $\eta > 0$. Further suppose that $\|\operatorname{grad} f\| \leq \eta$ on the whole of D. Then $p = (\bar{p}, f(\bar{p}))$ and $q = (\bar{q}, f(\bar{q}))$ from the graph of f satisfy the estimate

$$\frac{1}{\sqrt{1+\eta^2}} \|p - q\| \leq \|\bar{p} - \bar{q}\| \leq \|p - q\|.$$

Proof The inequality

$$\|\bar{p} - \bar{q}\| \leq \|p - q\|$$

is clear. From the estimate

$$\begin{aligned}|f(\bar{p}) - f(\bar{q})| &= \left| \int_0^1 \frac{d}{dt} f(t\bar{p} + (1-t)\bar{q}) dt \right| \\ &= \left| \int_0^1 \langle \operatorname{grad} f(t\bar{p} + (1-t)\bar{q}), \bar{p} - \bar{q} \rangle \, dt \right| \\ &\leq \int_0^1 |\langle \operatorname{grad} f(t\bar{p} + (1-t)\bar{q}), \bar{p} - \bar{q} \rangle| \, dt \\ &\leq \int_0^1 \| \operatorname{grad} f(t\bar{p} + (1-t)\bar{q}) \| \cdot \|\bar{p} - \bar{q}\| dt \\ &\leq \eta \cdot \|\bar{p} - \bar{q}\|,\end{aligned}$$

it follows that

$$\|p - q\|^2 = \|\bar{p} - \bar{q}\|^2 + (f(\bar{p}) - f(\bar{q}))^2 \leq (1 + \eta^2)\|\bar{p} - \bar{q}\|^2$$

and hence

$$\frac{1}{1+\eta^2}\|p - q\|^2 \leq \|\bar{p} - \bar{q}\|^2. \qquad \square$$

If $q \in S$ is a point on the surface S, then (after an application of a Euclidean motion) we can write S near q as the graph of a function $f : D_R \to \mathbb{R}$ by corollary 3.6.16, where D_R is the disc of radius R with centre $(0,0)^\top$, and $f(0,0) = 0$ as well as $\operatorname{grad} f(0,0) = (0,0)^\top$. For fixed η satisfying

$$0 < \eta \leq \frac{\sqrt{5}}{2},$$

say, by reducing R we can achieve that

$$\| \operatorname{grad} f \| \leq \eta$$

on the whole of D_R. This radius R depends on q, but if S is compact, then we can in a way similar to that used in the preceding lemmas argue that there exists an $R > 0$ that works for all $q \in S$. For $\bar{p}, \bar{q} \in D_R$ we then have

$$\frac{2}{3}\|p - q\| \leq \sqrt{\frac{1}{1+\eta^2}}\|p - q\| \leq \|\bar{p} - \bar{q}\| \leq \|p - q\| \tag{6.4}$$

by lemma 6.2.14.

We now begin to think about which triangles we want to use for the triangulation. Let A be an ε-configuration on a compact regular surface S, with ε

so small that we can apply all of the above lemmas. Further, let ε be so small that for every point $q \in S$ the part $S \cap B(q, \varepsilon)$ of the surface is contained in the graph over $D_\varepsilon \subset T_qS$, so that (6.4) applies.

Let $M \subset A$ be a maximal subset of common neighbours. Maximality means in this case that M is not contained in a proper superset that also consists of common neighbours. By lemma 6.2.12 the set $\bigcap_{p \in M} \mathrm{St}(p) \cap S$ consists of exactly one element:

$$\bigcap_{p \in M} \mathrm{St}(p) \cap S = \{q\}.$$

The points $p \in M$ all have the same distance $r := \|p - q\|$ from q, which by lemma 6.2.7 satisfies $r \leq \varepsilon$. By (6.4), the orthogonal projections on T_qS then satisfy

$$\frac{2r}{3} \leq \|\bar{p} - \bar{q}\| \leq r.$$

The points $\bar{p}$ therefore lie in one ring-shaped area with inner radius $2r/3$, outer radius r and centre $\bar{q}$. By lemma 6.2.9, distinct points $p, p' \in M$ satisfy

$$\|p - p'\| \geq \varepsilon \geq r$$

and thus, again by (6.4),

$$\|\bar{p} - \bar{p}'\| \geq \frac{2r}{3}.$$

Hence the point $\bar{p}' - \bar{p}$ cannot lie on the segment between $(2r/3\|\tilde{p} - \tilde{q}\|)(\bar{p} - \bar{q})$ and $(r/\|\tilde{p} - \tilde{q}\|)(\bar{p} - \bar{q})$, since this only has length $r/3$.

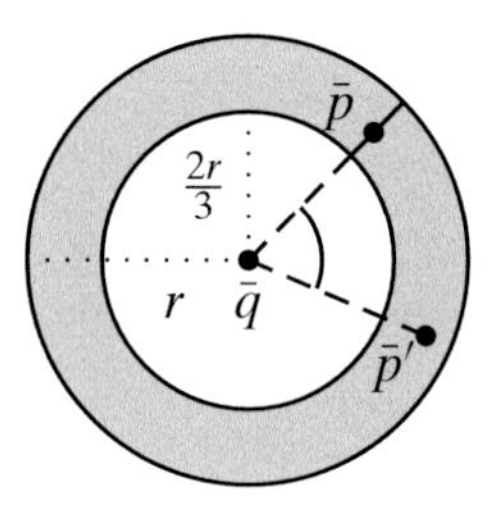

Hence the angle $\angle(\bar{p}, \bar{q}, \bar{p}')$ is positive, i.e. the unit vectors $(\bar{p} - \bar{q})/(\|\bar{p} - \bar{q}\|)$ and $(\bar{p}' - \bar{q})/(\|\bar{p}' - \bar{q}\|)$ do *not* agree. We now order the points $p \in M$ by traversing the unit circle once and enumerating the vectors $(\bar{p} - \bar{q})/(\|\bar{p} - \bar{q}\|)$ as we pass them. At which point we begin and in which direction we traverse the unit circle does not matter.

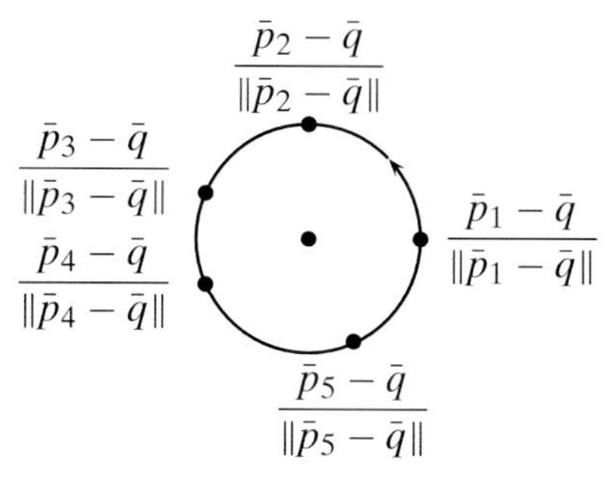

Using the order $M = \{p_1, \ldots, p_k\}$ we can now state which triangles we use for the triangulation: $\Delta(p_1, p_2, p_3), \Delta(p_1, p_3, p_4), \ldots, \Delta(p_1, p_{k-1}, p_k)$.

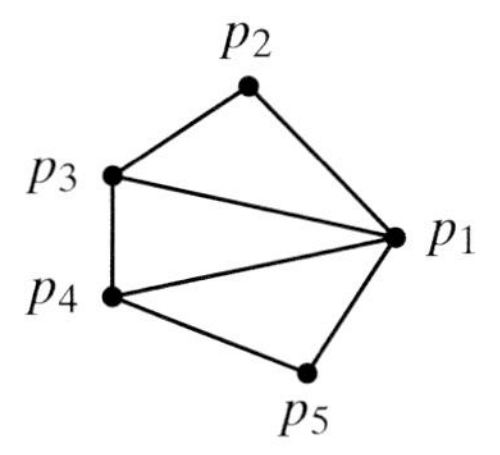

Now the triangulation is constructed as follows. Partition the ε-configuration A into maximal subsets of common neighbours. By corollary 6.2.13, any two different such maximal subsets have at most two points in common. For each maximal subset of common neighbours with at least three points perform the above construction.

Lemma 6.2.15 *Let $S \subset \mathbb{R}^3$ be a compact orientable regular surface. Then there exists a positive ρ_3 such that for every ε-configuration A on S with $\varepsilon \in (0, \rho_3]$ the resulting set of triangles forms a polyhedron.*

Proof Choose $\rho_3 > 0$ so small that all previous lemmas apply. Let Δ and Δ' be two triangles obtained by the above construction. We have to show that they intersect at exactly one edge, or at exactly one vertex, or not at all. If all their vertices belong to one maximal subset of common neighbours, then this is clear by construction. Suppose that the vertices p_1, p_2 and p_3 of Δ belong to one maximal subset M of common neighbours and let q be the point such that $\mathrm{St}(p_1) \cap \mathrm{St}(p_2) \cap \mathrm{St}(p_2) \cap S = \{q\}$, see lemma 6.2.12. Then all points in M have equal distance from q, hence lie on a sphere about q. If at least one vertex of Δ' does not belong to M, then it lies outside this sphere. At most two vertices of Δ' then lie on the sphere, therefore the triangles must be disjoint or have one or two vertices in common.

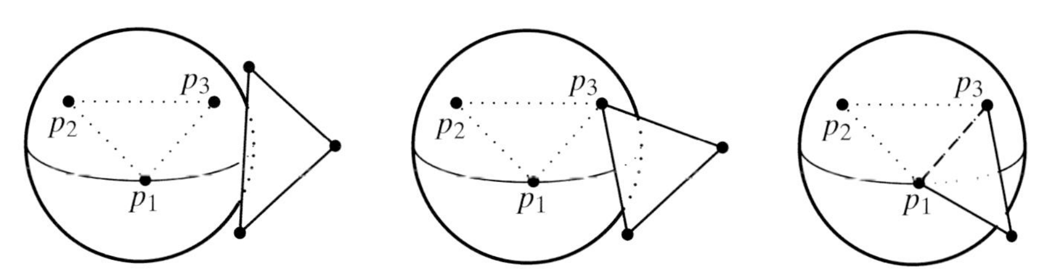

The lemma follows. □

If we write the surface locally as a graph of a function f with $\|\operatorname{grad} f\| \leq 1$ say, then for not too large ε the polyhedron $|X|$ is the graph of a piecewise linear function. If p, p' and p'' are the vertices of a triangle of X, then this function is affine linear over the triangle with the projected vertices $\bar{p}, \bar{p}'$ and $\bar{p}''$.

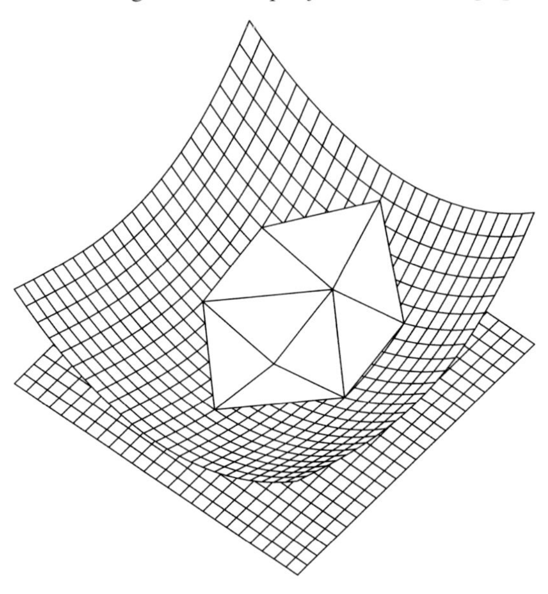

In particular, each point in $|X|$ has a neighbourhood in $|X|$ which is homeomorphic to an open disc.

Lemma 6.2.16 *Let S be a compact orientable regular surface. Then there exists a $\rho_4 > 0$ such that for every ε-configuration A with $\varepsilon \in (0, \rho_4]$ there is a polyhedron X whose vertices are points in A, such that*

$$\Phi := \mathscr{P}|_{|X|} : |X| \to S$$

is bijective.

Proof (a) We first show injectivity. If there were a sequence of ε_i-configurations A_i, $\varepsilon_i \searrow 0$, and points $x_i \neq y_i$ in the corresponding polyhedra with $\Phi(x_i) = \Phi(y_i) = q$, then we would have $\|x_i - y_i\| \to 0$ as $i \to \infty$. Hence we may assume that x_i and y_i both lie in a part of the surface which we may write as a graph of a function f with $f(p) = 0$ and $\|\operatorname{grad} f\| \leq 1$ for some $p \in S$. Since the normal vector $N(p)$ is projected to 0, $\overline{N}(p) = 0$, we may furthermore assume that $\|\overline{N}\| \leq \frac{1}{4}$.

Now we write $x_i = q + t \cdot N(q)$ and $y_i = q + t' \cdot N(q)$. Using lemma 6.2.14 we see that

$$|t - t'| = \|x_i - y_i\| \leq \sqrt{2}\,\|\bar{x}_i - \bar{y}_i\| = \sqrt{2}\,\|(t - t')\overline{N}(q)\| \leq \frac{\sqrt{2}}{4}\,|t - t'|,$$

hence $t = t'$ and $x_i = y_i$.

(b) Now we show surjectivity. Since $|X|$ is compact, so is $\Phi(|X|)$. Hence $\Phi(|X|)$ is a closed subset of S. Since bijective continuous maps between compact Hausdorff spaces are homeomorphisms we know that Φ maps $|X|$ homeomorphically onto its image. Since each point of $|X|$ has a neighbourhood homeomorphic to an open disc, the same holds for the image $\Phi(|X|)$. Therefore $\Phi(|X|)$ is also an open subset of S. Assuming without loss of generality that S is connected (otherwise apply the argument to each connected component seperately) we conclude $\Phi(|X|) = S$. □

We summarise:

Theorem 6.2.17 *Let S be a compact orientable regular surface. Then there exists a $\rho_0 > 0$ such that for every ε-configuration A with $\varepsilon \in (0, \rho_0]$ there is a polyhedron X whose vertices are points in A, such that*

$$\Phi := \mathscr{P}|_{|X|} : |X| \to S$$

defines a triangulation. In particular, S has triangulations.

Remark The assumption that the surface is orientable is, in fact, unnecessary. Compact regular surfaces in $\mathbb{R}^3$ are actually automatically orientable. This is the case because such surfaces divide $\mathbb{R}^3$ into two parts, the inside and the outside of the surface. At every point of the surface one of the two unit normal vectors points inside and the other points outside. The outer and the inner unit normal field are both smooth, and the surface is thus orientable.

Can we triangulate surfaces that are not compact? As our definitions imply that the geometric realisation of a polyhedron is always a finite union of compact triangles, it is itself compact. It can therefore only be homeomorphic to a compact surface. However, one could instead of finitely many triangles allow countably many. Then a more careful discussion than the one from the proof of theorem 6.2.17 will show that non-compact surfaces can also be triangulated. The details are left to the (ambitious) reader.

As a corollary of the fact that compact regular surfaces have triangulations, we can now come back to a remark about integration theory on surfaces, which we made in section 3.7.

Corollary 6.2.18 *Let S be a compact orientable regular surface. Then there exist three local parametrisations (U_i, F_i, V_i) of S, which cover S entirely:*

$$S \subset V_1 \cup V_2 \cup V_3.$$

Proof Let $\Phi : |X| \to S$ be a triangulation, $X = \{\Delta_1, \dots, \Delta_k\}$. For every triangle we choose a congruence to a triangle in the plane:

$$L_j : \Delta_j \overset{\cong}{\to} \Delta'_j \subset \mathbb{R}^2.$$

Composing L_j with a translation if necessary, we can assume that the triangles Δ'_j are pairwise disjoint. The first local parametrisation is defined on the union of the interiors of the plane triangles $U_1 := \bigcup_{j=1}^{k} \mathring{\Delta}'_j$ and on every open triangle it is given by $F_1|_{\mathring{\Delta}'_j} = \Phi \circ L_j^{-1}$. This first parametrisation already covers the entire surface except for the vertices and edges.

For every edge $v_1, \dots, v_\ell$ we choose a local parametrisation $G_n : D_n \to S$ with $G_n(0, n) = v_n$. The domain D_n must be contained in the unit disc in $\mathbb{R}^2$ at the point $(0, n)^\top$. This can be achieved for an arbitrary local parametrisation at v_n by first translating the pre-image of v_n to the point $(0, n)^\top$ and then restricting the parametrisation to a small disc centred at that point. The second local parametrisation is defined on $U_2 := \bigcup_{n=1}^{\ell} D_n$ and is for each D_n given by G_n. This second parametrisation contains all vertices of the triangulation in its image.

Fermi coordinates along the edges (except for little neighbourhoods of the edges) give local parametrisations that can be made disjoint for different edges by restriction and translation. These local parametrisations are then pieced together to a third one, which covers the rest of S. □

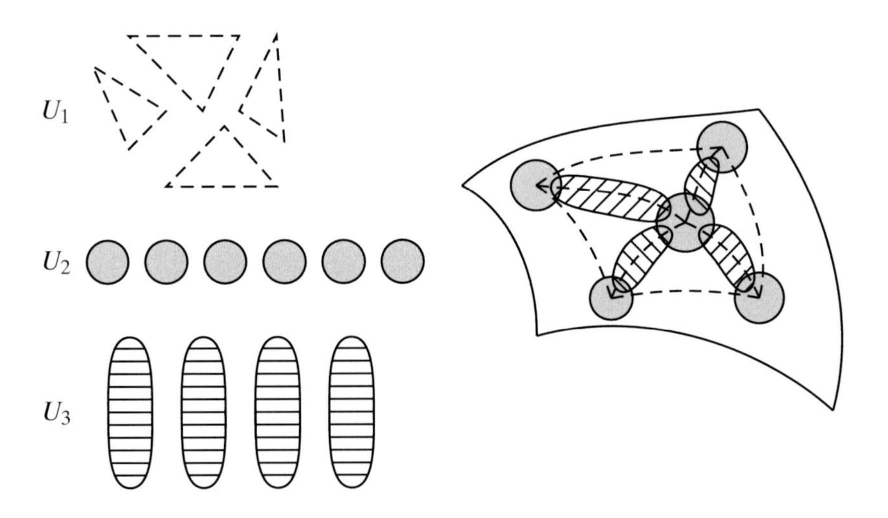

As mentioned above, we can drop the conditions of orientability and compactness. It is crucial that the surface can be triangulated. Countably many bounded subsets of the plane (e.g. the Δ'_j) can be made disjoint by translations as well. Corollary 6.2.18 therefore holds for all regular surfaces.

6.3 The Gauss–Bonnet theorem

In this section we will reach the climax of this book, the Gauss–Bonnet theorem, which relates the Euler–Poincaré characteristic to the Gauss curvature. Before arriving there, we need to go through some technical matters.

Definition 6.3.1 A triangulation $\Phi : |X| \to S$ is called ***smooth along the edges*** if any two triangles $\Delta_1, \Delta_2 \in X$ with a common edge satisfy the following: if $\Psi : \Delta_1 \cup \Delta_2 \to \Delta_1' \cup \Delta_2' \subset \mathbb{R}^2$ is a homeomorphism which maps Δ_1 congruently to the triangle Δ_1', analogously Δ_2 congruently to Δ_2', then the restriction of

$$\Phi \circ \Psi^{-1} : \Delta_1' \cup \Delta_2' \to S$$

to the interior of $\Delta_1' \cup \Delta_2'$ is a diffeomorphism onto its image.

By definition, $\Phi \circ \Psi^{-1}$ is automatically a diffeomorphism on the interior of Δ_1' or of Δ_2'. It is only on the common edge of Δ_1' and Δ_2' that $\Phi \circ \Psi^{-1}$ might not be smooth.

Exercise 6.5 Let $\Phi : |X| \to S$ be a triangulation of the compact regular surface S. Show that Φ can be turned into a triangulation $\hat{\Phi} : |X| \to S$ that is based on the same polyhedron and that is smooth along the edges.

After all these preparations we now reach the climax of this last chapter.

Theorem 6.3.2 (Gauss–Bonnet theorem) *Let S be a compact regular surface with Riemannian metric. Let $K : S \to \mathbb{R}$ be the Gauss curvature and dA the surface element. Let $\Phi : |X| \to S$ be a triangulation. Then*

$$\int_S K\, dA = 2\pi \chi(X).$$

Proof We know by theorem 5.2.7 that the number $\int_S K\, dA$ does not depend on the Riemannian metric. We therefore forget about the given Riemannian metric and construct a new one, which is tailored to the triangulation. We prove the Gauss–Bonnet theorem for this Riemannian metric. We assume that the triangulation is smooth along the edges.

Φ is a diffeomorphism onto its image on the interior of each triangle $\Delta \in X$. We can therefore pull back the Euclidean metric on $\mathring{\Delta}$ to $\Phi(\mathring{\Delta}) \subset S$ using Φ^{-1}. We thus obtain a Riemannian metric on S without the edges and vertices. As the triangulation is smooth along the edges, the metric extends smoothly to S minus the vertices. It follows by the definition 4.4.3 of the pulled-back metric that Φ is an isometry on the interiors of the triangles. Hence the Gauss curvature of this Riemannian metric vanishes. We have therefore found a

Riemannian metric on $S - \bigcup_e \Phi(e)$ with $K \equiv 0$, where the union is taken over all edges e of X. If we could extend this metric smoothly to the vertices, then the Gauss curvature of this extension would for reasons of continuity vanish as well, and we would have a Riemannian metric on S with $K \equiv 0$. But this will in general not be possible, since, for example, on the sphere $S = S^2$, we already know $\int_{S^2} K\,dA = 4\pi$.

Let us therefore investigate more closely where the extension to the vertices goes wrong. For this purpose let e be a vertex of X and let $\Delta \in X$ be a triangle which has e as a vertex. If X_1, X_2 form an orthonormal basis of the plane E spanned by Δ, more precisely $E = \{e + u^1 X_1 + u^2 X_2 |\, u^1, u^2 \in \mathbb{R}\}$, then $(r, \varphi)^\top \mapsto e + r\cos(\varphi)X_1 + r\sin(\varphi)X_2$ defines polar coordinates at e for this plane. The part of a small r_0-disc around e that lies in the interior of the triangle Δ is parametrised by the parameters $0 < r < r_0$ and $\varphi_0 < \varphi < \varphi_0 + \alpha$, where α is the interior angle of Δ at the vertex e. If the orthonormal basis X_1, X_2 is rotated, then only φ_0 changes. Composing with Φ gives coordinates of the image of this set in S. Let us fix e and parametrise, choosing suitable orthonormal bases, all triangles $\Delta_1, \ldots, \Delta_k$ that have e as a vertex, proceeding vertex by vertex. Then $0 < r < r_0$ and $\varphi_0 < \varphi < \varphi_0 + \alpha_1 + \cdots + \alpha_k$ parametrise a neighbourhood of e minus an edge beginning at e (that corresponds to the value $\varphi = \varphi_0$). Here α_j is always the interior angle of Δ_j at the vertex e. As the metric on $S - \bigcup_e \Phi(e)$ was defined in such a way that Φ is an isometry, the metric has in these coordinates the form

$$\left(g_{ij}\right)_{ij} = \begin{pmatrix} 1 & 0 \\ 0 & r^2 \end{pmatrix}.$$

If we now had $\alpha_1 + \cdots + \alpha_k = 2\pi$, then we would be dealing with polar coordinates of the plane with Euclidean metric, which would extend smoothly to the centre. It is generally not possible to extend the metric in such a way because the angle defect $\mathrm{def}(e) = 2\pi - \sum_{j=1}^k \alpha_j$ might be non-trivial.

To remove this defect, we first change the coordinates slightly by modifying the angle coordinate. We use the abbreviation $\alpha := \sum_{j=1}^k \alpha_j$ and define

$$\varphi := \frac{2\pi}{\alpha}(\varphi - \varphi_0).$$

Then φ is still from $(0, 2\pi)$ and r from $(0, r_0)$. In the coordinates (r, φ) the metric has the form

$$\begin{pmatrix} 1 & 0 \\ 0 & \left(\frac{\alpha}{2\pi} r\right)^2 \end{pmatrix}.$$

After this small alteration of the coordinates we are using, we now modify the metric near e. We choose r_1 between 0 and r_0 and a smooth function $\rho : \mathbb{R} \to \mathbb{R}$ with the properties $\rho(r) = 1$ for $r \leq r_1$, $\rho(r) = 0$ for $r \geq r_0$ and $0 \leq \rho \leq 1$ on $[r_1, r_0]$. We define the modified metric in the coordinates (r, φ) by

$$(\tilde{g}_{ij})_{ij} := \begin{pmatrix} 1 & 0 \\ 0 & \left(\rho(r) \cdot r + (1 - \rho(r))\frac{\alpha}{2\pi} r\right)^2 \end{pmatrix}.$$

This new Riemannian metric agrees with the old one for $r \geq r_0$. The metric has therefore only changed near $\Phi(e)$. For $r < r_1$ on the other hand the metric now has the form of the Euclidean metric in polar coordinates. We can therefore extend it smoothly to $\Phi(e)$. The Gauss curvature vanishes on the r_1-disc around $\Phi(e)$ as well. Only in the area $r_1 \leq r \leq r_0$ might we have accidentally introduced a non-vanishing curvature. We can easily calculate the integral of the curvature using lemma 4.6.12:

$$\begin{aligned}
\int_{\{r \leq r_0\}} K \, dA &= 2\pi - \int_0^{2\pi} \frac{d}{dr}\Big|_{r=r_0} \left(\rho(r) \cdot r + (1 - \rho(r))\frac{\alpha}{2\pi} r\right) d\varphi \\
&= 2\pi \left(1 - \dot{\rho}(r_0) r_0 - \rho(r_0) + \dot{\rho}(r_0)\frac{\alpha}{2\pi} r_0 - (1 - \rho(r_0))\frac{\alpha}{2\pi}\right) \\
&= 2\pi \left(1 - 0 \cdot r_0 - 0 + 0 \cdot \frac{\alpha}{2\pi} r_0 - (1 - 0)\frac{\alpha}{2\pi}\right) \\
&= 2\pi - \alpha = \operatorname{def}(e).
\end{aligned}$$

The deformation of the metric that was necessary to extend it smoothly to the vertex e gave us a curvature in the ring-shaped area $r_1 \leq r \leq r_0$, and the integral of this curvature is exactly the angle defect.

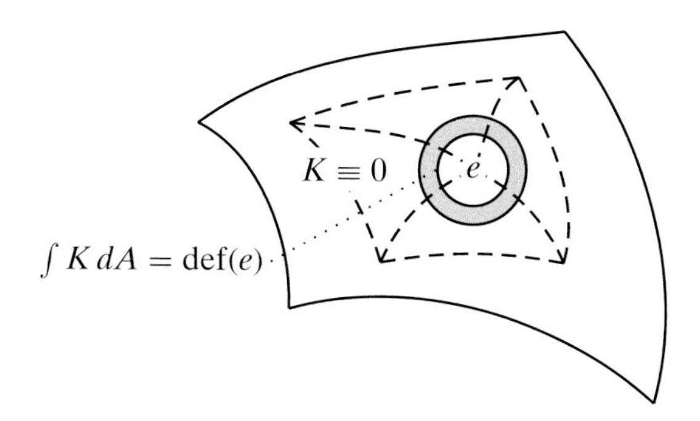

The statement of the theorem now follows by the Gauss–Bonnet formula for polyhedra (theorem 6.1.5) by summing over all vertices.

$$\int_S K \, dA = \sum_e \operatorname{def}(e) = 2\pi \chi(X). \qquad \square$$

Corollary 6.3.3 *Let S be a compact regular surface. Then the Euler–Poincaré characteristic of S does not depend on the choice of the triangulation. Definition 6.2.2 therefore makes sense.*

Exercise 6.6 The Euler–Poincaré characteristic of a torus is by the Gauss–Bonnet theorem equal to 0, since the torus has a Riemannian metric with vanishing Gauss curvature. Verify this using a triangulation.

6.4 Outlook

Having reached the end of this book, we have only got to know the elements of differential geometry. The reader may want to consult other textbooks on curves and surfaces such as [6, 9, 14, 23]. Differential geometry has developed enormously and has countless interconnections to other areas, not only in mathematics. We want to conclude this book with some notes about such developments and applications.

Knot theory We have seen in section 2.3 that knots must curve quite a lot. Knot theory, which mainly deals with the question of when two curves are ambient isotopic, has now developed into its own branch of topology, with some amazing connections to physics and other areas. Introductions can be found in, for example, [15, 20, 27].

Immersions The reader may have noticed that regular surfaces and regular curves were defined in very different ways. Curves are equivalence classes of parametrised curves, i.e. of certain maps. Regular surfaces on the other hand are subsets of $\mathbb{R}^3$ with certain properties. A notion of surfaces that is closer to our concept of curves is the one of an ***immersed surface***. Such an ***immersion*** is a map $F : S \to \mathbb{R}^3$, where S is a regular surface and d_pF has full rank for all $p \in S$. Two immersions F_1 and F_2 are equivalent if there exists a diffeomorphism $\Phi : S \to S$ with $F_2 = F_1 \circ \Phi$. An immersed surface is then an equivalence class of immersions. As we do not require immersions to be injective, immersed surfaces can have self-intersections, just like curves.

We have already met some immersed surfaces, e.g. Enneper's surface from example 3.8.12, which we then turned into regular surfaces by restricting the domain of the immersion. A large part of the surface theory discussed can easily be extended to immersed surfaces.

We can construct interesting non-orientable immersed surfaces that cannot be realised in $\mathbb{R}^3$ without self-intersections. Examples for this are the ***projective plane*** (or ***crosscap***) and the ***Klein bottle*** [12, 1.89]. In the first example the domain of the immersion is a sphere, in the second one a torus.

Classification of compact surfaces We have seen that the sphere and the ellipsoid are diffeomorphic, while the sphere and the torus are not. We say a bit loosely that the sphere and the ellipsoid are topologically equivalent, the sphere and the torus, on the other hand, are topologically different. But how many topologically different compact surfaces are there?

The answer to this question is remarkably simple. The Euler–Poincaré characteristic tells us whether two given compact surfaces are topologically equivalent or not. More precisely, let S_1 and S_2 be two connected compact orientable surfaces. Then the following are equivalent:

- S_1 and S_2 are diffeomorphic.
- S_1 and S_2 are homeomorphic.
- $\chi(S_1) = \chi(S_2)$.

Possible values for the Euler–Poincaré characteristic are the numbers $2, 0, -2, -4, \ldots$ [6].

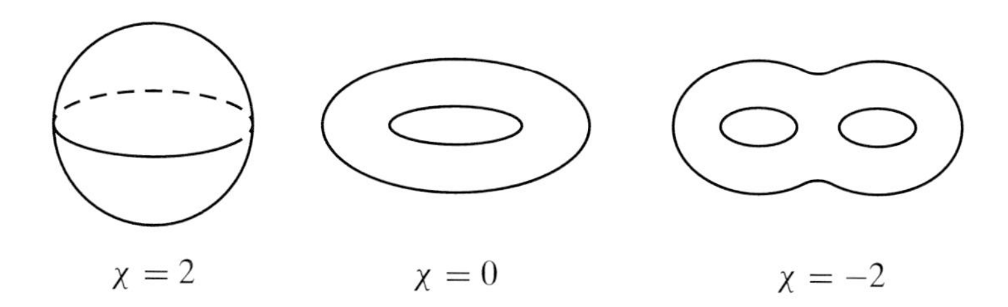

There is a similar classification for non-orientable compact surfaces. We have seen that the sphere and the torus have Riemannian metrics with constant Gauss curvature. This is also the case for the other compact surfaces. By the Gauss–Bonnet theorem a negative Euler–Poincaré characteristic implies that the Gauss curvature is also negative. Theorem 3.6.17 tells us that such a metric can never be the first fundamental form.

Minimal surfaces We have spent a considerable amount of time studying minimal surfaces, but an investigation of minimal surfaces would, done properly, require an entire course. In particular, conformal parametrisations could not be considered due to lack of space. See [1, 19, 26] for more.

Abstract manifolds Having passed from the first fundamental form to general Riemannian metrics, we have in a way taken the regular surfaces out of $\mathbb{R}^3$. They are still subsets of $\mathbb{R}^3$, but the relative position of the surface in space is no longer reflected by the Riemannian metric in any way. It is therefore in this context more natural to define surfaces with Riemannian metrics in a more abstract way, not as subsets of Euclidean space, but as abstract topological spaces with certain properties and additional structures. We have here refrained from taking this approach in order to keep the formal effort as small as possible, and because certain constructions, such as the tangent plane, are less intuitive.

Those concepts are nevertheless indispensable for a more detailed study of differential geometry [5, 16, 35]. The object of study is not only the two-dimensional case, but also the higher-dimensional "manifolds", sometimes even infinite-dimensional ones. A beautiful book on three-dimensional geometry is [32].

Riemannian geometry Further developments in the theory of the inner geometry of abstract manifolds lead to Riemannian geometry. A part of our discussion translates without any changes, e.g. the theory of geodesics. But there are very new phenomena as well. Recall that for surfaces it was possible to calculate the Gauss curvature from the Riemann curvature tensor and vice versa. This is no longer possible in higher dimensions for the corresponding generalisations. The curvature tensor carries more information than the scalar curvature (one of the generalisations of the Gauss curvature). There is additionally the concept of the Ricci curvature, which lies between the two others. See, for example, [10, 12, 29] for an introduction.

Lorentzian geometry and general relativity Einstein used Riemannian geometry in his gravitation theory, the theory of general relativity. But a modification needs to be made here. Instead of positive-definite scalar products we consider non-degenerate bilinear forms of index 1 on the tangent spaces. This is because we are not modelling curved space, but curved space-time. This variant of Riemannian geometry is called Lorentzian geometry [3, 25]. Einstein's field equation relates the geometric quantities, in particular the above-mentioned Ricci curvature, to physical quantities. Very good textbooks are [31, 34], see also [22] for an extensive introduction.

Ricci flow Fix a compact regular surface S and a Riemannian metric g_0 on S. Now look at the equation

$$\frac{\partial}{\partial t} g(t) = 2(k - K(t))g(t), \tag{6.5}$$

subject to the initial condition

$$g(0) = g_0.$$

Here $g(t)$ denotes a family of Riemannian metrics on S depending on a parameter $t \in [0, \infty)$, $K(t)$ is the Gauss curvature with respect to $g(t)$ and k is its mean, $k = \int_S K(t)\, dA(t)/\mathrm{area}(S, g(t))$. Equation (6.5) is known as the *normalized Ricci flow equation*.

It turns out that for any g_0 there is a unique solution and it is defined for all $t \in [0, \infty)$. Observe that if g_0 has constant Gauss curvature, then the right hand side of (6.5) vanishes, so the solution is constant in t, $g(t) \equiv g_0$. In general, it turns out that there is a limit metric $g_\infty = \lim_{t\to\infty} g(t)$ and g_∞ has constant Gauss curvature. Moreover, g_∞ is conformally equivalent to g_0, i.e. there is a positive function $f : S \to \mathbb{R}$ such that $g_\infty = f \cdot g_0$. See [8] together with [7].

This shows that for any Riemannian metric g_0 on S there is a conformally equivalent metric with constant Gauss curvature. This is the geometric version of a classical theorem, the *uniformatisation theorem*.

A lot of research has been done and is still being done on Ricci flow on higher-dimensional manifolds. The term "Ricci flow" comes from the fact that a formulation of (6.5) in higher dimensions requires the notion of Ricci curvature mentioned above. By far the most spectacular result was obtained in dimension 3. After many failed attempts by others, Grisha Perelman was able to prove the Poincaré conjecture which had been open for more than a hundred years. It says that any compact connected three-dimensional manifold in which every closed curve can be shrunk to a point must be homeomorphic to the three-dimensional sphere S^3. The corresponding statement in dimension 2 is an easy consequence of the classification of compact surfaces mentioned earlier. In fact, Perelman even proved a stronger conjecture due to William Thurston. See, for example, [24] for an introduction to these deep results.

Appendix A Hints for solutions to (most) exercises

1.1 Let L be the straight line containing p, q, r and s. Choose a point t not contained in L (I_4). Let u be a point with $t \in \overline{qu}$ (A_3). Then u is not contained in L. Put $L' := L(r,t)$.

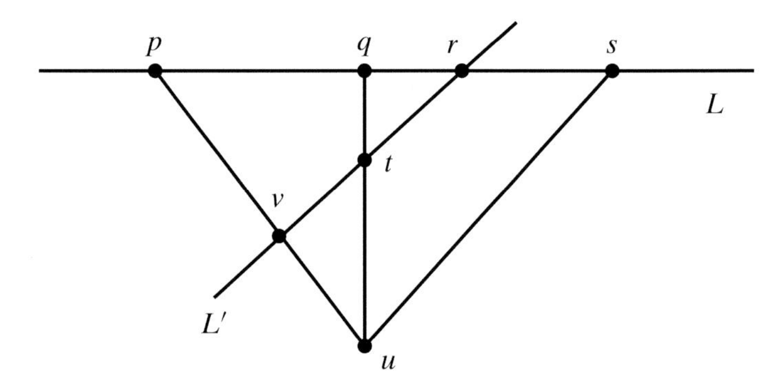

L' does not contain any of the points p, q, s since otherwise $L = L'$ and hence $t \in L$. Moreover, L' does not contain u since otherwise $L' = L(t,u)$ and hence $q \in L'$. Now L' enters the triangle qsu at r and leaves it at t. By axiom A_5 L' does not intersect $\overline{us}$. Moreover, L' enters the triangle pqu at t and does not intersect the side $\overline{pq}$. By axiom A_5 it intersects $\overline{pu}$ at a point v. Now L' enters the triangle pus at v and does not intersect $\overline{us}$. By axiom A_5 it intersects $\overline{ps}$ at r since r is the only point on L' also contained in L. This proves that r lies between p and s.

1.3 Reflexivity follows from axiom A_1 and symmetry is axiom A_2. Transitivity: let $q_1, q_2, q_3 \in L$, $q_i \neq p$, such that $p \notin \overline{q_1q_2}$ and $p \notin \overline{q_2q_3}$. We have to show $p \notin \overline{q_1q_3}$.

Suppose $p \in \overline{q_1q_3}$. Choose $r \notin L$. By theorem 1.1.1 we can find $s \in \overline{rq_1}$. The straight line $L' := L(p,s)$ intersects L at p only since otherwise $L' = L$, hence $s \in L$, hence $r \in L$, a contradiction.

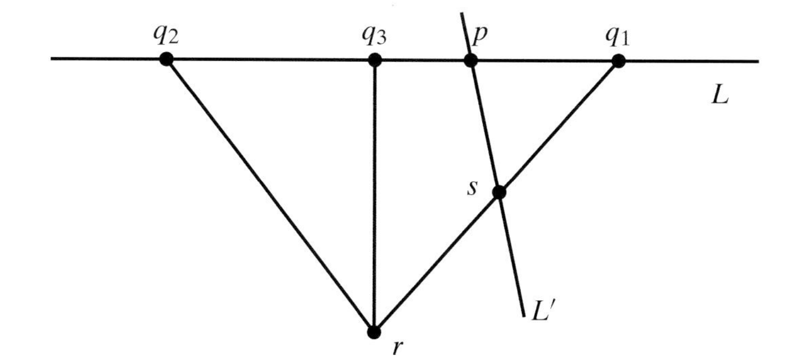

Moreover, $r \notin L'$ since otherwise $L' = L(s,r)$, thus $q_1 \in L'$, hence $L' = L$. Now L' enters the triangle q_1q_2r at s and intersects L only at p. Since by assumption $p \notin \overline{q_1q_2}$ axiom A_5 tells us that L' intersects $\overline{rq_2}$. Therefore, L' enters the triangle rq_2q_3 through $\overline{rq_2}$ and does not intersect $\overline{q_2q_3}$. By axiom A_5 L' intersects $\overline{rq_3}$. Thus L' intersects all three sides of the triangle q_1q_3r, a contradiction to axiom A_5.

1.4 By axiom A_3, there are at least two sides. The assumption that there are at least three sides will lead to a contradiction. Let $q_1, q_2, q_3 \in L$ be pairwise distinct points such that $p \in \overline{q_iq_j}$ for all $i \neq j$. Choose $r \notin L$ and let $s \in \overline{q_1r}$. The straight line $L' := L(p,s)$ does not contain r and no point of L other than p. Since L' intersects the sides $\overline{q_1q_2}$ and $\overline{q_1r}$ of the triangle q_1q_2r it does not intersect $\overline{q_2r}$. Similarly, it does not intersect $\overline{q_3r}$. Thus L' intersects the triangle q_2q_3r only at the side $\overline{q_2q_3}$, namely at p, a contradiction.

1.5 Reflexivity follows from axiom A_1 and symmetry is axiom A_2. Transitivity: suppose $\overline{q_1q_2}$ and $\overline{q_2q_3}$ do not intersect L. We have to show that $\overline{q_1q_3}$ does not intersect L either. In the case q_1, q_2, and q_3 do not lie on one line this follows from axiom A_5. If they lie on one line, then the problem reduces to that of exercise 1.3.

1.6 Let L be a straight line. Choose $q \notin L$ and $p \in L$. Choose a point r so that $p \in \overline{rq}$. Then $r \notin L$, since otherwise $q \in L$. Hence r and q represent two different sides.

Now we derive a contradiction from the assumption that q_1, q_2 and q_3 represent three pairwise different sides. If the three points do not lie on one line, then L intersects the triangle $q_1q_2q_3$ in all three sides contradicting axiom A_5. If they lie on one line, then the problem reduces to that of exercise 1.4.

1.9

$$\begin{aligned} F_{A,b}(F_{B,c}(x)) &= F_{A,b}(Bx + c) = A(Bx + c) + b \\ &= ABx + (Ac + b) = F_{AB,Ac+b}(x). \end{aligned}$$

Moreover,

$$F_{A,b} \circ F_{A^{-1},-A^{-1}b} = F_{AA^{-1},A(-A^{-1}b)+b} = F_{\mathrm{Id},0} = \mathrm{Id}$$

and similarly $F_{A^{-1},-A^{-1}b} \circ F_{A,b} = \mathrm{Id}$.

1.10 We assume without loss of generality that $L = L' = \{(x,0)^\top \mid x \in \mathbb{R}\}$. It is easy to see that two points $(x,y)^\top, (x',y')^\top \in \mathbb{R} - L$ lie on the same side of L if and only if y and y' have the same sign. Now consider the three points $p = (x,y)$, $q = (x',y')$ and $r = (x'',y'') \in \mathbb{R} - L$. If L intersects the line segment $\overline{pq}$, i.e. if p and q lie on different sides of L, then y and y' have opposite signs. Hence the sign of y'' must coincide with the sign of either y or y'. If y'' and y have the same sign, i.e. if y'' and y' have opposite signs, then L intersects the line segment $\overline{qr}$. In the other case it intersects $\overline{pr}$.

1.11 Axiom K_1: Choose $A \in \mathrm{SO}(2)$ such that $A(q-p) = \lambda \cdot (r_1 - p_1)$, $\lambda > 0$. Put $F(x) := Ax + (p_1 - Ap)$. Then $F \in \mathrm{E}(2)$, $F(p) = p_1$, and

$$F(q) - p_1 = Aq + p_1 - Ap - p_1 = A(q-p) = \lambda \cdot (r_1 - p_1).$$

Hence $q_1 := F(q)$ lies on the line through p_1 and q_1 on the same side as r_1 (because $\lambda > 0$).

Axiom K_2: This follows from the fact that $\mathrm{E}(2)$ is a group.

Axiom K_3: One shows that two line segments $\overline{pq}$ and $\overline{p'q'}$ in the Cartesian plane are congruent if and only if they have equal lengths, i.e. $\|p - q\| = \|p' - q'\|$. Then axiom K_3 is clear.

Axiom K_4: This follows from the fact that $\mathrm{E}(2)$ is a group.

Axiom K_5: There is a unique $A \in \mathrm{O}(2)$ such that $A(p - q) = \lambda \cdot (p_1 - q_1)$, $\lambda > 0$, and the determinants of the 2×2 matrices $(A(p-q), A(r-q))$ and $(p_1 - q_1, s_1 - q_1)$ have equal sign. Put $b := q_1 - Aq$ and $F(x) := Ax + b$. Then $F \in \mathrm{E}(2)$ and $r_1 := F(r)$ does the job.

Axiom K_6: Since $\angle(q,p,r) \equiv \angle(q_1,p_1,r_1)$, there exists $F \in \mathrm{E}(2)$ such that $F(p) = p_1$, $F(q) - F(p) = \lambda \cdot (q_1 - p_1)$ and $F(r) - F(p) = \mu \cdot (r_1 - p_1)$, $\lambda, \mu > 0$. From $\overline{pq} = \overline{p_1q_1}$ we conclude $\lambda = 1$ and similarly from $\overline{pr} = \overline{p_1r_1}$ we derive $\mu = 1$. Thus $F(p) = p_1$, $F(q) = q_1$, and $F(r) = r_1$. This implies all congruences.

1.14 Choose the old points p_1 and q_1 as in the proof of theorem 1.2.3. Now let q_2 be the midpoint of the line segment $\overline{p_1q_1}$. Observe that q_2 is again an old point. Let q_3 be the midpoint of the line segment $\overline{p_1q_2}$ or $\overline{q_2q_1}$ which contains n. Proceed inductively to define a Cauchy sequence $(q_k)_k$. By completeness of $\mathbb{R}^2$, this Cauchy sequence has a limit p_2. Now continue as in the proof of theorem 1.2.3.

1.15 If $\angle(y,x,z) = \angle(y',x',z')$, then $x = x'$, $y - x = \lambda \cdot (y' - x)$, and $z - x = \mu \cdot (z' - x)$ (or $y - x = \lambda \cdot (z' - x)$ and $z - x = \mu \cdot (y' - x)$, which one can treat similarly) with $\lambda, \mu > 0$. It follows that

$$\frac{\langle y - x, z - x\rangle}{\|y - x\| \cdot \|z - x\|} = \frac{\langle \lambda(y' - x), \mu(z' - x)\rangle}{\|\lambda(y' - x')\| \cdot \|\mu(z' - x')\|} = \frac{\langle y' - x', z' - x'\rangle}{\|y' - x'\| \cdot \|z' - x'\|}.$$

1.16 If $F \in \mathrm{E}(2)$, $F(x) = Ax + b$, then

$$\frac{\langle F(y) - F(x), F(z) - F(x)\rangle}{\|F(y) - F(x)\| \cdot \|F(z) - F(x)\|} = \frac{\langle A(y - x), A(z - x)\rangle}{\|A(y - x)\| \cdot \|A(z - x)\|} = \frac{\langle y - x, z - x\rangle}{\|y - x\| \cdot \|z - x\|}.$$

Hence congruent angles have the same interior angle. For the converse observe (using axiom K_5) that each angle with interior angle γ is congruent to the angle $\angle((1,0)^\top, (0,0)^\top, (\cos(\gamma), \sin(\gamma))^\top)$. Hence two angles with the same interior angle are congruent to the same angle, thus congruent to each other.

1.17 We first compute

$$\frac{1 - \cos(2t)}{1 + \cos(2t)} = \frac{1 - (\cos^2(t) - \sin^2(t))}{1 + (\cos^2(t) - \sin^2(t))} = \frac{2\sin^2(t)}{2\cos^2(t)} = \tan^2(t).$$

Now the cosine rule yields

$$\tan^2\left(\frac{\alpha}{2}\right) = \frac{1 - \cos(\alpha)}{1 + \cos(\alpha)} = \frac{1 - \dfrac{-a^2 + b^2 + c^2}{2bc}}{1 + \dfrac{-a^2 + b^2 + c^2}{2bc}}$$
$$= \frac{(a + b - c)(a - b + c)}{(-a + b + c)(a + b + c)}.$$

1.18 Axioms I_1, I_3 and I_4 are valid. Axiom I_2 fails because pairs of antipodal points p and $-p$ are contained in infinitely many great circles.

1.19 Let $p, q \in S^2$ be two distinct points on the sphere. If they are not antipodal, $p \neq -q$, then the line segment $\overline{pq}$ would be the unique great circle containing p and q with these two points removed. If they are antipodal, $p = -q$, then the line segment would be $\overline{pq} = S^2 - \{p, q\}$. Axioms A_1, A_2 and

A3 are obviously valid while axiom A4 certainly fails. Axiom A5 fails because L will intersect all three sides of a triangle.

2.1 For $c(t) = \begin{pmatrix} \sin(t) \\ \cos(t) + \ln\tan(t/2) \end{pmatrix}$ one easily computes

$$\dot{c}(t) = \begin{pmatrix} \cos(t) \\ -\sin(t) + 1/\sin(t) \end{pmatrix}.$$

Now fix t. The tangent of c at $c(t)$ is the straight line parametrised by

$$s \mapsto c(t) + s \cdot \dot{c}(t) = \begin{pmatrix} \sin(t) \\ \cos(t) + \ln\tan(t/2) \end{pmatrix} + s \cdot \begin{pmatrix} \cos(t) \\ -\sin(t) + 1/\sin t \end{pmatrix}.$$

One sees that this line intersects the y-axis for $s_0 = -\tan(t)$. Now

$$s_0 \cdot \dot{c}(t) = \begin{pmatrix} -\sin(t) \\ -\cos(t) \end{pmatrix}$$

indeed has lenght 1 for each t.

2.2 For $F(x) = Ax + b$ with $A \in O(2)$ one has

$$(F \circ c)^{\cdot}(t) = A\dot{c}(t).$$

Matrices in O(2) preserve the length of vectors.

2.4 Let $c : \mathbb{R} \to \mathbb{R}^n$ be a periodic parametrisation with period L. Let ψ and $\varphi = \psi^{-1}$ be the parameter transformations constructed in the proof of proposition 2.1.13. By lemma 2.1.14 it is sufficient to show that $\tilde{c} := c \circ \varphi$ is periodic. Let $\Lambda := L[c|_{[0,L]}]$ be the length of the curve when restricted to one period interval. Then we have for all $s \in \mathbb{R}$

$$\psi(s + L) - \psi(s) = \int_s^{s+L} \|\dot{c}(t)\| \, dt = \Lambda$$

and therefore (substituting $t = \psi(s)$)

$$\varphi(t + \Lambda) - \varphi(t) = \varphi(\psi(s) + \Lambda) - \varphi(\psi(s)) = s + L - s = L$$

for all $t \in \mathbb{R}$. Hence

$$\tilde{c}(t + \Lambda) = c(\varphi(t + \Lambda)) = c(\varphi(t) + L) = c(\varphi(t)) = \tilde{c}(t).$$

2.5 (a) We parametrise the curve such that the centre at time t is located at $(t,1)^\top$. Hence t is the length of the segment of the x-axis which has been touched by the circle of radius 1. The arc-length parametrisation of the unit circle starting at $(0,-1)^\top$ at $t=0$ is given by $t \mapsto -(\sin(t),\cos(t))^\top$. Thus

$$c(t) = \begin{pmatrix} t \\ 1 \end{pmatrix} - r \cdot \begin{pmatrix} \sin(t) \\ \cos(t) \end{pmatrix} = \begin{pmatrix} t - r\sin(t) \\ 1 - r\cos(t) \end{pmatrix}.$$

(b)

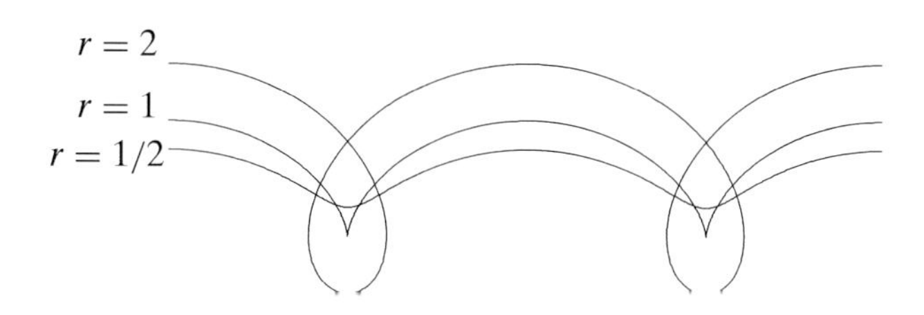

(c) From $\dot{c}(t) = \begin{pmatrix} 1 - r\cos(t) \\ r\sin(t) \end{pmatrix}$ we see that $\dot{c}(t) = (0,0)^\top$ if and only if $t \in \mathbb{Z}\cdot\pi$ and $0 = 1 \mp r$. Thus c is regular if and only if $r \neq 1$.

(d) One computes using the substitution $t = 2u$

$$\begin{aligned}
\int_0^{2\pi} \|\dot{c}(t)\|\, dt &= \int_0^{2\pi} \left\| \begin{pmatrix} 1 - \cos(t) \\ \sin(t) \end{pmatrix} \right\| dt = \int_0^{2\pi} \sqrt{2 - 2\cos(t)}\, dt \\
&= 2\sqrt{2} \int_0^{\pi} \sqrt{1 - \cos(2u)}\, du \\
&= 2\sqrt{2} \int_0^{\pi} \sqrt{1 - \cos^2(u) + \sin^2(u)}\, du \\
&= 2\sqrt{2} \int_0^{\pi} \sqrt{2\sin^2(u)}\, du = 4 \int_0^{\pi} \sin(u)\, du = 8.
\end{aligned}$$

2.6 (a) Let c be a curve connecting x and y which contains a point z that does not lie on the line segment $\overline{xy}$. Then the triangle inequality shows that any polygon P having x, z and y as vertices must have length $L[P] \geq \|x - z\| + \|z - y\| > \|x - y\|$. Now choose a sequence of polygons approximating c which contain x, z, and y as vertices. Proposition 2.1.18 then says $L[c] \geq \|x - z\| + \|z - y\| > \|x - y\|$.

(b) Without loss of generality assume $x = (0,0,\ldots,0)^\top$ and $y = (y^1,0,\ldots,0)^\top$. Let $c : [a,b] \to \mathbb{R}^n$ be a parametrised curve connecting x and y. Then

$$L[c] = \int_a^b \|\dot{c}(t)\| \, dt \geq \int_a^b |\dot{c}^1(t)| \, dt \geq \int_a^b \dot{c}^1(t) \, dt$$
$$= c^1(b) - c^1(a) = y^1 = \|x - y\|.$$

Thus there is no curve connecting x and y which is shorter than the line segment. If we have equality, then we must have $\|\dot{c}(t)\| = |\dot{c}^1(t)|$ for all t, thus $\dot{c}^2 = \cdots = \dot{c}^n = 0$. Hence $c^2 = \cdots = c^n = 0$ and c is a parametrisation of the line segment.

2.7 When restricted to $(0, 2\pi)$ the first parametrised curve is injective while the second is not.

2.8 Let I be the parameter interval of c_1 and J that of c_2. Let $C := c_1(I) = c_2(J)$ be the trace of the curves. By assumption $c_1 : I \to C$ and $c_2 : J \to C$ are bijective so that $\varphi := (c_2)^{-1} \circ c_1$ is defined. It remains to prove smoothness of φ (interchanging the roles of c_1 and c_2 will then also show smoothness of φ^{-1}). Let $t_0 \in I$ and put $s_0 := \varphi(t_0) \in J$. Since c_2 is regular, $\dot{c}_2(s_0)$ does not vanish. Without loss of generality $\dot{c}_2^1(s_0) \neq 0$. Thus the function c_2^1 has a smooth inverse $(c_2^1)^{-1}$ near s_0. Then $\varphi = (c_2^1)^{-1} \circ c_1^1$ is smooth near t_0 as well.

2.9 For $\tilde{c} := F \circ c$ we have $\dot{\tilde{c}} = A \cdot \dot{c}$ and $\ddot{\tilde{c}} = A \cdot \ddot{c}$. Hence

$$\tilde{n} = \begin{pmatrix} 0 & -1 \\ 1 & 0 \end{pmatrix} \cdot A \cdot \dot{c} = A \cdot \begin{pmatrix} 0 & -1 \\ 1 & 0 \end{pmatrix} \cdot \dot{c} = An$$

and therefore

$$\ddot{\tilde{c}} = A \cdot \ddot{c} = A\kappa n = \kappa A n = \kappa \tilde{n}.$$

This shows $\tilde{\kappa} = \kappa$.

If F reverses the orientation, then A anticommutes with $\begin{pmatrix} 0 & -1 \\ 1 & 0 \end{pmatrix}$ so that $\tilde{\kappa} = -\kappa$.

2.10 It is easy to see that the function κ defined in this exercise coincides with the curvature if the curve is parametrised by arc-length. Let $\tilde{c} = c \circ \varphi$ be an orientation-preserving reparametrisation. Then $\dot{\tilde{c}} = \dot{\varphi} \cdot (\dot{c} \circ \varphi)$ and $\ddot{\tilde{c}} = \ddot{\varphi} \cdot (\dot{c} \circ \varphi) + \dot{\varphi}^2 \cdot (\ddot{c} \circ \varphi)$. Thus

$$\det(\dot{\tilde{c}}, \ddot{\tilde{c}}) = \det(\dot{\varphi} \cdot (\dot{c} \circ \varphi), \ddot{\varphi} \cdot (\dot{c} \circ \varphi) + \dot{\varphi}^2 \cdot (\ddot{c} \circ \varphi) = \dot{\varphi}^3 \det(\dot{c} \circ \varphi, \ddot{c} \circ \varphi)$$

and

$$\|\dot{\tilde{c}}\|^3 = \dot{\varphi}^3 \|\dot{c} \circ \varphi\|^3.$$

This shows $\tilde{\kappa} = \kappa \circ \varphi$.

2.11 The function $t \mapsto \|c(t)\|^2$ has a maximum at $t = t_0$. Hence $(d/dt)|_{t=t_0} \|c(t)\|^2 = 0$ and $(d^2/dt^2)|_{t=t_0}\|c(t)\|^2 \leq 0$. The first condition implies $\ddot{c}(t_0) = \alpha \cdot c(t_0)$ and the second $\alpha \leq -1/R^2$. Therefore $|\kappa(t_0)| = \|\ddot{c}(t_0)\| = |\alpha| \cdot \|c(t_0)\| \geq 1/R^2 \cdot R = 1/R$.

2.12 A regular parametrisation is given by $c : (0, \infty) \to \mathbb{R}$, $c(t) = (t^2, t^3)^\top$. It extends to a non-regular parametrisation on $[0, \infty)$. We compute the length using the substitution $u = 4 + 9t^2$:

$$
\begin{aligned}
L[c|_{[0,T]}] &= \int_0^T \sqrt{(2t)^2 + (3t^2)^2}\, dt = \int_0^T t\sqrt{4 + 9t^2} \\
&= \int_4^{4+9T^2} \sqrt{u}\, \frac{du}{18} = \frac{1}{18}\left[\frac{2}{3}u^{3/2}\right]_{u=4}^{4+9T^2} \\
&= \frac{1}{27}\left((4 + 9T^2)^{3/2} - 8\right) = \frac{1}{27}\left((4 + 9x)^{3/2} - 8\right),
\end{aligned}
$$

where $c(T) = (x, y)$. We compute the curvature using exercise 2.10:

$$
\begin{aligned}
\kappa(t) &= \frac{1}{\left((2t)^2 + (3t^2)^2\right)^{3/2}} \det\begin{pmatrix} 2t & 2 \\ 3t^2 & 6t \end{pmatrix} = \frac{12t^2 - 6t^2}{t^3(4 + 9t^2)^{3/2}} \\
&= \frac{6}{t(4 + 9t^2)^{3/2}}.
\end{aligned}
$$

This shows that

$$
\lim_{t \searrow 0} \kappa(t) = \infty \quad \text{and} \quad \lim_{t \to \infty} \kappa(t) = 0,
$$

hence κ takes all values from $(0, \infty)$.

2.13 For the length we have

$$
L[c|_{[0,T]}] = \sqrt{\pi} \int_0^T \sqrt{\cos(\pi t^2/2)^2 + \sin(\pi t^2/2)^2}\, dt = \sqrt{\pi}\, T.
$$

The curve is parametrised proportional to arc-length with $\|\dot{c}\| = \sqrt{\pi}$. Hence we get for the curvature

$$
\begin{aligned}
|\kappa(t)| &= \frac{1}{\pi}\|\ddot{c}(t)\| = \frac{1}{\sqrt{\pi}}\sqrt{\left(-\sin(\pi t^2/2) \cdot t\pi\right)^2 + \left(\cos(\pi t^2/2) \cdot t\pi\right)^2} \\
&= \frac{1}{\sqrt{\pi}}\sqrt{t^2\pi^2} = \sqrt{\pi}\, t.
\end{aligned}
$$

2.14 The osculating circle $\tilde{c}$ is constructed in such a way that it contains $c(t_0)$. Without loss of generality $\tilde{c}(t_0) = c(t_0)$ and $\tilde{c}$ is parametrised by arc-length.

Since both $\dot{c}(t_0)$ and $\dot{\tilde{c}}(t_0)$ are perpendicular to $n(t_0)$ we have $\dot{c}(t_0) = \pm\dot{\tilde{c}}(t_0)$. With the right choice of orientation of $\tilde{c}$ we have $\dot{c}(t_0) = \dot{\tilde{c}}(t_0)$. Denote the centre of $\tilde{c}$ by $M := c(t_0) + (1/\kappa(t_0))n(t_0)$. Since the radius of $\tilde{c}$ is $r = 1/|\kappa(t_0)|$ we have

$$\ddot{\tilde{c}}(t_0) = \frac{1}{r}\frac{M - \tilde{c}(t_0)}{\|M - \tilde{c}(t_0)\|} = |\kappa(t_0)| \cdot \frac{(1/\kappa(t_0))n(t_0)}{\|(1/\kappa(t_0))n(t_0)\|} = \kappa(t_0)n(t_0) = \ddot{c}(t_0).$$

2.15 We parametrise the parabola by $c(x) = (x, x^2)^\top$. Using exercise 2.10 one easily computes $\kappa(x) = 2/(1+4x^2)^{3/2}$ and $n(x) = (1/\sqrt{1+4x^2})(-2x, 1)^\top$. Thus the centre of the osculating circle is given by

$$M(x) = c(x) + \frac{1}{\kappa(x)}n(x) = \begin{pmatrix} -4x^3 \\ \frac{1}{2} + 3x^2 \end{pmatrix}.$$

We conclude

$$\{M(x) \mid x \in \mathbb{R}\} = \left\{ \begin{pmatrix} X \\ Y \end{pmatrix} \in \mathbb{R}^2 \,\middle|\, \left(Y - \frac{1}{2}\right)^3 = 27\left(\frac{X}{4}\right)^2 \right\}.$$

2.16 We compute

$$\tilde{v}(t) = \frac{d}{dt}F(c(t)) = \frac{d}{dt}(A \cdot c(t) + p) = A \cdot \dot{c}(t) = Av(t)$$

and

$$\ddot{\tilde{c}}(t) = \dot{\tilde{v}}(t) = A\dot{v}(t) = A\ddot{c}(t).$$

Hence

$$\tilde{\kappa}(t) = \|\ddot{\tilde{c}}(t)\| = \|A\ddot{c}(t)\| = \|\ddot{c}(t)\| = \kappa(t),$$

because A is orthogonal and thus preserves lengths of vectors. This also shows $\tilde{n}(t) = An(t)$. Since A is orthogonal and orientation-preserving the unique vector completing $Av(t)$ and $An(t)$ as a positively oriented orthonormal basis is given by $Ab(t)$. Thus $\tilde{b}(t) = Ab(t)$. For the torsion we get, using once more that A is orthogonal,

$$\tilde{\tau}(t) = \left\langle \dot{\tilde{n}}(t), \tilde{b}(t) \right\rangle = \langle A\dot{n}(t), Ab(t)\rangle = \langle \dot{n}(t), b(t)\rangle = \tau(t).$$

If F and hence A is orientation-reversing, then similar reasoning yields $\tilde{b}(t) = -Ab(t)$ and therefore $\tilde{\tau}(t) = -\tau(t)$ while still $\tilde{\kappa}(t) = \kappa(t)$.

2.17 The proof is analogous to that of theorem 2.3.9 where one replaces (2.15) by

$$\frac{d}{dt}(c, v, n) = (c, v, n) \cdot \begin{pmatrix} 0 & 0 & 0 \\ 1 & 0 & -\kappa \\ 0 & \kappa & 0 \end{pmatrix}$$

and (2.16) by

$$\frac{d}{dt}\begin{pmatrix} \langle v, v\rangle \\ \langle n, n\rangle \\ \langle n, v\rangle \end{pmatrix} = \begin{pmatrix} 0 & 0 & 2\kappa \\ 0 & 0 & -2\kappa \\ -\kappa & \kappa & 0 \end{pmatrix} \cdot \begin{pmatrix} \langle v, v\rangle \\ \langle n, n\rangle \\ \langle n, v\rangle \end{pmatrix}.$$

2.18 First observe that

$$\mu(c, -e) = |\{\text{local minima of the function } \mathbb{R} \to \mathbb{R},\ t \mapsto \langle c(t), e\rangle \text{ in } [0, L)\}|.$$

The assertion follows because between any two local maxima there is a local minimum and vice versa.

2.19 *Any simple closed space curve is ambient isotopic to itself:* Put $\Phi(t, x) := x$ for all $x \in \mathbb{R}^3$ and all $t \in [0, 1]$.

If c_1 is ambient isotopic to c_2, then c_2 is ambient isotopic to c_1: Let Φ be an ambient isotopy deforming c_1 into c_2. Then Ψ deforms c_2 into c_1 where $\Psi(t, \cdot) = \Phi(t, \cdot)^{-1}$.

If c_1 is ambient isotopic to c_2 and c_2 is ambient isotopic to c_3, then c_1 is ambient isotopic to c_3: If Φ_1 deforms c_1 into c_2 and Φ_2 deforms c_2 into c_3, then Φ deforms c_1 into c_3 where

$$\Phi(t, x) = \begin{cases} \Phi_1(2t, x), & \text{if } t \in [0, 1/2], \\ \Phi_2(2t - 1, \Phi_1(1, x)), & \text{if } t \in [1/2, 1]. \end{cases}$$

3.1 There are many ways to cover S^2 by two coordinate neighbourhoods. We use the *stereographic projection*. Put $V_1 := \mathbb{R}^3 \setminus \{(0, 0, 1)^\top\}$, $U_1 = \mathbb{R}^2$, and $F_1(u, v) := (1/(1 + u^2 + v^2))(2u, 2v, u^2 + v^2 - 1)^\top$. One easily checks that $F_1(U_1) = S^2 \setminus \{(0, 0, 1)^\top\}$ with inverse map

$$F_1^{-1} : S^2 \setminus \{(0, 0, 1)^\top\} \to \mathbb{R}^2, \quad F_1^{-1}(x, y, z) = \frac{1}{1 - z}(x, y)^\top.$$

In particular, both F_1 and F_1^{-1} are continuous which verifies condition (i) in definition 3.1.1. Condition (ii) is also easily checked.

This local parametrisation covers all of S^2 except one single point, namely $(0,0,1)$. So we need a second local parametrisation. For example, we may take $V_2 := \mathbb{R}^3 \setminus \{(0,0,-1)^\top\}$, $U_2 = \mathbb{R}^2$, and $F_2(u,v) := (1/1+u^2+v^2)(2u,2v,1-u^2-v^2)^\top$.

3.2 Any local parametrisation (U,F,V) of S gives rise to a local parametrisation $(F^{-1}(W), F|_{F^{-1}(W)}, V\cap W)$ of $S\cap W$. These local parametrisations cover $S\cap W$.

3.3 For any point of $p\in S$ there is an open neighbourhood V of p such that $S\cap V$ is a regular surface. Hence there is a local parametrisation (U,F,V') of $S\cap V$ with $p\in V'$. Now (U,F,V') is also a local parametrisation of S, thus S can be covered by local parametrisations.

3.4 Let $f: V\to\mathbb{R}^3$ be smooth. Proposition 3.1.9 tells us that $F_2^{-1}\circ f$ is smooth for any local parametrisation (U_2,F_2,V_2) of S_2. Proposition 3.1.11 now implies that $F_2^{-1}\circ f\circ F_1$ is smooth for any local parametrisation (U_1,F_1,V_1) of S_1.

3.5

$$\frac{\partial F_3^+}{\partial x} = \begin{pmatrix} 1 \\ 0 \\ -\dfrac{x}{\sqrt{1-x^2-y^2}} \end{pmatrix}, \quad \frac{\partial F_3^+}{\partial y} = \begin{pmatrix} 0 \\ 1 \\ -\dfrac{y}{\sqrt{1-x^2-y^2}} \end{pmatrix},$$

$$(g_{ij})_{ij} = \begin{pmatrix} \dfrac{1-y^2}{1-x^2-y^2} & \dfrac{xy}{1-x^2-y^2} \\ \dfrac{xy}{1-x^2-y^2} & \dfrac{1-x^2}{1-x^2-y^2} \end{pmatrix}$$

3.6 For $F(x,y) = (x,y,f(x,y))$ we get

$$\frac{\partial F}{\partial x} = \begin{pmatrix} 1 \\ 0 \\ \dfrac{\partial f}{\partial x} \end{pmatrix}, \quad \frac{\partial F}{\partial y} = \begin{pmatrix} 0 \\ 1 \\ \dfrac{\partial f}{\partial y} \end{pmatrix},$$

$$(g_{ij})_{ij} = \begin{pmatrix} 1+\left(\dfrac{\partial f}{\partial x}\right)^2 & \dfrac{\partial f}{\partial x}\dfrac{\partial f}{\partial y} \\ \dfrac{\partial f}{\partial x}\dfrac{\partial f}{\partial y} & 1+\left(\dfrac{\partial f}{\partial y}\right)^2 \end{pmatrix}.$$

3.7

$$\frac{\partial F}{\partial \varphi} = \begin{pmatrix} -r\sin(\varphi) \\ r\cos(\varphi) \\ 0 \end{pmatrix}, \quad \frac{\partial F}{\partial r} = \begin{pmatrix} \cos(\varphi) \\ \sin(\varphi) \\ 1 \end{pmatrix},$$

$$(g_{ij})_{ij} = \begin{pmatrix} r^2 & 0 \\ 0 & 2 \end{pmatrix}.$$

3.8 Let (U, F, V) be a local parametrisation of S. Now

$$\tilde{N} := \frac{\dfrac{\partial F}{\partial x} \times \dfrac{\partial F}{\partial y}}{\left\| \dfrac{\partial F}{\partial x} \times \dfrac{\partial F}{\partial y} \right\|}$$

is a smooth unit normal field on $S \cap V$. Hence we have $N = f \cdot \tilde{N}$ with a function $f : S \cap V \to \{-1, 1\}$. Such an f is continuous if and only if it is locally constant if and only if it is smooth.

3.9 Near p write S locally as $S = \{x \in \mathbb{R}^3 \mid f_1(x) = 0\}$, where f_1 has non-vanishing gradient. Put $Y := N(p) \times X \in T_pS$. Then $E + p = \{q \in \mathbb{R}^3 \mid \langle q - p, Y\rangle = 0\}$. Put $f_2(q) := \langle q - p, Y\rangle$. Then, near p, $S \cap (E + p) = \{x \in \mathbb{R}^3 \mid f_1(x) = f_2(x) = 0\}$. Since $\operatorname{grad} f_1(p)$ and Y are orthogonal and non-zero, $D(f_1, f_2)(p)$ has maximal rank 2. Without loss of generality, let the second and the third row of this 3×2 matrix be linearly independent. Then, by the implicit function theorem, there exist smooth functions g_1 and g_2 such that

$$S \cap (E + p) = \left\{ (t, g_1(t), g_2(t))^\top \mid t \in (-\varepsilon, \varepsilon) \right\}$$

near p.

3.10 $(d/dt)N(c(t)) = dN(\dot{c}(t)) = -W_{c(t)}(\dot{c}(t))$.

3.11 According to corollary 3.6.16 near p S is the graph of a function f on the tangent plane. Translating p to $(0, 0, 0)^\top$ and rotating S such that T_pS becomes the plane spanned by e_1 and e_2 this function is of the form

$$f(u^1, u^2) = \frac{1}{2} \sum_{ij} h_{ij}(0, 0) u^i u^j + \varphi(u),$$

where $\varphi(u) = O(\|u\|^3)$, i.e. $|\varphi(u)| \le C \cdot \|u\|^3$ for some constant C and all u in a neighbourhood of $(0, 0)$. Now rotate T_pS about the origin such that e_1 and e_2 become an eigenbasis of the Weingarten map at $p = (0, 0, 0)^\top$.

Table A.1 *Graphs with vanishing Gauss curvature at the origin*

$f \equiv 0$	$S = T_pS$
$f(x,y) = y^2$	S lies on one side of T_pS and intersects T_pS in a line
$f(x,y) = x^4 + y^4$	S lies on one side of T_pS and $S \cap T_pS = \{p\}$
$f(x,y) = x^4 - y^4$	S intersects both sides of T_pS

Then we have

$$f(u^1, u^2) = \tfrac{1}{2}\left(\kappa_1(0,0)(u^1)^2 + \kappa_2(0,0)(u^2)^2\right) + \varphi(u).$$

(a) Let $K(p) > 0$. Then $\kappa_1(0,0)$ and $\kappa_2(0,0)$ are both positive or both negative. Without loss of generality, assume that they are both positive, the other case being similar. Put $k := \min\{\kappa_1(0,0), \kappa_2(0,0)\} > 0$. Now we have for all $u \neq (0,0)^\top$ in the domain of f with $\|u\| \leq k/4C$

$$f(u) \geq \frac{k}{2}\|u\|^2 + \varphi(u) \geq \frac{k}{2}\|u\|^2 - C\|u\|^3 \geq \frac{k}{4}\|u\|^2 > 0.$$

Hence this part of S lies on one side of T_pS.

(b) Let $K(p) < 0$. Then the two principal curvatures have opposite sign, say $\kappa_1(0,0) < 0$ and $\kappa_2(0,0) > 0$. An argument similar to the one in (a) shows that near p the function f is negative along the e_1-axis and positive along the e_2-axis. Hence each neighbourhood of p in S meets both sides of the tangent plane.

(c) Nothing general can be said in the case $K(p) = 0$ as can be seen by the following examples. Let S be the graph of the function $f : \mathbb{R}^2 \to \mathbb{R}$. Then at $p = (0,0,0)^\top$ the Gauss curvature vanishes; curves with vanishing Gauss curvature are listed in table A.1.

3.12 (a) Let $N \in S^2$. Let E be the orthogonal complement of N. Since S is compact the affine plane $E_t := E + t \cdot N$ will not intersect S provided t is large enough. Now decrease t until the affine plane E_t touches S for the first time. Then E is the tangent plane of S at this intersection point p, hence N is the exterior unit normal vector to S at p.

(b) Suppose there is a point $p \in S$ such that $K(p) < 0$. By exercise 3.11(b) there are points of S on both sides of the affine tangent plane $T_pS + p$. The construction in (a) above yields a point $q \in S$ such that $N(q) = N(p)$. By construction S lies entirely on one side of the affine tangent plane $T_qS + q$. Therefore $q \neq p$ and the Gauss map is not injective.

(c) For the pre-image p of N constructed in (a) the surface S lies entirely on one side of its tangent plane. Hence the Gauss curvature at this point cannot be negative.

3.13 Since f is integrable in the sense of definition 3.7.4 we can write f in the form $f = \tilde{f}_1 + \cdots + \tilde{f}_l$, where each $\tilde{f}_j$ is integrable in the sense of definition 3.7.1. Since $\chi = \chi_{V_j - (V_1 \cup \cdots \cup V_{j-1})}$ is integrable and bounded, the function $f_i = \chi \cdot \tilde{f}_1 + \cdots + \chi \cdot \tilde{f}_l$ is integrable as well.

3.14 The problem is that the familiar linearity of the integral of functions defined on $\mathbb{R}^n$ gives us linearity for the integral of functions on a surface a priori only if they are supported in a common range of a local parametrisation. Now let $f = f_1 + \cdots + f_k$ be a decomposition as in definition 3.7.4. Let $\chi_j = \chi_{V_j - (V_1 \cup \cdots \cup V_{j-1})}$, $\tilde{f}_j = \chi_j \cdot f$ and $f = \tilde{f}_1 + \cdots + \tilde{f}_l$ as in the remark after definition 3.7.4. It suffices to show

$$\sum_{i=1}^{k} \int_S f_i \, dA = \sum_{j=1}^{l} \int_S \tilde{f}_j \, dA .$$

Using $\sum_j \chi_j = 1$ and the linearity of the integral for functions supported in a common range of a local parametrisation we get

$$\begin{aligned} \sum_{i=1}^{k} \int_S f_i \, dA &= \sum_{i=1}^{k} \int_S \sum_{j=1}^{l} \chi_j f_i \, dA = \sum_{i,j} \int_S \chi_j f_i \, dA \\ &= \sum_{j=1}^{l} \int_S \sum_{i=1}^{k} \chi_j f_i \, dA = \sum_{j=1}^{l} \int_S \chi_j f \, dA . \end{aligned}$$

3.16 At a point $p \in S$, where $K(p) > 0$, the determinant of the differential of the Gauss map is non-zero. By the inverse function theorem there exist neighbourhoods U of p in S and V of $N(p)$ in S^2 such that the Gauss map is a diffeomorphism from U to V. The transformation formula for the integral implies

$$\int_U K \, dA = \int_V dA = A[V].$$

By exercise 3.12 (c) we know that the Gauss map $S_+ \to S^2$ is onto. By Sard's theorem the set $S_0 := \{x \in S \mid K(x) = 0\}$ is mapped to a null set in S^2. Choose (countably many) open subsets $U_j \subset S_+ \setminus S_0$ and $V_j \subset S^2$ such that

- the Gauss map is a diffeomorphism from U_j to V_j,
- the V_j are pairwise disjoint,
- the V_j cover S^2 up to a null set.

Since the V_j are pairwise disjoint, so are the U_j. We conclude

$$\int_{S_+} K\,dA \geq \sum_j \int_{U_j} K\,dA = \sum_j A[V_j] = A[S^2] = 4\pi.$$

3.17 From $K = \det(h_{ij})/\det(g_{ij}) = -h_{12}^2/\det(g_{ij})$, we see that $K(F(t,s)) = 0$ if and only if

$$0 = h_{12}(t,s) = \left\langle \frac{\partial^2 F}{\partial s \partial t}(t,s), N(F(t,s)) \right\rangle = \langle \dot{v}(t), N(F(t,s)) \rangle,$$

i.e. if and only if $\dot{v}(t) \in T_{F(t,s)}S$. Since $T_{F(t,s)}S$ is spanned by $(\partial F/\partial t)(t,s) = \dot{c}(t) + s\dot{v}(t)$ and $(\partial F/\partial s)(t,s) = v(t)$, the claim follows.

3.18 For the cylinder $\dot{v}(t) = 0$ and for the cone $\dot{v}(t) = -\dot{c}(t)$. In both cases $\dot{c}(t), v(t)$ and $\dot{v}(t)$ are linearly dependent and exercise 3.17 applies.

3.19 For the hyperboloid of revolution we have

$$\dot{c}(t) = \begin{pmatrix} -\sin(t) \\ \cos(t) \\ 0 \end{pmatrix}, \quad v(t) = \begin{pmatrix} -\sin(t) \\ \cos(t) \\ 1 \end{pmatrix}, \quad \dot{v}(t) = \begin{pmatrix} -\cos(t) \\ -\sin(t) \\ 0 \end{pmatrix}.$$

Thus

$$\det(\dot{c}(t), v(t), \dot{v}(t)) = \det \begin{pmatrix} -\sin(t) & -\sin(t) & -\cos(t) \\ \cos(t) & \cos(t) & -\sin(t) \\ 0 & 1 & 0 \end{pmatrix} = -1 \neq 0,$$

hence $\dot{c}(t), v(t)$ and $\dot{v}(t)$ are linearly independent and exercise 3.17 applies.

For the hyperbolic paraboloid the corresponding computation yields

$$\det(\dot{c}(t), v(t), \dot{v}(t)) = \frac{1}{1+t^2} \neq 0,$$

while we get for the Möbius strip

$$\det(\dot{c}(t), v(t), \dot{v}(t)) = -\tfrac{1}{2} \neq 0.$$

3.20 First partial derivatives of F:

$$\frac{\partial F}{\partial u^1}\left(u^1,u^2\right)=\begin{pmatrix}1-(u^1)^2+(u^2)^2\\ 2\,u^2\,u^1\\ 2\,u^1\end{pmatrix},\quad \frac{\partial F}{\partial u^2}\left(u^1,u^2\right)=\begin{pmatrix}2\,u^2\,u^1\\ 1-(u^2)^2+(u^1)^2\\ -2\,u^2\end{pmatrix}.$$

First fundamental form:

$$(g_{ij})=(1+(u^1)^2+(u^2)^2)^2\begin{pmatrix}1&0\\0&1\end{pmatrix}.$$

Its inverse:

$$(g^{ij})=(1+(u^1)^2+(u^2)^2)^{-2}\begin{pmatrix}1&0\\0&1\end{pmatrix}.$$

Unit normal field:

$$N=\frac{1}{1+(u^1)^2+(u^2)^2}\cdot\begin{pmatrix}-2u^1\\ 2u^2\\ 1-(u^1)^2-(u^2)^2\end{pmatrix}.$$

Second partial derivatives of F:

$$\frac{\partial^2 F}{\partial(u^1)^2}=\begin{pmatrix}-2\,u^1\\ 2\,u^2\\ 2\end{pmatrix},\quad \frac{\partial^2 F}{\partial u^2\partial u^1}=\begin{pmatrix}2\,u^2\\ 2\,u^1\\ 0\end{pmatrix},\quad \frac{\partial^2 F}{\partial(u^2)^2}=\begin{pmatrix}2\,u^1\\ -2\,u^2\\ -2\end{pmatrix}.$$

Second fundamental form:

$$(h_{ij})=\begin{pmatrix}2&0\\0&-2\end{pmatrix}.$$

Weingarten map:

$$\left(w_i^j\right)=(1+(u^1)^2+(u^2)^2)^{-2}\begin{pmatrix}2&0\\0&-2\end{pmatrix}.$$

Mean curvature:

$$H=\tfrac{1}{2}\,\mathrm{Trace}(W)=0.$$

3.21 First partial derivatives of F:

$$\frac{\partial F}{\partial u^1}\left(u^1,u^2\right)=\begin{pmatrix}\sinh(u^1)\cos(u^2)\\ \sinh(u^1)\sin(u^2)\\ 1\end{pmatrix},\quad \frac{\partial F}{\partial u^2}\left(u^1,u^2\right)=\begin{pmatrix}-\cosh(u^1)\sin(u^2)\\ \cosh(u^1)\cos(u^2)\\ 0\end{pmatrix}.$$

First fundamental form:

$$(g_{ij})=\begin{pmatrix}\cosh^2(u^1) & 0\\ 0 & \cosh^2(u^1)\end{pmatrix}.$$

Its inverse:

$$(g^{ij})=\begin{pmatrix}\cosh^{-2}(u^1) & 0\\ 0 & \cosh^{-2}(u^1)\end{pmatrix}.$$

Unit normal field:

$$N=\frac{1}{\cosh(u^1)}\begin{pmatrix}-\cos(u^2)\\ -\sin(u^2)\\ \sinh(u^1)\end{pmatrix}.$$

Second partial derivatives of F:

$$\frac{\partial^2 F}{\partial (u^1)^2}\left(u^1,u^2\right)=\begin{pmatrix}\cosh(u^1)\cos(u^2)\\ \cosh(u^1)\sin(u^2)\\ 0\end{pmatrix},$$

$$\frac{\partial^2 F}{\partial u^2\partial u^1}\left(u^1,u^2\right)=\begin{pmatrix}-\sinh(u^1)\sin(u^2)\\ \sinh(u^1)\cos(u^2)\\ 0\end{pmatrix},$$

$$\frac{\partial^2 F}{\partial (u^2)^2}\left(u^1,u^2\right)=\begin{pmatrix}-\cosh(u^1)\cos(u^2)\\ -\cosh(u^1)\sin(u^2)\\ 0\end{pmatrix}.$$

Second fundamental form:

$$(h_{ij})=\begin{pmatrix}-1 & 0\\ 0 & 1\end{pmatrix}.$$

Weingarten map:

$$\left(w_i^j\right) = \begin{pmatrix} -\cosh^{-2}(u^1) & 0 \\ 0 & \cosh^{-2}(u^1) \end{pmatrix}.$$

Mean curvature:

$$H = \tfrac{1}{2}\,\mathrm{Trace}(W) = 0.$$

3.22 First partial derivatives of F:

$$\frac{\partial F}{\partial u^1}\left(u^1,u^2\right) = \begin{pmatrix} \sin\left(u^2\right) \\ -\cos\left(u^2\right) \\ 0 \end{pmatrix}, \quad \frac{\partial F}{\partial u^2}\left(u^1,u^2\right) = \begin{pmatrix} u^1 \cos\left(u^2\right) \\ u^1 \sin\left(u^2\right) \\ 1 \end{pmatrix}.$$

First fundamental form:

$$(g_{ij}) = \begin{pmatrix} 1 & 0 \\ 0 & 1+(u^1)^2 \end{pmatrix}.$$

Its inverse:

$$(g^{ij}) = \begin{pmatrix} 1 & 0 \\ 0 & \left(1+(u^1)^2\right)^{-1} \end{pmatrix}.$$

Unit normal field:

$$N = \frac{1}{\sqrt{1+(u^1)^2}} \begin{pmatrix} -\cos(u^2) \\ -\sin(u^2) \\ u^1 \end{pmatrix}.$$

Second partial derivatives of F:

$$\frac{\partial^2 F}{\partial (u^1)^2}\left(u^1,u^2\right) = \begin{pmatrix} 0 \\ 0 \\ 0 \end{pmatrix}, \quad \frac{\partial^2 F}{\partial u^2 \partial u^1}\left(u^1,u^2\right) = \begin{pmatrix} \cos\left(u^2\right) \\ \sin\left(u^2\right) \\ 0 \end{pmatrix},$$

$$\frac{\partial^2 F}{\partial (u^2)^2}\left(u^1,u^2\right) = \begin{pmatrix} -u^1 \sin\left(u^2\right) \\ u^1 \cos\left(u^2\right) \\ 0 \end{pmatrix}.$$

Second fundamental form:

$$(h_{ij}) = \frac{1}{\sqrt{1+(u^1)^2}} \begin{pmatrix} 0 & -1 \\ -1 & 0 \end{pmatrix}.$$

Weingarten map:

$$\left(w_i^j\right) = \begin{pmatrix} 0 & -\left(1+(u^1)^2\right)^{-3/2} \\ -\dfrac{1}{\sqrt{1+(u^1)^2}} & 0 \end{pmatrix}.$$

Mean curvature:

$$H = \tfrac{1}{2}\,\mathrm{Trace}(W) = 0.$$

The helicoid is a ruled surface because $F(u^1, u^2) = c(u^2) + u^1 \cdot v(u^2)$, where $c(u^2) = (0, 0, u^2)^\top$ and $v(u^2) = (\sin(u^2), -\cos(u^2), 0)^\top$. Clearly, $\dot{c}(u^2) = (0, 0, 1)^\top$ and $v(u^2)$ are linearly independent.

3.23 The principal curvatures are the eigenvalues of the the Weingarten map and hence the zeros of its characteristic polynomial. The characteristic polynomial of any 2×2 matrix A is given by

$$\chi_A(\lambda) = \lambda^2 - \lambda \cdot \mathrm{Trace}(A) + \det(A).$$

3.24 The graph of φ has the following parametrisation:

$$F(x, y) = \begin{pmatrix} x \\ y \\ \varphi(x, y) \end{pmatrix}.$$

Its partial derivatives:

$$\frac{\partial F}{\partial x} = \begin{pmatrix} 1 \\ 0 \\ \dfrac{\partial \varphi}{\partial x} \end{pmatrix}, \quad \frac{\partial F}{\partial y} = \begin{pmatrix} 0 \\ 1 \\ \dfrac{\partial \varphi}{\partial y} \end{pmatrix}.$$

First fundamental form:

$$(g_{ij}) = \begin{pmatrix} 1 + \left(\frac{\partial \varphi}{\partial x}\right)^2 & \frac{\partial \varphi}{\partial x}\frac{\partial \varphi}{\partial y} \\ \frac{\partial \varphi}{\partial x}\frac{\partial \varphi}{\partial y} & 1 + \left(\frac{\partial \varphi}{\partial y}\right)^2 \end{pmatrix}.$$

Its inverse:

$$(g^{ij}) = \frac{1}{1 + \|\operatorname{grad}\varphi\|^2} \begin{pmatrix} 1 + \left(\frac{\partial \varphi}{\partial y}\right)^2 & -\frac{\partial \varphi}{\partial x}\frac{\partial \varphi}{\partial y} \\ -\frac{\partial \varphi}{\partial x}\frac{\partial \varphi}{\partial y} & 1 + \left(\frac{\partial \varphi}{\partial x}\right)^2 \end{pmatrix}.$$

Unit normal field:

$$N = \frac{1}{\sqrt{1 + \|\operatorname{grad}\varphi\|^2}} \begin{pmatrix} -\frac{\partial \varphi}{\partial x} \\ -\frac{\partial \varphi}{\partial y} \\ 1 \end{pmatrix}.$$

Second partial derivatives of the parametrisation:

$$\frac{\partial^2 F}{\partial x^2} = \begin{pmatrix} 0 \\ 0 \\ \frac{\partial^2 \varphi}{\partial x^2} \end{pmatrix}, \quad \frac{\partial^2 F}{\partial y \partial x} = \begin{pmatrix} 0 \\ 0 \\ \frac{\partial^2 \varphi}{\partial x \partial y} \end{pmatrix}, \quad \frac{\partial^2 F}{\partial y^2} = \begin{pmatrix} 0 \\ 0 \\ \frac{\partial^2 \varphi}{\partial y^2} \end{pmatrix}.$$

Second fundamental form:

$$(h_{ij}) = \frac{1}{\sqrt{1 + \|\operatorname{grad}\varphi\|^2}} \begin{pmatrix} \frac{\partial^2 \varphi}{\partial x^2} & \frac{\partial^2 \varphi}{\partial x \partial y} \\ \frac{\partial^2 \varphi}{\partial x \partial y} & \frac{\partial^2 \varphi}{\partial y^2} \end{pmatrix}.$$

Weingarten map:

$$\left(1+\|\operatorname{grad}\varphi\|^2\right)^{3/2}\cdot\left(w_i^j\right)=$$

$$\begin{pmatrix} \frac{\partial^2\varphi}{\partial x^2}\cdot\left(1+\left(\frac{\partial\varphi}{\partial y}\right)^2\right)-\frac{\partial^2\varphi}{\partial x\partial y}\cdot\frac{\partial\varphi}{\partial x}\cdot\frac{\partial\varphi}{\partial y} & -\frac{\partial^2\varphi}{\partial x^2}\cdot\frac{\partial\varphi}{\partial x}\cdot\frac{\partial\varphi}{\partial y}+\frac{\partial^2\varphi}{\partial x\partial y}\cdot\left(1+\left(\frac{\partial\varphi}{\partial x}\right)^2\right) \\ \frac{\partial^2\varphi}{\partial x\partial y}\cdot\left(1+\left(\frac{\partial\varphi}{\partial y}\right)^2\right)-\frac{\partial^2\varphi}{\partial y^2}\cdot\frac{\partial\varphi}{\partial x}\cdot\frac{\partial\varphi}{\partial y} & -\frac{\partial^2\varphi}{\partial x\partial y}\cdot\frac{\partial\varphi}{\partial x}\cdot\frac{\partial\varphi}{\partial y}+\frac{\partial^2\varphi}{\partial y^2}\cdot\left(1+\left(\frac{\partial\varphi}{\partial x}\right)^2\right) \end{pmatrix}.$$

The claim follows.

3.25 From the above computation of the Weingarten map we see

$$K=\frac{\frac{\partial^2\varphi}{\partial x^2}\cdot\frac{\partial^2\varphi}{\partial y^2}-\left(\frac{\partial^2\varphi}{\partial x\partial y}\right)^2}{\left(1+\left(\frac{\partial\varphi}{\partial y}\right)^2+\left(\frac{\partial\varphi}{\partial x}\right)^2\right)^2}=\frac{\det(\operatorname{Hess}\varphi)}{(1+\|\operatorname{grad}\varphi\|^2)^2}.$$

Now any symmetric real 2×2 matrix has positive determinant if and only if it is (positive or negative) definite (both eigenvalues positive or both eigenvalues negative) while it has negative determinant if and only if it is indefinite and non-degenerate (one eigenvalue positive and one negative).

3.26 Use exercise 3.24.

3.27 (a) After the parameter transformation $v^1=u^1/R$, $v^2=u^2$, the rescaled catenoid is given by the parametrisation $G_{(R)}(v^1,v^2)=R\cdot(\cosh(v^1)\cos(v^2),\cosh(v^1)\sin(v^2),v^1)^\top = R\cdot G_{(1)}(v^1,v^2)$. By exercise 3.21 we know that $G_{(1)}$ parametrises a minimal surface. One easily sees that we have for the first fundamental form $g_{(R)ij}=R^2\cdot g_{(1)ij}$, for its inverse $g_{(R)}^{ij}=R^{-2}\cdot g_{(1)}^{ij}$, for the normal field $N_{(R)}=N_{(1)}$, for the second fundamental form $h_{(R)ij}=R\cdot h_{(1)ij}$, and hence for the Weingarten map $w_{(R)}{}_i^j=R^{-1}\cdot w_{(1)}{}_i^j$. Thus $\mathscr{H}_{(R)}=R^{-1}\cdot\mathscr{H}_{(1)}=(0,0,0)^\top$.

(b) At height h the rescaled catenoid describes a circle of radius $R\cdot\cosh(h/R)$. Hence $h=R\operatorname{arcosh}(1/R)$.

(c) There is an upper bound for h because

$$\lim_{R\searrow 0}R\operatorname{arcosh}(1/R)=\lim_{R\nearrow 1}R\operatorname{arcosh}(1/R)=0.$$

3.28 First partial derivatives of the parametrisation:

$$\frac{\partial F}{\partial u^1}(u^1,u^2)=\begin{pmatrix}\sin(\alpha)\sinh(u^1)\cos(u^2)+\cos(\alpha)\cosh(u^1)\sin(u^2)\\ \sin(\alpha)\sinh(u^1)\sin(u^2)-\cos(\alpha)\cosh(u^1)\cos(u^2)\\ \sin(\alpha)\end{pmatrix},$$

$$\frac{\partial F}{\partial u^2}(u^1,u^2) = \begin{pmatrix} -\sin(\alpha)\cosh(u^1)\sin(u^2) + \cos(\alpha)\sinh(u^1)\cos(u^2) \\ \sin(\alpha)\cosh(u^1)\cos(u^2) + \cos(\alpha)\sinh(u^1)\sin(u^2) \\ \cos(\alpha) \end{pmatrix}.$$

First fundamental form:

$$(g_{ij}) = \begin{pmatrix} \cosh^2(u^1) & 0 \\ 0 & \cosh^2(u^1) \end{pmatrix}.$$

Its inverse:

$$(g^{ij}) = \begin{pmatrix} \cosh^{-2}(u^1) & 0 \\ 0 & \cosh^{-2}(u^1) \end{pmatrix}.$$

Unit normal field:

$$N = \frac{1}{\cosh(u^1)} \begin{pmatrix} -\cos(u^2) \\ -\sin(u^2) \\ \sinh(u^1) \end{pmatrix}.$$

Second partial derivatives of the parametrisation:

$$\frac{\partial^2 F}{\partial (u^1)^2}(u^1,u^2) = \begin{pmatrix} \sin(\alpha)\cosh(u^1)\cos(u^2) + \cos(\alpha)\sinh(u^1)\sin(u^2) \\ \sin(\alpha)\cosh(u^1)\sin(u^2) - \cos(\alpha)\sinh(u^1)\cos(u^2) \\ 0 \end{pmatrix}.$$

$$\frac{\partial^2 F}{\partial u^2 \partial u^1}(u^1,u^2) = \begin{pmatrix} -\sin(\alpha)\sinh(u^1)\sin\left(u^2\right) + \cos(\alpha)\cosh(u^1)\cos(u^2) \\ \sin(\alpha)\sinh(u^1)\cos(u^2) + \cos(\alpha)\cosh(u^1)\sin(u^2) \\ 0 \end{pmatrix},$$

$$\frac{\partial^2 F}{\partial (u^2)^2}(u^1,u^2) = \begin{pmatrix} -\sin(\alpha)\cosh(u^1)\cos(u^2) - \cos(\alpha)\sinh(u^1)\sin(u^2) \\ -\sin(\alpha)\cosh(u^1)\sin(u^2) + \cos(\alpha)\sinh(u^1)\cos(u^2) \\ 0 \end{pmatrix}.$$

Second fundamental form:

$$(h_{ij}) = \begin{pmatrix} -\sin(\alpha) & -\cos(\alpha) \\ -\cos(\alpha) & \sin(\alpha) \end{pmatrix}.$$

Weingarten map:

$$\left(w_i^j\right) = \cosh^{-2}(u^1) \begin{pmatrix} -\sin(\alpha) & -\cos(\alpha) \\ -\cos(\alpha) & \sin(\alpha) \end{pmatrix}.$$

We conclude $H=0$ for all α. For $\alpha=\pi/2$ we get the parametrisation of the catenoid as in example 3.8.13. For $\alpha = 0$ we get the parametrisation of the helicoid as in example 3.8.14 after the parameter transformation $v^1 = \sinh(u^1)$, $v^2 = u^2$.

3.29 One can easily check that in both (a) and (b) the ruled surface is a minimal surface. Conversely, suppose that it is minimal. From the parametrisation $F(t,s) = c(t) + sv(t)$ we see that

$$\frac{\partial F}{\partial t}(t,s) = \dot{c}(t) + s\dot{v}(t), \quad \frac{\partial F}{\partial s}(t,s) = v(t).$$

Hence we have for the first and second fundamental forms

$$(g_{ij}) = \begin{pmatrix} g_{11} & 0 \\ 0 & 1 \end{pmatrix}, \quad (h_{ij}) = \begin{pmatrix} h_{11} & h_{12} \\ h_{21} & 0 \end{pmatrix}.$$

Thus

$$\left(w_i^j\right) = \begin{pmatrix} h_{11}/g_{11} & 0 \\ 0 & 0 \end{pmatrix}$$

and therefore minimality means that $h_{11} = 0$, in other words $N \perp \partial^2 F/\partial t^2$, i.e.

$$\begin{aligned} 0 &= \langle \ddot{c} + s\ddot{v}, \dot{c} \times v + s\dot{v} \times v \rangle \\ &= \langle \ddot{c}, \dot{c} \times v \rangle + s\left(\langle \ddot{v}, \dot{c} \times v \rangle + \langle \ddot{c}, \dot{v} \times v \rangle\right) + s^2 \langle \ddot{v}, \dot{v} \times v \rangle. \end{aligned}$$

Thus

(i) $\langle \ddot{c}, \dot{c} \times v \rangle = 0$,
(ii) $\langle \ddot{v}, \dot{c} \times v \rangle + \langle \ddot{c}, \dot{v} \times v \rangle = 0$,
(iii) $\langle \ddot{v}, \dot{v} \times v \rangle = 0$.

If v is constant, then (ii) and (iii) hold automatically while (i) implies $\ddot{c}(t) = \alpha(t) \cdot v$ because both v and $\ddot{c}$ are perpendicular to $\dot{c}$. Integrating twice yields $\dot{c}(t) = \beta(t) \cdot v + w$ and $c(t) = \gamma(t) \cdot v + tw + u$. Thus c parametrises a straight line and the surface is part of a plane.

Let us now look at those values of t for which $\dot{v}(t) \neq (0,0,0)^\top$. From (iii) we conclude that $\ddot{v}(t) = \alpha(t)v(t) + \beta(t)\dot{v}(t)$ and thus

$$\frac{d}{dt}(\dot{v} \times v) = \ddot{v} \times v + \dot{v} \times \dot{v} = \beta\dot{v} \times v.$$

The unique solution to this ordinary differential equation for $\dot{v} \times v$ with initial value $\dot{v}(t_0) \times v(t_0)$ at $t = t_0$ can be written down explicitly:

$$\dot{v}(t) \times v(t) = \exp\left(\int_{t_0}^{t} \beta(s)ds\right) \cdot \dot{v}(t_0) \times v(t_0).$$

Hence $\dot{v}(t) \times v(t)$ are linearly dependent for all t, i.e. $v(t)$ and $\dot{v}(t)$ are contained in a fixed plane. After applying a Euclidean motion we may assume that this is the x–y plane. Since $\|v\| = 1$, the curve v parametrises part of the unit circle in the plane, i.e. we can write

$$v(t) = \begin{pmatrix} \sin(\Omega(t)) \\ -\cos(\Omega(t)) \\ 0 \end{pmatrix}$$

for some smooth function Ω. Now (i) yields $\ddot{c}(t) = \alpha(t)v(t)$, while (ii) gives $\ddot{v}(t) = \beta(t)\dot{c}(t) + \gamma(t)v(t)$. On the other hand,

$$\ddot{v}(t) = \ddot{\Omega}(t) \begin{pmatrix} \cos(\Omega(t)) \\ \sin(\Omega(t)) \\ 0 \end{pmatrix} - \dot{\Omega}^2(t) \cdot v(t).$$

Comparing we find

$$\beta(t)\dot{c}(t) = \ddot{\Omega}(t) \begin{pmatrix} \cos(\Omega(t)) \\ \sin(\Omega(t)) \\ 0 \end{pmatrix}.$$

If $\ddot{\Omega}(t) \neq 0$, then $\dot{c}(t) = \pm(\cos(\Omega(t)), \sin(\Omega(t)), 0)^\top$, thus c lies in an affine plane parallel to the x–y plane. Then the ruled surface is contained in this affine plane. Let therefore $\ddot{\Omega} \equiv 0$. Then $\Omega(t) = \omega t + t_0$. After applying a rotation we may assume $t_0 = 0$. Since $\dot{c}$ is perpendicular to v, we can write

$$\dot{c}(t) = \beta(t) \begin{pmatrix} \cos(\omega t) \\ \sin(\omega t) \\ 0 \end{pmatrix} + \gamma(t) \begin{pmatrix} 0 \\ 0 \\ 1 \end{pmatrix} \tag{A.1}$$

with $\beta^2 + \gamma^2 = 1$. Thus

$$\ddot{c}(t) = \dot{\beta}(t) \begin{pmatrix} \cos(\omega t) \\ \sin(\omega t) \\ 0 \end{pmatrix} + \beta(t)\omega \begin{pmatrix} -\sin(\omega t) \\ \cos(\omega t) \\ 0 \end{pmatrix} + \dot{\gamma}(t) \begin{pmatrix} 0 \\ 0 \\ 1 \end{pmatrix}.$$

Comparing this to $\ddot{c}(t) = \alpha(t)v(t)$, we find $\dot{\beta} = \dot{\gamma} = 0$ and $\beta = -\alpha$. Putting $\beta =: A$, integrating (A.1), and a last translation concludes the proof.

3.30 From the formula for the minimal curvature we see that a surface of revolution is minimal if and only if

$$\ddot{r} = \frac{1+\dot{r}^2}{r}$$

holds. Now $r(t) = c_1 \cosh(t + c_2)/c_1$ solves this differential equation and for suitable choices of $c_1 > 0$ and c_2 realises all initial values $r(t_0) = c_1 \cosh(t_0 + c_2)/c_1 > 0$ and $\dot{r}(t_0) = \sinh(t_0 + c_2)/c_1$.

3.31 We observe that for the third component $u(t) = \cos(t) + \ln(\tan(t/2))$ of F we have

$$\frac{du}{dt} = \frac{\cos^2(t)}{\sin(t)} > 0,$$

so that it is strictly monotonically increasing. Denote the inverse function by $t(u)$. Then

$$\begin{aligned} r &= \sin(t(u)), \\ \frac{dr}{du} &= \cos(t(u)) \cdot \frac{dt}{du} = \frac{\sin(t(u))}{\cos(t(u))} = \tan(t(u)), \\ \frac{d^2r}{du^2} &= \frac{1}{\cos^2(t(u))} \cdot \frac{dt}{du} = \frac{\sin(t(u))}{\cos^4(t(u))}. \end{aligned}$$

Hence

$$K = -\frac{\ddot{r}(t)}{r(t)(1+\dot{r}(t)^2)^2} = -\frac{\sin(t)/\cos^4(t)}{\sin(t)(1+\tan^2(t))^2} = -1.$$

3.32 Let c be parametrised on the interval I. Then

$$\begin{aligned} \int_S K\,dA &= -\int_I \int_0^{2\pi} \frac{1}{r} \frac{\kappa(t)\cos(\varphi)}{1 - r\cos(\varphi)\kappa(t)} r(1 - r\cos(\varphi)\kappa(t))\,d\varphi dt \\ &= -\int_I \int_0^{2\pi} \kappa(t)\cos(\varphi)\,d\varphi dt = 0. \end{aligned}$$

3.33 Choose $r > 0$ so small that $r < 1/\kappa(t)$ for all $t \in I$. Let S be the tubular surface around c. We only need to consider t for which $\kappa(t) \neq 0$, i.e. $\kappa(t) > 0$. From

$$K = -\frac{1}{r} \frac{\kappa(t)\cos(\varphi)}{1 - r\cos(\varphi)\kappa(t)}$$

we see that $K > 0$ if and only if $\cos(\varphi) < 0$, i.e. $\varphi \in (\pi/2, 3\pi/2)$. Using exercise 3.16 we find

$$4\pi \le \int_{\{K\ge 0\}} K\,dA = \int_I \int_{\pi/2}^{3\pi/2} \kappa(t)\cos(\varphi)\,d\varphi\,dt = 2\kappa(c).$$

This proof of Fenchel's estimate is due to Konrad Voss [33].

4.1 The standard vectors e_1 and e_2 form an orthonormal basis of T_pS_1 for any $p = (x, y, 0)^\top$. We compute

$$\begin{aligned} d_pf(e_1) &= \frac{d}{dt}\Big|_{t=0} f(p + te_1) \\ &= \frac{d}{dt}\Big|_{t=0} f(x+t, y, 0) \\ &= \frac{\partial f}{\partial x}(x, y, 0) \\ &= (-\sin(x), \cos(x), 0)^\top \end{aligned}$$

and similarly

$$d_pf(e_2) = \frac{\partial f}{\partial y}(x, y, 0) = (0, 0, 1)^\top.$$

Since $d_pf(e_1)$ and $d_pf(e_2)$ are again orthonormal, d_pf is a linear isometry.

4.2 At $p = (x, y, 0)^\top$ we compute

$$d_pf(e_1) = \frac{\partial f}{\partial x}(x, y, 0) = \frac{1}{2(x^2+y^2)^{3/2}} \begin{pmatrix} x(x^2+3y^2) \\ 2y^3 \\ \sqrt{3}x(x^2+y^2) \end{pmatrix},$$

$$d_pf(e_2) = \frac{\partial f}{\partial y}(x, y, 0) = \frac{1}{2(x^2+y^2)^{3/2}} \begin{pmatrix} -y(3x^2+y^2) \\ 2x^3 \\ \sqrt{3}y(x^2+y^2) \end{pmatrix}.$$

One can easily check that $d_pf(e_1)$ and $d_pf(e_2)$ are again orthonormal.

4.3 If we denote the coefficient function of the first fundamental form of F by g_{ij} and the ones of $f \circ F$ by $\tilde{g}_{ij}$, then, using the chain rule and the fact that df is a linear isometry, we get

$$\begin{aligned} \tilde{g}_{ij} &= \left\langle \frac{\partial(f\circ F)}{\partial u^i}, \frac{\partial(f\circ F)}{\partial u^j} \right\rangle = \langle D(f\circ F)(e_i), D(f\circ F)(e_j)\rangle \\ &= \langle df(DF(e_i)), df(DF(e_j))\rangle = \langle DF(e_i), DF(e_j)\rangle = g_{ij}. \end{aligned}$$

4.4 This follows from $d_q(f^{-1}) = (d_p f)^{-1}$, where $f(p) = q$ and the fact that the inverse of a linear isometry is again a linear isometry.

4.5 Given a local isometry $f : S_1 \to S_2$ we can restrict it to an isometry of a neighbourhood U of any given point $p \in S_1$ onto its image $V = f(U)$. By exercise 4.4 $(f|_U)^{-1} : V \to U$ is also an isometry. Since f is onto we can find such a neighbourhood V for any given point $q \in S_2$.

This last argument fails if f is not assumed surjective. For example, let S_2 be a surface having elliptic points and hyperbolic points (such as the torus). Let $S_1 \subset S_2$ be the open subset of elliptic points. Then the embedding $f : S_1 \hookrightarrow S_2$ is a local isometry. But no hyperbolic point in S_2 can have a neighbourhood which is isometric to an open subset of S_1 because isometries preserve Gauss curvature as we shall see (see Theorem 4.3.8).

4.6 $d_p f = D_p F|_{T_p S} = A|_{T_p S}$ is a linear isometry.

4.7 Take a Euclidean motion $F : \mathbb{R}^3 \to \mathbb{R}^3$ such that $F(E_1) = E_2$. Now exercise 4.6 applies.

4.8 Look at the curve $c : \mathbb{R} \to S$, $c(t) = (\cos(t+t_0), \sin(t+t_0), z)^\top$, where t_0 is chosen such that $c(0) = (x, y, z)^\top$. Then $\dot{c}(0) = X(x, y, z)$ and hence

$$\begin{aligned} \partial_{X(x,y,z)} f_1 &= d_{(x,y,z)} f_1(X(x,y,z)) = \left.\frac{d}{dt}\right|_{t=0} f_1(c(t)) = \left.\frac{d}{dt}\right|_{t=0} \cos(t+t_0) \\ &= -\sin(t_0) = -y \end{aligned}$$

and similarly

$$\partial_{X(x,y,z)} f_2 = x, \qquad \partial_{X(x,y,z)} f_3 = 0.$$

For Y we can take the curve $c(t) = (x, y, z+t)^\top$. Then one gets

$$\partial_{Y(x,y,z)} f_1 = \partial_{Y(x,y,z)} f_2 = 0, \qquad \partial_{Y(x,y,z)} f_2 = 1.$$

4.9 Compute

$$\begin{aligned} \partial_X(\partial_Y f) &= \partial_X\left(\sum_j \eta_j \frac{\partial(f\circ F)}{\partial u^j}\right) \\ &= \sum_{ij} \xi_i \frac{\partial}{\partial u^i}\left(\eta_j \frac{\partial(f\circ F)}{\partial u^j}\right) \\ &= \sum_{ij} \xi_i \left(\frac{\partial \eta_j}{\partial u^i}\frac{\partial(f\circ F)}{\partial u^j} + \eta_j \frac{\partial^2(f\circ F)}{\partial u^i u^j}\right) \end{aligned}$$

and similarly

$$\partial_Y(\partial_X f) = \sum_{ij} \eta_j \left(\frac{\partial \xi_i}{\partial u^j} \frac{\partial (f \circ F)}{\partial u^i} + \xi_i \frac{\partial^2 (f \circ F)}{\partial u^j u^i} \right).$$

By Schwarz's theorem,

$$\frac{\partial^2 (f \circ F)}{\partial u^i u^j} = \frac{\partial^2 (f \circ F)}{\partial u^j u^i}$$

and hence

$$\begin{aligned}
\partial_X(\partial_Y f) - \partial_Y(\partial_X f) &= \sum_{ij} \left(\xi_i \frac{\partial \eta_j}{\partial u^i} \frac{\partial (f \circ F)}{\partial u^j} - \eta_j \frac{\partial \xi_i}{\partial u^j} \frac{\partial (f \circ F)}{\partial u^i} \right) \\
&= \sum_{ij} \left(\xi_i \frac{\partial \eta_j}{\partial u^i} \frac{\partial (f \circ F)}{\partial u^j} - \eta_i \frac{\partial \xi_j}{\partial u^i} \frac{\partial (f \circ F)}{\partial u^j} \right) \\
&= \sum_{ij} \left(\xi_i \frac{\partial \eta_j}{\partial u^i} - \eta_i \frac{\partial \xi_j}{\partial u^i} \right) \frac{\partial (f \circ F)}{\partial u^j} \\
&= \partial_Z f.
\end{aligned}$$

4.10 From $\ddot{c}(t) = (-\cos(t)\cos(\theta), -\sin(t)\cos(\theta), 0)^\top$ and $N(c(t)) = -c(t) = (-\cos(t)\cos(\theta), -\sin(t)\cos(\theta), -\sin(\theta))^\top$ we get

$$\begin{aligned}
\frac{\nabla}{dt}\dot{c}(t) &= \ddot{c}(t) - \langle \ddot{c}(t), N(c(t)) \rangle N(c(t)) \\
&= \begin{pmatrix} -\cos(t)\cos(\theta) \\ -\sin(t)\cos(\theta) \\ 0 \end{pmatrix} - \cos^2(\theta) \begin{pmatrix} -\cos(t)\cos(\theta) \\ -\sin(t)\cos(\theta) \\ -\sin(\theta) \end{pmatrix} \\
&= \begin{pmatrix} -\cos(t)\cos(\theta)\sin^2(\theta) \\ -\sin(t)\cos(\theta)\sin^2(\theta) \\ -\sin(\theta)\cos^2(\theta) \end{pmatrix},
\end{aligned}$$

which vanishes if and only if $\theta = 0$.

4.12 The Gauss equation implies

$$\begin{aligned}
\sum_{ijk} g^{jk} R^i_{ijk} &= \sum_{ijk} g^{jk} \left(h_{jk} w^i_i - h_{ik} w^i_j \right) = \sum_{ij} \left(w^j_j w^i_i - w^j_i w^i_j \right) \\
&= \mathrm{Trace}(W)^2 - \mathrm{Trace}(W^2)
\end{aligned}$$

$$= (\kappa_1 + \kappa_2)^2 - \left(\kappa_1^2 + \kappa_2^2\right)$$
$$= 2\kappa_1\kappa_2$$
$$= 2K.$$

4.13 Covariant differentiation of vector fields along curves is now defined by (4.2):

$$\frac{\nabla}{dt}v = \sum_{k=1}^{2}\left(\dot{\xi}^k + \sum_{i,j=1}^{2}\Gamma_{ij}^k(\tilde{c})\xi^i\dot{\tilde{c}}^j\right)\frac{\partial F}{\partial u^k}(\tilde{c}),$$

where $v(t) = \xi^1(t)(\partial F/\partial u^1)(\tilde{c}(t)) + \xi^2(t)(\partial F/\partial u^2)(\tilde{c}(t))$ and $\tilde{c} = F^{-1} \circ c$. Statements (a) and (b) of lemma 4.2.12 follow immediately. For (c), writing $w(t) = \eta^1(t)(\partial F/\partial u^1)(\tilde{c}(t)) + \eta^2(t)(\partial F/\partial u^2)(\tilde{c}(t))$, we compute

$$\frac{d}{dt}I(v,w) = \frac{d}{dt}\left(\sum_{ij} g_{ij}\xi^i\eta^j\right)$$
$$= \sum_{ij}\left(\sum_k \frac{\partial g_{ij}}{\partial u^k}\dot{\tilde{c}}^k\xi^i\eta^j + g_{ij}\dot{\xi}^i\eta^j + g_{ij}\xi^i\dot{\eta}^j\right)$$

and

$$g\left(\frac{\nabla}{dt}v, w\right) = g\left(\sum_i \dot{\xi}^i\frac{\partial F}{\partial u^i} + \sum_{ikm}\Gamma_{ik}^m\xi^i\dot{\tilde{c}}^k\frac{\partial F}{\partial u^m}, \sum_j \eta^j\frac{\partial F}{\partial u^j}\right)$$
$$= \sum_{ij}\left(\dot{\xi}^i\eta^j g_{ij} + \sum_{km}\Gamma_{ik}^m\xi^i\dot{\tilde{c}}^k\eta^j g_{mj}\right)$$

and similarly

$$g\left(v, \frac{\nabla}{dt}w\right) = \sum_{ij}\left(\xi^i\dot{\eta}^j g_{ij} + \sum_{km}\Gamma_{jk}^m\xi^i\dot{\tilde{c}}^k\eta^j g_{mi}\right).$$

Hence, by (4.3),

$$g\left(\frac{\nabla}{dt}v, w\right) + g\left(v, \frac{\nabla}{dt}w\right)$$
$$= \sum_{ij}\left(\dot{\xi}^i\eta^j g_{ij} + \xi^i\dot{\eta}^j g_{ij} + \sum_{km}\left(\Gamma_{ik}^m g_{mj} + \Gamma_{jk}^m g_{mi}\right)\xi^i\dot{\tilde{c}}^k\eta^j\right)$$
$$= \sum_{ij}\left(\dot{\xi}^i\eta^j g_{ij} + \xi^i\dot{\eta}^j g_{ij} + \sum_k \frac{\partial g_{ij}}{\partial u^k}\xi^i\dot{\tilde{c}}^k\eta^j\right).$$

Note that (4.3) follows directly from (4.7) and therefore holds for general Riemannian metrics. This proves (c). Assertion (d) is again easily checked. Checking lemma 4.2.17 is similar.

4.14 Computing the left hand side of the equation in local coordinates one sees that the terms containing Christoffel symbols cancel due to their symmetry in the lower indices. What remains is the local expression for the Lie bracket.

4.15 The proof of lemma 4.3.10 no longer applies because it uses the Gauss equation and hence the second fundamental form and Weingarten map, which we do not have for general Riemannian metrics. Assertion (a), however, is a direct consequence of the definition of R:

$$R(v,w)x = \nabla^2_{v,w}x - \nabla^2_{w,v}x.$$

For (b), using exercise 4.13, we compute

$$\begin{aligned} g\left(\nabla^2_{v,w}x, y\right) &= g(\nabla_v \nabla_w x, y) - g(\nabla_{\nabla_v w} x, y) \\ &= \partial_v g(\nabla_w x, y) - g(\nabla_w x, \nabla_v y) - \partial_{\nabla_v w} g(x,y) + g(x, \nabla_{\nabla_v w} y) \\ &= \partial_v \partial_w g(x,y) - \partial_v g(x, \nabla_w y) - \partial_w g(x, \nabla_v y) + g(x, \nabla_w \nabla_v y) \\ &\quad - \partial_{\nabla_v w} g(x,y) + g(x, \nabla_{\nabla_v w} y) \end{aligned}$$

and similarly

$$\begin{aligned} g\left(\nabla^2_{w,v}x, y\right) &= \partial_w \partial_v g(x,y) - \partial_w g(x, \nabla_v y) - \partial_v g(x, \nabla_w y) + g(x, \nabla_v \nabla_w y) \\ &\quad - \partial_{\nabla_w v} g(x,y) + g(x, \nabla_{\nabla_w v} y). \end{aligned}$$

Thus, by exercise 4.14,

$$\begin{aligned} g(R(v,w)x, y) &= g\left(\nabla^2_{v,w}x, y\right) - g\left(\nabla^2_{w,v}x, y\right) \\ &= \partial_{[v,w]} g(x,y) + g(x, \nabla_w \nabla_v y) - g(x, \nabla_v \nabla_w y) - \partial_{\nabla_v w - \nabla_w v} g(x,y) \\ &\quad + g(x, \nabla_{\nabla_v w} y) - g(x, \nabla_{\nabla_w v} y) \\ &= -g(x, R(v,w)y) \\ &= -g(R(v,w)y, x). \end{aligned}$$

This shows (b). Before we continue with the proof of lemma 4.3.10 we observe that the proof of lemma 4.3.11 remains valid for general Riemannian metrics without change. As in the proof of lemma 4.3.11 one sees that (a) and (b) of lemma 4.3.10 imply that

$$g(R(v,w)x,y) = (v^1w^2 - v^2w^1)(x^1y^2 - x^2y^1)g(R(e_1,e_2)e_1,e_2),$$

where e_1, e_2 are a fixed basis of T_pS and $v = v^1e_1 + v^2e_2$ and similarly of the other tangent vectors. This directly implies (c). For (d), we calculate

$$\begin{aligned} &g(R(v,w)x + R(x,v)w + R(w,x)v, y) \\ &\quad = \big((v^1w^2 - v^2w^1)(x^1y^2 - x^2y^1) + (x^1v^2 - x^2v^1)(w^1y^2 - w^2y^1) \\ &\qquad + (w^1x^2 - w^2x^1)(v^1y^2 - v^2y^1)\big)g(R(e_1,e_2)e_1,e_2) \\ &\quad = 0. \end{aligned}$$

4.16 Straightforward computation using the formulae from lemma 4.2.14 (for Γ_{ij}^k), lemma 4.3.5 (for R_{ijk}^ℓ) and exercise 4.12.

4.17 Invariance of the length under reparametrisations can be shown exactly as in lemma 2.1.16 (replacing the Euclidean norm by the Riemannian norm). For the energy, we see that even a simple rescaling of the parametrisation does indeed change it. Put $\tilde{c}(t) = c(\alpha t)$. Then

$$\begin{aligned} E[\tilde{c}] &= \frac{1}{2}\int_a^b g(\dot{\tilde{c}}(t), \dot{\tilde{c}}(t))\,dt = \frac{\alpha^2}{2}\int_a^b g(\dot{c}(\alpha t), \dot{c}(\alpha t))\,dt \\ &= \frac{\alpha^2}{2}\int_{\alpha a}^{\alpha b} g(\dot{c}(s), \dot{c}(s))\,\frac{ds}{\alpha} = \alpha E[c]. \end{aligned}$$

4.18 If $F : U \to S_1$ is a local parametrisation of S_1, then $f \circ F : U \to S_2$ is one of S_2 and the components of the Riemannian metrics $g_{ij} : U \to \mathbb{R}$ are the same for both. Hence the Christoffel symbols are also the same and the claim follows.

4.19 From $\dot{\tilde{c}}(t) = \alpha\dot{c}(\alpha t + \beta)$ we get

$$\frac{\nabla}{dt}\dot{\tilde{c}}(t) = \alpha\frac{\nabla}{dt}(\dot{c}(\alpha t + \beta)) = \alpha^2\left(\frac{\nabla}{dt}\dot{c}\right)(\alpha t + \beta) = 0.$$

4.20 We know from example 4.5.9 that the equator is a geodesic when parametrised by arc-length. By exercise 4.19, this is also true if the equator is parametrised proportional to arc-length. Any other great circle is the isometric image of the equator and hence, by exercise 4.18, also a geodesic when parametrised proportional to arc-length. Simply choose $A \in \mathrm{O}(3)$ such that it maps the x–y plane to the plane E whose intersection with S^2 is the given great circle.

4.21 From exercise 4.19 we know that $\tilde{c}$ is a geodesic. Since clearly $\tilde{c}(0) = c(0) = p$ and $\dot{\tilde{c}}(0) = \delta\dot{c}(0) = \delta v$, it is the unique geodesic with these initial values.

4.22 (a) Let $c : [a,b] \to S$ be a curve, $c(s) = F(t(s), \varphi(s))$. We compare its length to that of a line of longitude, $\tilde{c}(t) = F(t, \varphi_0)$, $t \in [t(a), t(b)]$. Here we assume without loss of generality that $t(a) < t(b)$. Since $\partial F/\partial t$ and $\partial F/\partial\varphi$ are perpendicular, we get

$$
\begin{aligned}
L[c] &= \int_a^b \left\| \frac{d}{ds} F(t(s), \varphi(s)) \right\| ds \\
&= \int_a^b \left\| \frac{\partial F}{\partial t}(t(s), \varphi(s)) t'(s) + \frac{\partial F}{\partial \varphi}(t(s), \varphi(s)) \varphi'(s) \right\| ds \\
&\geq \int_a^b \left\| \frac{\partial F}{\partial t}(t(s), \varphi(s)) t'(s) \right\| ds \\
&= \int_a^b \left\| \frac{\partial F}{\partial t}(t(s), \varphi_0) \right\| \cdot |t'(s)|\, ds \\
&\geq \int_a^b \left\| \frac{\partial F}{\partial t}(t(s), \varphi_0) \right\| \cdot t'(s)\, ds \\
&= \int_a^b \left\| \frac{\partial F}{\partial t}(t, \varphi_0) \right\| dt \\
&= L[\tilde{c}].
\end{aligned}
$$

(b) The line of latitude $\varphi \mapsto F(t_0, \varphi)$ is a geodesic if and only if $\nabla_w w = 0$, where $w = \partial F/\partial\varphi$. From the proof of theorem 4.5.13 we know that $\nabla_w w = \alpha v$, $v = \partial F/\partial t$. Hence the line of latitude is a geodesic if and only if $\langle \partial^2 F/\partial\varphi^2, v \rangle = 0$. We calculate

$$
\begin{aligned}
\left\langle \frac{\partial^2 F}{\partial \varphi^2}, \frac{\partial F}{\partial t} \right\rangle &= \frac{\partial}{\partial \varphi} \underbrace{\left\langle \frac{\partial F}{\partial \varphi}, \frac{\partial F}{\partial t} \right\rangle}_{=0} - \left\langle \frac{\partial F}{\partial \varphi}, \frac{\partial^2 F}{\partial \varphi \partial t} \right\rangle \\
&= -\frac{1}{2} \frac{\partial}{\partial t} \left\langle \frac{\partial F}{\partial \varphi}, \frac{\partial F}{\partial \varphi} \right\rangle \\
&= -\frac{1}{2} \frac{\partial}{\partial t} (r(t)^2) \\
&= -r(t_0) \dot{r}(t_0).
\end{aligned}
$$

Thus the line of latitude is a geodesic if and only if $\dot{r}(t_0) = 0$.

4.23 Choose an orientation-preserving parameter transformation φ such that $\hat{c}(s) - c(\varphi(s))$ is parametrised by arc-length. Then $1 = \|\hat{c}'(s)\|_g = \|\dot{c}(\varphi(s))\varphi'(s)\|_g$, hence $\varphi'(s) = \|\dot{c}(\varphi(s))\|_g^{-1}$. Thus

$$\begin{aligned} \kappa_g &= g\left(\frac{\nabla}{ds}\hat{c}', n\right) = g\left(\frac{\nabla}{ds}(\dot{c}\cdot\varphi'), n\right) \\ &= g\left(\left(\frac{\nabla}{dt}\dot{c}\right)\cdot(\varphi')^2 + \dot{c}\cdot\varphi'', n\right) \\ &= g\left(\frac{\nabla}{dt}\dot{c}, n\right)\cdot(\varphi')^2 = \frac{g\left(\frac{\nabla}{dt}\dot{c}, n\right)}{g(\dot{c},\dot{c})}. \end{aligned}$$

4.24 The first column in the matrix corresponds to the definition of κ_g. For the second we first observe that

$$0 = \frac{d}{dt}\langle n, n\rangle = 2\left\langle \frac{\nabla}{dt}n, n\right\rangle$$

and hence $(\nabla/dt)n(t) = \alpha(t)\dot{c}(t)$. From

$$0 = \frac{d}{dt}\langle n, \dot{c}\rangle = \left\langle \frac{\nabla}{dt}n, \dot{c}\right\rangle + \left\langle n, \frac{\nabla}{dt}\dot{c}\right\rangle = \alpha + \kappa_g.$$

we conclude that $\alpha = -\kappa_g$ and the claim follows.

4.25 Since the curves $s \mapsto F(t,s) = \exp_{c(t)}(sn(t))$ are geodesics we have

$$0 = \nabla_{\frac{\partial F}{\partial s}}\frac{\partial F}{\partial s} = \Gamma_{22}^1(t,s)\frac{\partial F}{\partial t} + \Gamma_{22}^2(t,s)\frac{\partial F}{\partial s},$$

hence $\Gamma_{22}^1(t,s) = \Gamma_{22}^2(t,s) = 0$. From

$$\begin{aligned} \kappa_g(t)\cdot n(t) &= \frac{\nabla}{dt}\dot{c}(t) = \nabla_{\frac{\partial F}{\partial t}}\frac{\partial F}{\partial t} \\ &= \Gamma_{11}^1(t,0)\frac{\partial F}{\partial t} + \Gamma_{11}^2(t,0)\frac{\partial F}{\partial s} = \Gamma_{11}^1(t,0)\dot{c}(t) + \Gamma_{11}^2(t,0)n(t), \end{aligned}$$

we conclude $\Gamma_{11}^1(t,0) = 0$ and $\Gamma_{11}^2(t,0) = \kappa_g(t)$. Similarly, using the Frenet equation $(\nabla/dt)n = -\kappa_g\dot{c}$ (see exercise 4.24) we get $\Gamma_{12}^1(t,0) = 0$ and $\Gamma_{12}^2(t,0) = -\kappa_g(t)$. The remaining Christoffel symbols are determined because of symmetry in the lower indices.

4.26 If $F : U \to S$ is a local parametrisation of S_1, then $f \circ F : U \to S_2$ is local parametrisation of S_2. With respect to these parametrisations both surfaces have the same Christoffel symbols. This shows that v is parallel if and only if

$f \circ v$ is parallel. For given $v_0 \in T_pS_1$, let v be the unique parallel vector field on S_1 along c with $v(t_0) = v_0$. Then

$$d_qf(P_c(v_0)) = d_qf(v(t_1)) = (df \circ v)(t_1) = P_{f\circ c}((df \circ v)(t_0)) = P_{f\circ c}(d_pf(v_0)).$$

4.28 In the case of a constant curve the local differential equations (4.15) for a parallel vector field reduce to $\dot{\xi}^k = 0$. Hence the vector field is parallel if and only if it is constant.

4.29 By exercise 4.22 the lines of longitude are geodesics, hence $\partial F/\partial\varphi$ is the variation vector field of a geodesic variation.

4.30 For each $p \in \mathbb{M}_{-1}$ the surface $\mathbb{M}_{-1}$ lies on one side of the affine tangent plane $T_p\mathbb{M}_{-1}+p$. By exercise 3.11(b) it cannot have negative curvature, hence $K \geq 0$. To see that $K > 0$ one can apply exercise 3.25. The surface $\mathbb{M}_{-1}$ is the graph of the function $\varphi(x,y) = \sqrt{1+x^2+y^2}$ whose Hessian

$$\operatorname{Hess}\varphi(x,y) = \left(1+x^2+y^2\right)^{-3/2}\begin{pmatrix}1+y^2 & -xy\\ -xy & 1+x^2\end{pmatrix}$$

is positive definite.

4.31 Let $f:\mathbb{M}_\kappa \to \mathbb{M}_\kappa$ be an isometry and let $c(t) = \exp_p(tX)$ be the geodesic in $\mathbb{M}_\kappa$ with initial values $c(0) = p$ and $\dot{c}(0) = X$. Since isometries map geodesics to geodesics, $f \circ c$ is the geodesic with initial values $f(p)$ and $d_pf(X)$. Hence $f(c(t)) = \exp_{f(p)}(td_pf(X))$. In particular, for $t=1$ we get $f(\exp_p(X)) = \exp_{f(p)}(d_pf(X))$. On $\mathbb{M}_\kappa$ any two points p and q can be joined by a geodesic; just intersect $\mathbb{M}_\kappa$ with the plane containing p, q and 0. Thus $\exp_p : T_p\mathbb{M}_\kappa \to \mathbb{M}_\kappa$ is surjective. Hence for any $q \in \mathbb{M}_\kappa$ we can find an $X \in T_p\mathbb{M}_\kappa$ with $\exp_p(X) = q$. Then $f(q) = \exp_{f(p)}(d_pf(X))$. This shows that f is uniquely detemined by the point $f(p)$ and the linear map $d_pf : T_p\mathbb{M}_\kappa \to T_{f(p)}\mathbb{M}_\kappa$.

Now fix $p = (0,0,1)^\top \in \mathbb{M}_\kappa$. For any $q \in \mathbb{M}_\kappa$ we can find an $f \in G_\kappa$ with $f(p) = q$. Moreover, for any linear isometry $A: T_p\mathbb{M}_\kappa \to T_q\mathbb{M}_\kappa$ we can modify $f \in G_\kappa$ in such a way that $d_pf = A$ (by composing with a suitable $\tilde{f} \in G_\kappa$ with $\tilde{f}(p) = p$). Since the isometry f is determined by $f(p)$ and d_pf, the group G_κ contains all isometries.

4.32 Let H_C be the point where the perpendicular from C intersects the side c. Applying the sine rule to the triangle (A, C, H_C) yields $\mathfrak{s}_\kappa(h_c)/\sin(\alpha) = \mathfrak{s}_\kappa(b)/\sin(\pi/2) = \mathfrak{s}_\kappa(b)$ and similarly for the triangle (B, C, H_C).

4.33 By the sine rule we get

$$a^2 + b^2 - 2ab\cos(\gamma) = \frac{\sin^2(\alpha)}{\sin^2(\gamma)}c^2 + \frac{\sin^2(\beta)}{\sin^2(\gamma)}c^2 - 2\frac{\sin(\alpha)}{\sin(\gamma)}\frac{\sin(\beta)}{\sin(\gamma)}c^2\cos(\gamma)$$
$$= c^2 \cdot \frac{\sin^2(\alpha) + \sin^2(\beta) - 2\sin(\alpha)\sin(\beta)\cos(\gamma)}{\sin^2(\gamma)}. \tag{A.2}$$

From the cosine rule for angles, which is equivalent to $\alpha + \beta + \gamma = \pi$ as we have seen, we obtain

$$\begin{aligned}
&\sin^2(\alpha) + \sin^2(\beta) - 2\sin(\alpha)\sin(\beta)\cos(\gamma) \\
&\quad = \sin^2(\alpha) + \sin^2(\beta) + 2\sin(\alpha)\sin(\beta)(\cos(\alpha)\cos(\beta) - \sin(\alpha)\sin(\beta)) \\
&\quad = \sin^2(\alpha)(1 - \sin^2(\beta)) + \sin^2(\beta)(1 - \sin^2(\alpha)) + 2\sin(\alpha)\sin(\beta)\cos(\alpha)\cos(\beta) \\
&\quad = \sin^2(\alpha)\cos^2(\beta) + \sin^2(\beta)\cos^2(\alpha) + 2\sin(\alpha)\sin(\beta)\cos(\alpha)\cos(\beta) \\
&\quad = (\sin(\alpha)\cos(\beta) + \sin(\beta)\cos(\alpha))^2 \\
&\quad = \sin^2(\alpha + \beta) \\
&\quad = \sin^2(\pi - \alpha - \beta) \\
&\quad = \sin^2(\gamma).
\end{aligned}$$

Plugging this into (A.2) yields theorem 1.2.4.

4.34 For our construction h must take values in $(-1, 1)$. In fact, if $|h(0)| \geq 1$, then h remains outside $(-1, 1)$ on its maximal interval of definition. Now for $h(0) \in (-1, 1)$ the solution is of the form $h(x) = \tanh(x + x_0)$. Thus the chart differs from Mercator's projection only by a shift in the x-variable.

4.35 Suppose we have a local parametrisation of S^2 which is conformal and area-preserving. Then

$$(g_{ij}(u^1, u^2)) = c(u^1, u^2) \cdot \begin{pmatrix} 1 & 0 \\ 0 & 1 \end{pmatrix}.$$

This implies $dA = c(u^1, u^2)\, du^1\, du^2$ and since the chart is area-preserving we have $c = 1$, i.e.

$$(g_{ij}(u^1, u^2)) = \begin{pmatrix} 1 & 0 \\ 0 & 1 \end{pmatrix}.$$

This means that the local parametrisation is an isometry, which is impossible.

4.36 We compute the area in the Klein model. Using the formulae for the Riemannian metric coefficients, one easily computes the area element,

$$dA = (1 - x^2 - y^2)^{-3/2}\, dx\, dy.$$

We integrate over U by passing to polar coordinates.

$$\begin{aligned}\int_U (1 - x^2 - y^2)^{-3/2}\, dx\, dy &= \int_0^{2\pi} \int_0^1 (1 - r^2)^{-3/2}\, r\, dr\, d\theta \\ &= 2\pi \int_0^1 (1 - r^2)^{-3/2}\, r\, dr \\ &= 2\pi \left[(1 - r^2)^{-1/2} \right]_{r=0}^1 \\ &= \infty.\end{aligned}$$

4.37 From the metric coefficient matrix we deduce directly that the area element is given by

$$dA = \frac{du\, dv}{u^2}.$$

Integrating,

$$\begin{aligned}\int_\Delta \frac{du\, dv}{u^2} &= \int_{-1}^1 \int_{\sqrt{1-v^2}}^\infty u^{-2}\, du\, dv \\ &= \int_{-1}^1 \left[-u^{-1} \right]_{u=\sqrt{1-v^2}}^\infty dv \\ &= \int_{-1}^1 \frac{dv}{\sqrt{1 - v^2}} \\ &= [\arcsin(v)]_{v=-1}^1 \\ &= \pi.\end{aligned}$$

5.1 We choose $S_{\mathrm{reg}} = S^2$ and use the local parametrisation

$$F(\varphi, \theta) = \begin{pmatrix} \cos\theta \cdot \cos\varphi \\ \cos\theta \cdot \sin\varphi \\ \sin\theta \end{pmatrix}$$

with domain $U = (0, 2\pi) \times (-\pi/2, \pi/2)$. Then $F(\varphi, \theta)$ lies in the upper hemisphere if and only if $\theta \geq 0$. Similarly, we can take this parametrisation with domain $U' = (-\pi, \pi) \times (-\pi/2, \pi/2)$. These two local parametrisations cover all points of S except for $(0, 0, 1)^\top$. This point is an interior point and can be covered by the local parametrisation $\tilde{F} : \tilde{U} \to S \subset S^2$, $\tilde{F}(x, y) = (x, y, \sqrt{1 - x^2 - y^2})^\top$, $U = \{(x, y)^\top \in \mathbb{R}^2 \mid x^2 + y^2 < 1\}$.

5.2 (a) Use local parametrisations of the form $\tilde{F}(t,s) = c(t) + \left(s - \frac{1}{2}\right) v(t)$ and $\hat{F}(t,s) = c(t) + \left(\frac{1}{2} - s\right) v(t)$, $(t,s) \in (t_0, t_0 + L) \times \left(-\frac{1}{2}, \frac{1}{2}\right)$.

(b) The equivalence of (i) and (ii) is seen by looking at the normal field $N(t,s) = (\dot{c}(t) + s\dot{v}(t)) \times v(t)$ which closes up if and only if $v(t+L) = +v(t)$. The boundary is given by $\left\{c(t) + \frac{1}{2}v(t) \mid t \in \mathbb{R}\right\} \cup \left\{c(t) - \frac{1}{2}v(t) \mid t \in \mathbb{R}\right\}$. If $v(t+L) = +v(t)$, then this defines the disjoint union of two closed space curves. If $v(t+L) = -v(t)$, then this is one space curve. This shows the equivalence of (iii) and answers (c).

5.3 For the inner unit normal vector we have $(d_uF)^{-1}(v(p)) = (x,y)^\top$ with $y > 0$. Then $t \mapsto F(u + t(x,y)) = F((u^1, 0) + t(x,y)) = F(u^1 + tx, ty)$, $t \in [0, \varepsilon)$, yields the required curve because $ty \geq 0$ and hence $F(u^1 + tx, ty) \in S$.

For the outer unit normal vector there is no such curve because any curve $c : [0, \varepsilon) \to S$ starting at p can be written as $c(t) = F(x(t), y(t))$ with $y(t) \geq 0$ and $y(0) = 0$. Thus $\dot{y}(0) \geq 0$.

5.4 Reparametrise c_j by arc-length and use substitution in the integral.

5.5

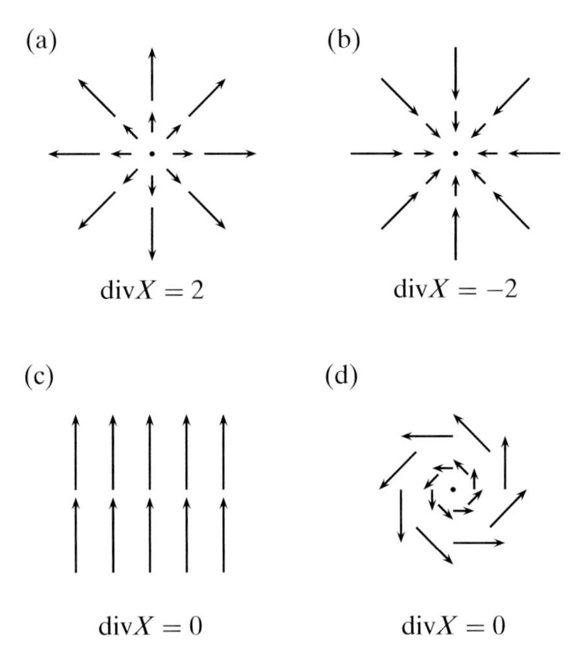

5.6 From $\nabla_Y(fX) = f\nabla_Y X + df(Y)X$ we have $\nabla_{\cdot}(fX) = f\nabla_{\cdot}X + df(\cdot)X$. Taking traces yields the claim.

5.7 (a) Apply the divergence theorem to the vector field $f_2 \cdot \operatorname{grad} f_1$ and use exercise 5.6.

(b) This follows from (a).

5.8 Let S be compact and let $f : S \to \mathbb{R}$ be harmonic. By exercise 5.7 we have

$$\int_S g(\operatorname{grad} f, \operatorname{grad} f)\, dA = -\int_S f \Delta f\, dA = 0.$$

Since the integrand $g(\operatorname{grad} f, \operatorname{grad} f)$ is non-negative this implies $g(\operatorname{grad} f, \operatorname{grad} f) \equiv 0$. Thus $\operatorname{grad} f \equiv (0,0,0)^\top$. Now let $p, q \in S$ be any two points. Choose a smooth curve $c : [0,1] \to S$ with $c(0) = p$ and $c(1) = q$. Then

$$f(q) - f(p) = \int_0^1 \frac{d}{dt} f(c(t))\, dt = \int_0^1 df(\dot{c}(t))\, dt = \int_0^1 g(\operatorname{grad} f(c(t)), \dot{c}(t))\, dt = 0.$$

Hence f is constant.

Note that this is no longer true if S is not connected because then f can take different values on the different connected components of S.

5.9 If F is a local parametrisation of S, then $F_t = F + tfN$ is a local parametrisation of the deformed surface. We get for the first fundamental form

$$\begin{aligned} g_{t,ij} &= \left\langle \frac{\partial F_t}{\partial u^i}, \frac{\partial F_t}{\partial u^j} \right\rangle \\ &= g_{ij} + t \cdot \left\{ \left\langle \frac{\partial F}{\partial u^i}, f \frac{\partial N}{\partial u^j} \right\rangle + \left\langle f \frac{\partial N}{\partial u^i}, \frac{\partial F}{\partial u^j} \right\rangle \right\} + O(t^2) \\ &= g_{ij} - 2tfII_{ij} + O(t^2) \end{aligned}$$

and hence

$$\dot{g} = -2fII.$$

Thus

$$\left.\frac{d}{dt}\right|_{t=0} A[S_t] = \frac{1}{2} \int_S \operatorname{Trace}(-2fII)\, dA = -2 \int_S fH\, dA = -2 \int_S \langle \Phi, \mathscr{H} \rangle\, dA.$$

6.1 Let p be a point in the interior of the tetrahedron. Then each straight half-line emanating from p will intersect the tetrahedron at exactly one point.

Thus the central projection with centre p will give a homeomorphism between the tetrahedron and the sphere with centre p.

6.2 Since all triangles are equilateral, all interior angles are equal to $\pi/3$. Thus for each vertex $\operatorname{def}(v) = 2\pi - 3\pi/3 = \pi$ and since the tetrahedron has four vertices $\sum_v \operatorname{def}(v) = 4\pi = 2\pi \cdot 2 = 2\pi \cdot \chi(X)$.

6.3 Using exercise 6.1 we see that S^2 can be triangulated by the tetrahedron. Hence its Euler–Poincaré characteristic is 2, see example 6.1.3.

6.4 Let $p_1, \ldots, p_k$ be common neighbours. Hence there exists $q \in S \cap \bigcap_{j=1}^k \operatorname{St}(p_j)$. From $q \in S \cap \operatorname{St}(p_j) \subset B(p_j, \varepsilon)$, we conclude $p_j \in B(q, \varepsilon)$ and hence $B(p_j, \varepsilon/2) \subset B(q, 3\varepsilon/2)$. Since the $\varepsilon/2$-balls about the points p_j are pairwise disjoint we have

$$k \cdot \frac{4\pi}{3} \cdot \left(\frac{\varepsilon}{2}\right)^3 = \sum_{j=1}^{k} \operatorname{vol}(B(p_j, \varepsilon/2)) \leq \operatorname{vol}(B(q, 3\varepsilon/2)) = \frac{4\pi}{3} \cdot \left(\frac{3\varepsilon}{2}\right)^3.$$

This shows $k \leq 3^3 = 27$.

6.5 Along each edge deform Φ in the interior of the dashed region (e.g. using Fermi coordinates) such that it becomes smooth along the edge.

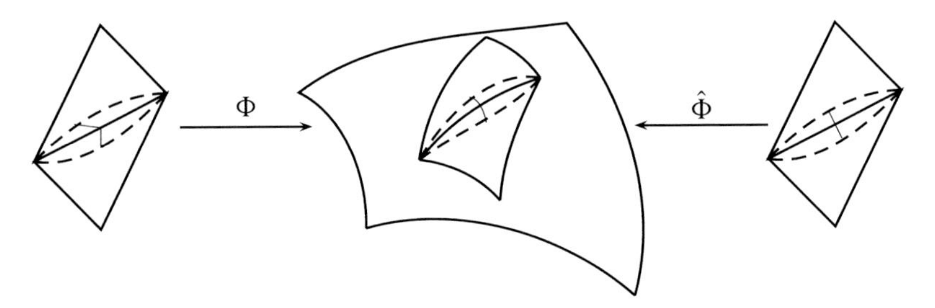

6.6 A torus can be obtained by identifying the opposite sides of a square. Thus a triangulation can be obtained as follows:

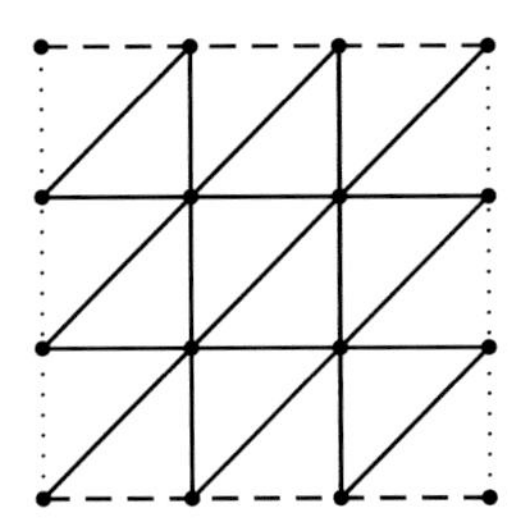

Making sure that we count each vertex and each edge which is identified only once we obtain 18 triangles, 27 edges, and 9 vertices. Hence the Euler–Poincaré characteristic is 0.

Appendix B Formulary

First fundamental form:

$$g_{ij} = \left\langle \frac{\partial F}{\partial u^i}, \frac{\partial F}{\partial u^j} \right\rangle.$$

Inverse of the first fundamental form:

$$\left(g^{ij}\right)_{ij} = \frac{1}{g_{11}g_{22} - g_{12}^2} \begin{pmatrix} g_{22} & -g_{12} \\ -g_{21} & g_{11} \end{pmatrix}.$$

Second fundamental form:

$$h_{ij} = \left\langle \frac{\partial^2 F}{\partial u^j \partial u^i}, N \right\rangle.$$

Weingarten map:

$$W\left(\frac{\partial F}{\partial u^i}\right) = \sum_j w_i^j \frac{\partial F}{\partial u^j},$$

$$h_{ij} = \sum_k w_i^k g_{kj}, \quad \sum_k h_{ik} g^{kj} = w_i^j.$$

Christoffel symbols:

$$\Gamma_{ij}^k = \frac{1}{2} \sum_m \left(\frac{\partial g_{jm}}{\partial u^i} + \frac{\partial g_{im}}{\partial u^j} - \frac{\partial g_{ij}}{\partial u^m} \right) g^{mk}.$$

Covariant derivative:

$$\frac{\nabla}{dt} v = \Pi_c(\dot{v}) = \dot{v} + \sum_{ijk} \Gamma_{ij}^k v^i \dot{c}^j \frac{\partial F}{\partial u^k}.$$

Riemann curvature tensor:

$$R(v,w)z = \nabla^2_{v,w}z - \nabla^2_{w,v}z = \nabla_v\nabla_w z - \nabla_w\nabla_v z - \nabla_{[v,w]}z$$
$$= \sum_{ijk\ell} R^{\ell}_{ijk} v^i w^j z^k \frac{\partial F}{\partial u^{\ell}},$$
$$R^{\ell}_{ijk} = \frac{\partial \Gamma^{\ell}_{kj}}{\partial u^i} - \frac{\partial \Gamma^{\ell}_{ki}}{\partial u^j} + \sum_m \left(\Gamma^{\ell}_{mi}\Gamma^m_{kj} - \Gamma^{\ell}_{mj}\Gamma^m_{ki}\right)$$
$$= h_{jk}w^{\ell}_i - h_{ik}w^{\ell}_j$$
$$= K \cdot \left(g_{jk}\delta^{\ell}_i - g_{ik}\delta^{\ell}_j\right).$$

Gauss curvature:

$$K = \frac{\det(h_{ij})}{\det(g_{ij})} = \frac{1}{2}\sum_{ijk} g^{jk} R^i_{ijk}.$$

Mean curvature:

$$H = \frac{\kappa_1 + \kappa_2}{2} = \frac{w^1_1 + w^2_2}{2} = \frac{1}{2}\sum_{ij} g^{ij} h_{ij}.$$

Divergence of a vector field:

$$\operatorname{div} X(p) = \operatorname{Trace}(Y_p \mapsto \nabla_{Y_p} X)$$
$$= \lim_{r\searrow 0} \frac{\int_{\partial \bar{D}(p,r)} g(X,\nu)ds}{A[\bar{D}(p,r)]}$$
$$= \sum_j \left(\frac{\partial \xi^j}{\partial u^j} + \sum_i \Gamma^j_{ij}\xi^i\right)$$
$$= \frac{1}{\sqrt{\det(g_{k\ell})}} \sum_j \frac{\partial}{\partial u^j}\left(\sqrt{\det(g_{k\ell})}\xi^j\right).$$

Divergence of a symmetric (2, 0) *tensor field:*

$$(\operatorname{div} b)^{\ell} = \sum_{ijk} g^{k\ell} g^{ij} \left(\frac{\partial b_{jk}}{\partial u^i} - \sum_{\alpha}\left(\Gamma^{\alpha}_{ij} b_{\alpha k} + \Gamma^{\alpha}_{ik} b_{\alpha j}\right)\right).$$

Euclidean trigonometry

Sine rule:

$$\frac{a}{\sin(\alpha)} = \frac{b}{\sin(\beta)} = \frac{c}{\sin(\gamma)}.$$

Cosine rule:

$$c^2 = a^2 + b^2 - 2ab\cos(\gamma).$$

Height formula:

$$h_c = b\sin(\alpha) = a\sin(\beta).$$

Sum of angles in a triangle:

$$\alpha + \beta + \gamma = \pi.$$

Spherical trigonometry

Sine rule:

$$\frac{\sin(a)}{\sin(\alpha)} = \frac{\sin(b)}{\sin(\beta)} = \frac{\sin(c)}{\sin(\gamma)}.$$

Cosine rule for sides:

$$\begin{aligned}
\cos(a) &= \cos(b)\cos(c) + \sin(b)\sin(c)\cos(\alpha),\\
\cos(b) &= \cos(a)\cos(c) + \sin(a)\sin(c)\cos(\beta),\\
\cos(c) &= \cos(a)\cos(b) + \sin(a)\sin(b)\cos(\gamma).
\end{aligned}$$

Cosine rule for angles:

$$\begin{aligned}
\cos(\alpha) &= \cos(a)\sin(\beta)\sin(\gamma) - \cos(\beta)\cos(\gamma),\\
\cos(\beta) &= \cos(b)\sin(\alpha)\sin(\gamma) - \cos(\alpha)\cos(\gamma),\\
\cos(\gamma) &= \cos(c)\sin(\alpha)\sin(\beta) - \cos(\alpha)\cos(\beta).
\end{aligned}$$

Height formula:

$$\sin(h_c) = \sin(b)\sin(\alpha) = \sin(a)\sin(\beta).$$

Sum of angles in a triangle:

$$\alpha + \beta + \gamma > \pi.$$

Hyperbolic trigonometry

Sine rule:

$$\frac{\sinh(a)}{\sin(\alpha)} = \frac{\sinh(b)}{\sin(\beta)} = \frac{\sinh(c)}{\sin(\gamma)}.$$

Cosine rule for sides:

$$\cosh(a) = \cosh(b)\cosh(c) - \sinh(b)\sinh(c)\cos(\alpha),$$
$$\cosh(b) = \cosh(a)\cosh(c) - \sinh(a)\sinh(c)\cos(\beta),$$
$$\cosh(c) = \cosh(a)\cosh(b) - \sinh(a)\sinh(b)\cos(\gamma).$$

Cosine rule for angles:

$$\cos(\alpha) = \cosh(a)\sin(\beta)\sin(\gamma) - \cos(\beta)\cos(\gamma),$$
$$\cos(\beta) = \cosh(b)\sin(\alpha)\sin(\gamma) - \cos(\alpha)\cos(\gamma),$$
$$\cos(\gamma) = \cosh(c)\sin(\alpha)\sin(\beta) - \cos(\alpha)\cos(\beta).$$

Height formula:

$$\sinh(h_c) = \sinh(b)\sin(\alpha) = \sinh(a)\sin(\beta).$$

Sum of angles in a triangle:

$$\alpha + \beta + \gamma < \pi.$$

Appendix C List of symbols

References

[1] F. R. Almgren. *Plateau's Problem. An Invitation to Varifold Geometry.* Providence, RI: American Mathematical Society revised edition, 2001

[2] R. Baer. The fundamental theorems of elementary geometry. *Trans. Amer. Math. Soc.* **56** (1944), 94–129

[3] J. K. Beem, P. E. Ehrlich, K. L. Easley. *Global Lorentzian Geometry.* New York, NY: Marcel Dekker second edition, 1996

[4] W. Blaschke. *Einführung in die Differentialgeometrie.* Berlin: Springer-Verlag, 1950

[5] R. L. Bishop, R. J. Crittenden. *Geometry of Manifolds.* Providence, RI: AMS Chelsea Publishing, 2001

[6] E. D. Bloch. *A First Course in Geometric Topology and Differential Geometry.* Boston, MA: Birkhäuser, 1997

[7] X. Chen, P. Lu, G. Tian. A note an uniformization of Riemann surfaces by Ricci flow. *Proc. Amer. Math. Soc.* **134** (2006), 3391–3393

[8] B. Chow, D. Knopf. *The Ricci Flow: An Introduction.* Providence, RI: American Mathematical Society, 2004

[9] M. P. do Carmo. *Differential Geometry of Curves and Surfaces.* Englewood Cliffs, NJ: Prentice-Hall, 1976

[10] M. P. do Carmo. *Riemannian Geometry.* Boston, MA: Birkhäuser, 1992

[11] I. Fáry. Sur la courbure totale d'une courbe gauche faisant un noeud. *Bull. Soc. Math. Fr.* **77** (1949), 128–138

[12] S. Gallot, D. Hulin, J. Lafontaine. *Riemannian Geometry.* Berlin: Springer-Verlag, 1987

[13] D. Hilbert. *Grundlagen der Geometrie.* Stuttgart: Teubner-Verlag, ninth edition, 1962

[14] A. Katok, V. Climenhaga. *Lectures on Surfaces.* Providence, RI: American Mathematical Society, 2008

[15] L. H. Kauffman. *Knots and Physics.* River Edge, NJ: World Scientific, third edition, 2001

[16] S. Kobayashi, K. Nomizu. *Foundations of Differential Geometry.* Vols. I and II. New York, NY: John Wiley & Sons, 1996

[17] S. Lang. *Linear Algebra*. New York: NY: Springer-Verlag, third edition, 1987

[18] S. Lang. *Undergraduate Analysis*. New York, NY: Springer-Verlag, second edition, 1997

[19] H. B. Lawson, Jr. *Lectures on Minimal Submanifolds*. Vol. I. Wilmington, DE: Publish or Perish, 1980

[20] C. Livingston. *Knot Theory*. Washington, DC: Mathematical Association of America, 1993

[21] J. Milnor. On the total curvature of knots. *Ann. Math.* **52** (1950), 248–257

[22] C. W. Misner, K. S. Thorne, J. A. Wheeler. *Gravitation*. San Francisco, CA: W. H. Freeman and Co., 1973

[23] S. Montiel, A. Ros. *Curves and Surfaces*. Providence, RI: American Mathematical Society, 2005

[24] J. Morgan, G. Tian. *Ricci Flow and the Poincaré Conjecture*. Providence, RI: American Mathematical Society, 2007

[25] B. O'Neill. *Semi-Riemannian Geometry. With Applications to Relativity*. New York, NY: Academic Press, 1983

[26] R. Osserman. *A survey of minimal surfaces*. New York, NY: Dover Publications, second edition, 1986

[27] V. V. Prasolov, A. B. Sossinsky. *Knots, Links, Braids and 3-manifolds*. Providence, RI: American Mathematical Society, 1997

[28] A. Robinson, J. Morrison, P. Muehrcke, J. Kimerling, S. Guptill. *Elements of Cartography*. New York, NY: John Wiley & Sons, sixth edition, 1995

[29] T. Sakai. *Riemannian Geometry*. Providence, RI: American Mathematical Society, 1996

[30] E. M. Schröder. *Geometrie euklidischer Ebenen*. Paderborn: Ferdinand Schoening, 1984

[31] H. Stephani. *General Relativity. An Introduction to the Theory of the Gravitational Field*. Cambridge: Cambridge University Press, 1982

[32] W. P. Thurston. *Three-Dimensional Geometry and Topology*. Vol. 1. Princeton, NJ: Princeton University Press, 1997

[33] K. Voss. Eine Bemerkung über die Totalkrümmung geschlossener Raumkurven. *Archiv für Mathematische Logik und Grundlagenforschung* **6** (1955), 259–263

[34] R. M. Wald. *General Relativity*. Chicago, IL: University of Chicago Press, 1984

[35] F. W. Warner. *Foundations of Differentiable Manifolds and Lie Groups*. New York, NY: Springer-Verlag, 1983

Index

Printed in the United States
by Baker & Taylor Publisher Services